Layers of Reality

Perception Study of a Synaesthete

Anna
Püschel

When I started art school, I did not know about synaesthesia. The fact that my emotions are coloured or that memories carry different tints did not appear as anything special to me.

It was just a few years ago that I realised that those associations of mine were not as common as I thought. Some of us can see music or taste words, while others see time in a geometrical shape or perceive numbers or letters as personalities. I found that this phenomenon has a name: synaesthesia—the merging of the senses.

I was relieved to hear that many synaesthetes are not aware that their reality looks different to those of others, just like me—and that, once they find out about their divergence in perception, they mostly state how much they appreciate the way their brain brings the world to them.
I started a project on synaesthesia because I do not perceive it only as a funny twist of evolution. For me, this topic also raises the fundamental question of the extent to which we as individuals share the same reality.

Layers of Reality is not a reference work, but a personal journey into my individual perception. My goal is to show that this phenomenon is not a disease, nor are synaesthetes making anything up. We are not sick and we do not need to be treated. Synaesthesia creates much controversy, as I learned during my research, and I do not aspire to solve all misconceptions or explain this phenomenon in every tiny detail. What I want is to make it accessible by offering a personal approach, and to help spread awareness about synaesthesia. By enriching my own findings with scientific approaches from different disciplines, I hope this book will answer a few questions and raise many more.

I feel we should all remember that what we perceive is not static, but a construct of our senses. It is our perception that shapes our world.

There is not one single reality.

An Introduction to Synaesthesia by Sam Beekhuizen 7

Outlines of the Perception Study 13

Chapter 1 Familiarity 19

Chapter 2 Motif 79

Chapter 3 Tonality 107

Chapter 4 Consistency 135

Chapter 5 Origins 193

Perception and Synaesthesia by Erika Schippmann 221

A Reflection on *Layers of Reality* by Olympia Colizoli 225

Visualising Synaesthesia 233

Coloured Emotions as Qualia by Alexandra Spaeth 249

An Introduction to Synaesthesia

Sam Beekhuizen, MSc
psychological researcher

Synaesthesia is a composition of the Greek words 'syn' — 'together' and 'aesthesia' — 'sensation'. It is a neurological condition in which the sensory experience in one sense (the inducing experience) involuntarily leads to a perceptual experience in another sense (the concurrent experience).

However, the concurrent does not always have to be an experience in another sense. It can also be in the same sensory modality or cognitive in nature.[1] A certain sound can be linked with the visual experience of a colour, or a number can be linked with the idea of a certain personality (for instance, '7' is cold but very intelligent).

The phenomenon of synaesthesia was first investigated by Georg Tobias Ludwig Sachs in 1812.[2] However, research into this interesting phenomenon progressed slowly for a long time. It received a new surge of interest from the scientific community in the last few decades; many different forms of synaesthesia have been characterised by now. There are some common characteristics that seem to apply to synaesthetic percepts. Firstly, the coupling between the inducing sensory stimulus and the following perceptual concurrent seems to be consistent over time.[3] Indeed, synaesthetes report that the links between the perceptual experiences are not random, but appear to them as normal, just as intuitive as regular sensory experiences. Secondly, the synaesthetic percepts are evoked involuntarily and automatically.[4] There is no conscious influence

1
Simner, J. Defining synesthesia. British Journal of Psychology, 103(1), 1–5 (2012).

2
Jewanski, D. & W., cited in Simner, J. (2012). Defining synesthesia 1, 1–15 (2009).

3
Eagleman, D. M., Kagan, A. D., Nelson, S. S., Sagaram, D. & Sarma, A. K. A standardized test battery for the study of synesthesia. Journal of Neuroscience Methods 159(1), 139–145 (2007).

4
Van der Veen, F. M., Aben, H. P., Smits, M. & Röder, C. H. Grapheme-color synesthesia interferes with color perception in a standard Stroop task. Neuroscience 258, 246–253 (2014).

that synaesthetes can exert on the percepts; they just happen to them. The spatial location of the synaesthetic experience, however, can differ between individuals.[5] Sometimes the synaesthetic percept is experienced as a natural association in the mind's eye, while for others the synaesthetic percept behaves as an actual perceptual experience coming from the outside world. The latter are named 'projector' synaesthetes, whereas the former are named 'associator' synaesthetes.

5

Cytowic, R. E. Synesthesia: A union of the senses. MIT Press (2002).

A very common form of synaesthesia is grapheme–colour synaesthesia, in which seeing letters or numbers evokes the perception of colour.[6] However, we all have certain associations, sometimes strange, between different shapes, colours, emotions and concepts. Think for a second about a much more common form of what could be called synaesthesia: the pairing of colours to certain days of the week. Or having a specific mental lay-out for the week. In people with genuine synaesthesia, however, the concurrents are less like mental imagery, but more like actual sensory experiences entering their consciousness from the outside world. It 'happens' to them, more than being 'formed' consciously.

6

Colizoli, O., Murre, J. M. & Rouw, R. Defining (trained) grapheme-color synesthesia. Frontiers in Human Neuroscience 8, 368 (2014).

To date, the origins of synaesthesia remain unclear. Research shows that, to a large extent, synaesthetic percepts entail real-world properties and mimic the directness and intensity of 'true' sensory experiences.[7] And, indeed, studies of the brain have shown that synaesthetic experiences show activity in the same parts

7

Nunn, J. A., Gregory, L. J., Brammer, M., Williams, S. C. R., Parslow, D. M., Morgan, M. J., Morris, R. G., Bullmore, E. T., Baron-Cohen, S. & Gray, J. A. Functional magnetic resonance imaging of synesthesia: Activation of V4/V8 by spoken words. Nature Neuroscience 5, 371–375 (2002).

of the brain as 'normal' sensory experiences. However, a complete account of the neural basis of synaesthesia deals with the difficult problem of the huge diversity of synaesthetic experiences. Each form of synaesthesia shows its own characteristics; even within the same form, there are many variations and individual differences.

Often, individuals that have one form of synaesthesia also show other forms. In *Layers of Reality*, Anna focuses on her emotion to colour synaesthesia, but she also experiences many other forms of perceptual divergence that, depending on the classification criteria chosen, would fall under synaesthesia: ordinal linguistic personification (the feeling that numbers or letters have a gender and/or personality), a slight grapheme–colour synaesthesia (letters appear coloured, but in her case only vowels are consistent), spatio–temporal synaesthesia (units of time such as the week and the year occupy a structure in space), personality–colour synaesthesia (feeling that specific colours belong to specific people), and colours and textures associated to strong sensory experiences such as orgasms or intense physical pain.

There are a number of general theories as to the neurological cause of synaesthesia.[8] Some focus on the existence of cross-activation between different perceptual and cognitive parts of the brain. It is assumed that, early during human development, different areas of the brain cortex are not yet specialised to process information from specific sensory modalities.

8
Spector, F. & Maurer, D. Synesthesia: A new approach to understanding the development of perception. Developmental Psychology 45, 175–189 (2009).

There are still a lot of cross-connections between different brain areas. During development, these are pruned and removed based on which connections prove themselves useful and efficient. Thus, every sensory modality develops into an increasingly efficient, specialised system in a specific location in the brain. According to these theories, the pruning of synaptic connections between neurons is not complete in synaesthesia. This allows different sensory modalities to influence each other.

Other theories assume that connections between brain areas that process different sensory modalities are common, but that these connections are usually inhibited by feedback from other brain areas. In synaesthesia, this feedback is not strong enough to inhibit these 'incorrect' connections or malfunctions, so the senses are able to influence each other in strange and interesting ways.

It is however unclear what the exact functional and structural differences are in the brains of synaesthetes compared to non-synaesthetes.

The fact that synaesthesia is a general term for such a large number of perceptual conditions makes it even more difficult to find a coherent definition.[9] Moreover, our views on reality are always individual. That is why synaesthesia is so difficult to capture in quantitative measurements.

9
Beekhuizen, S. H. Defining grapheme-color synesthesian (Master's thesis, University of Amsterdam, Amsterdam, The Netherlands, 2016).

As a comparison: how do you study people's perception of the colour 'red', or the way someone feels when studying a painting?

It is theorised that about 5% of the adult population has some form of synaesthesia,[10] but the phenomenon could be more prevalent than we think. Rather than an abnormal way of processing sensory stimuli, it is more like an exaggeration of the norm, depending on how the millions upon millions of neurons in our brains are wired. The senses do not necessarily operate as separate modules, but rather as a network of communicative pathways between different parts of our brains and minds.

This book is a dive into Anna's synaesthetic reality.

Enjoy.

10
Simner, J., Mulvenna, C., Sagiv, N., Tsakanikos, E., Witherby, S., Fraser, C., Scott, K. & Ward, J. (2006), cited in Spector, F. & Maurer, D. Synesthesia: A new approach to understanding the development of perception. Developmental Psychology 45, 175–189 (2009).

Outlines of the Perception Study

Anna Püschel
artist

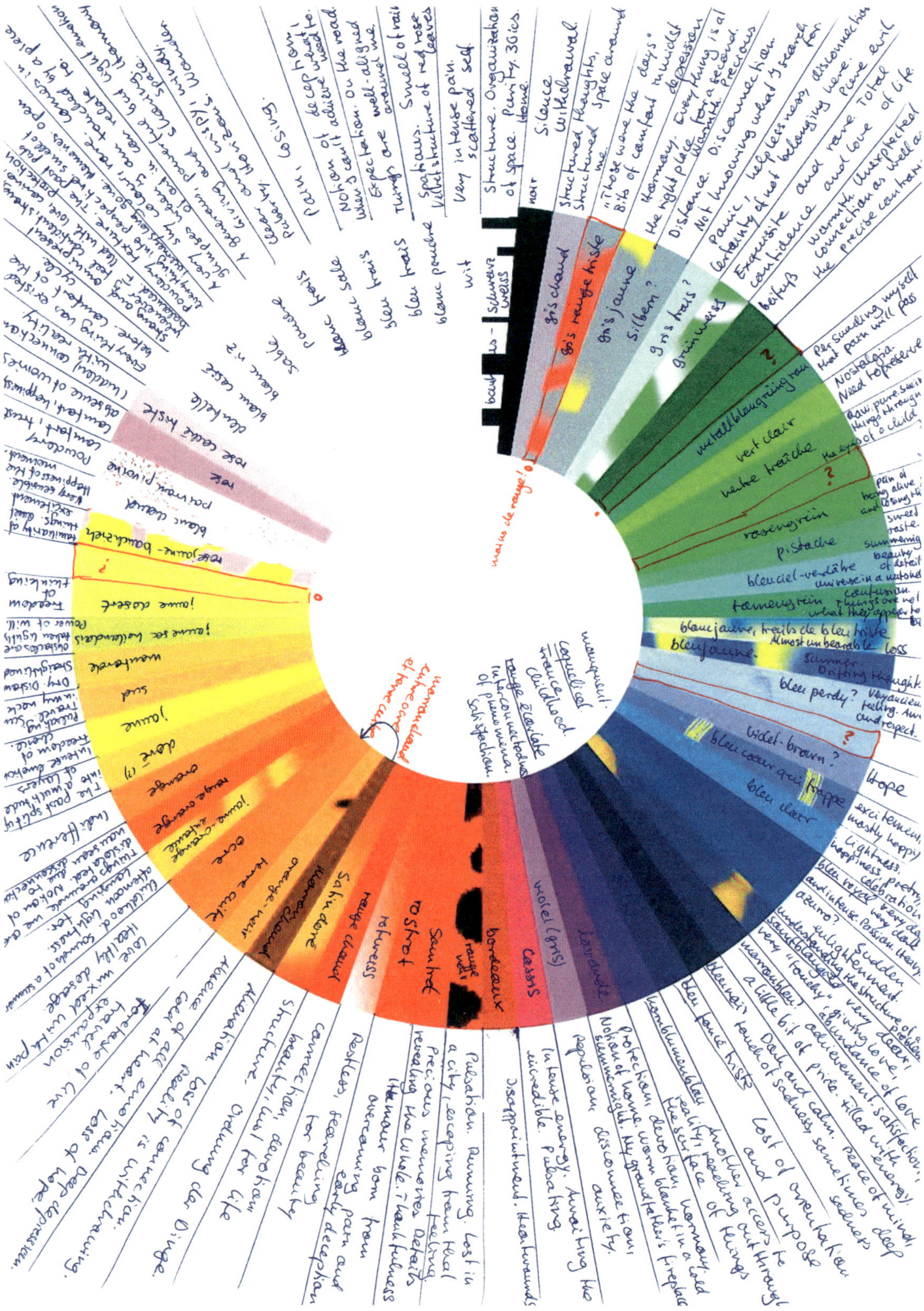
noir
Hope
bordeaux
Cassis
lavande
pistache
vert clair
orange
jaune
ocre
Sable

I started this project in order to understand what distinguished my perception and thus my reality from that of others. *Layers of Reality* began as an exploration of my condition in order to figure out how it works. Diving deep into a matter is the basis for my practice as a photographer and artist.

My research became an intimate journey guided by the wish to understand a complex phenomenon that draws that subtle line between my perception and those of others, and the desire to communicate about it.

I started with organising the colours of my emotions into a colour range. What do the colours mean, where do they come from, when do they appear?

What would their description be in terms of feelings? I am often asked which colour a specific feeling has—"which colour is love?", but that is not how I perceive emotions. They are first and foremost subtle, three-dimensional tints that I feel inside of me. I can describe verbally how I feel, what the colour means individually, but the colours are not clearly classified in such categories as hate, pain, or happiness.

Love, for instance, can have a magnificent variety of shades. It can be tinted by joy, excitement, satisfaction, a feeling of safety and trust when met—these would be blending into a warm and soft bath of ivory, white and sandy shades—but it can also be hard, dark blue, cold as concrete when deceived. To me, the colour of the feeling is a much more precise indicator of how I feel, with words often failing when I want to include all aspects of an emotion in a verbal description.

Intrigued by this variety of sensations and the seemingly unsurmountable task to break them down into an accessible record, I decided to investigate all questions that came up when I realised that my reality was different from that of most people around me. In order to make a representative investigation, I needed a database to base my research on. I collected 1500 images from my private archive, covering my photo-making lifespan and a wide range of subjects. That is why there are many snapshots and private images without much artistic value, but crucial for the research, since I wanted to examine my emotional response to images and therefore needed a large database including 'all of life'.

In my private photographs, I shoot to hold on to a feeling, a memory—that is why my own photographs often evoke the strongest colours: they touch a string in me. I see fragments of my life in them. A selection of 216 images from this archive were picked and randomised by the designers. These became the images I used for the research. The different steps I undertook are explained at the beginning of every chapter, each covering one particular question I asked myself about synaesthesia. I decided to work together with scientists who could give a context to my findings and place them within what is known today about this neurological phenomenon.

This is how I want to give a first impression of what synaesthesia can be like, make it accessible through a personal approach, and show how fascinating this phenomenon is.

Chapter 1
Familiarity

To investigate which photographs trigger a synaesthetic colour response, I compared a selection of 216 images from my own photographic archive to similar 'anonymous' pictures picked from Google as a neutral reference. All images were mixed, randomised and renamed. I noted which colours I felt when looking at each image individually. My own photographs are on the left; their anonymous counterpart is on the right side of each page.

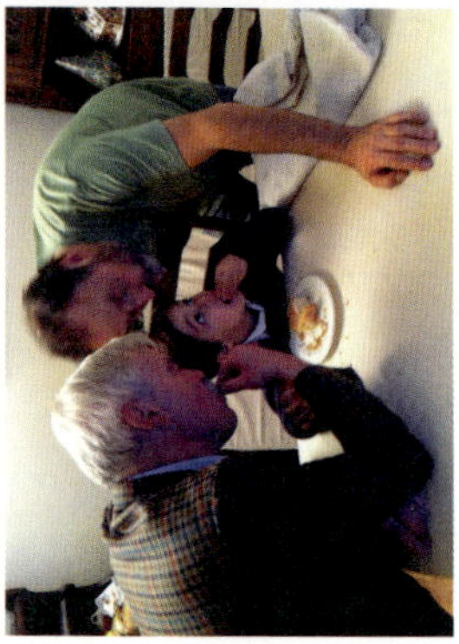
0004P

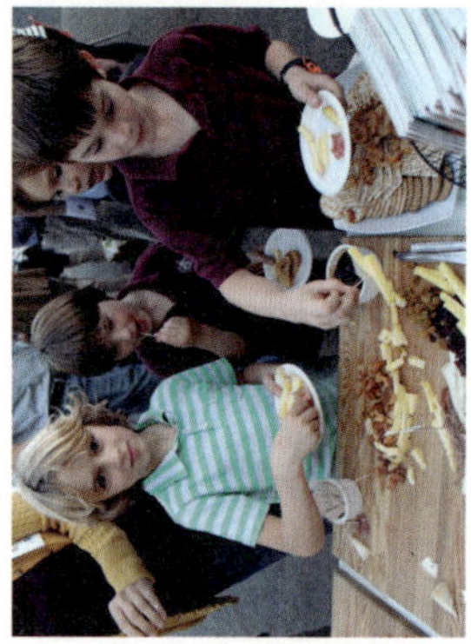
0004G

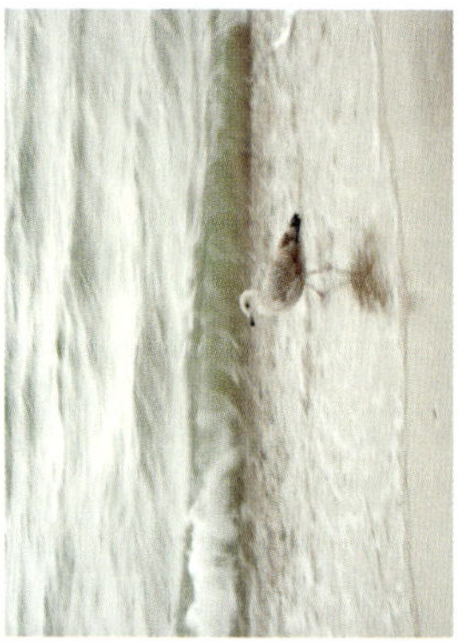
0014P

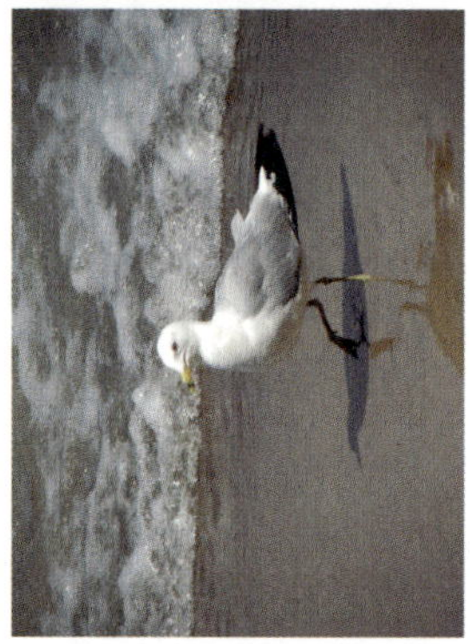
0014G

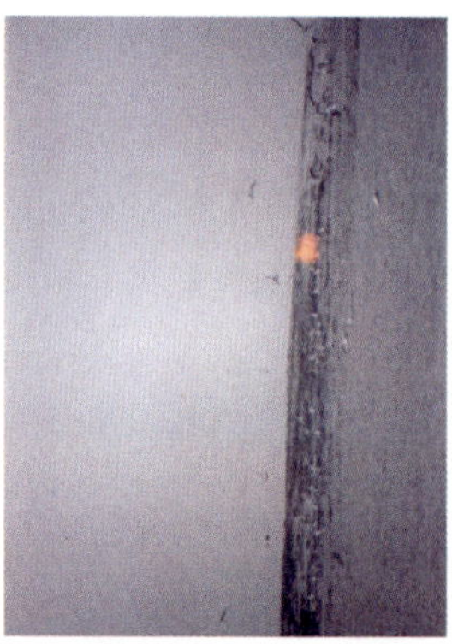
0015P

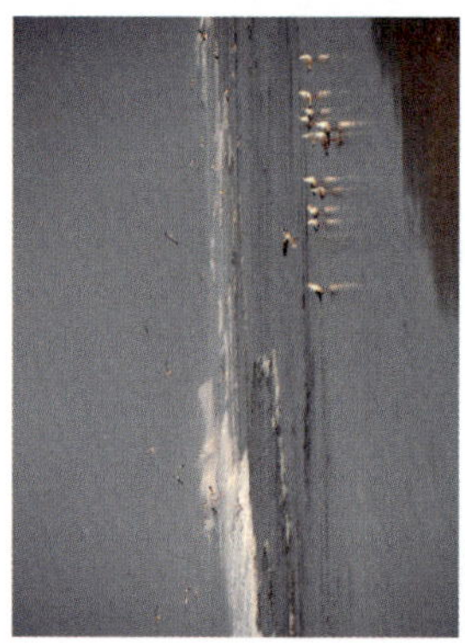
0015G

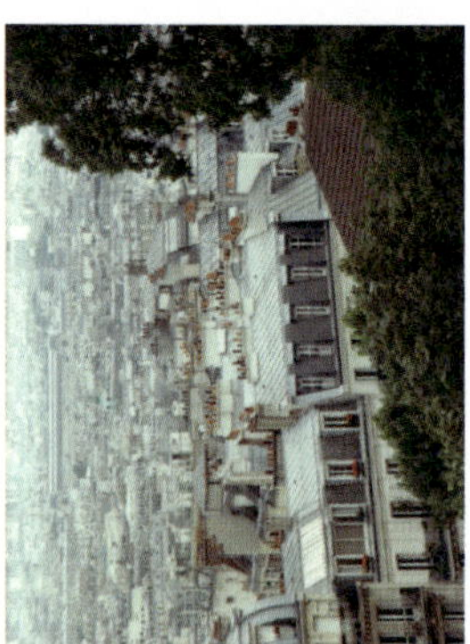
0019P

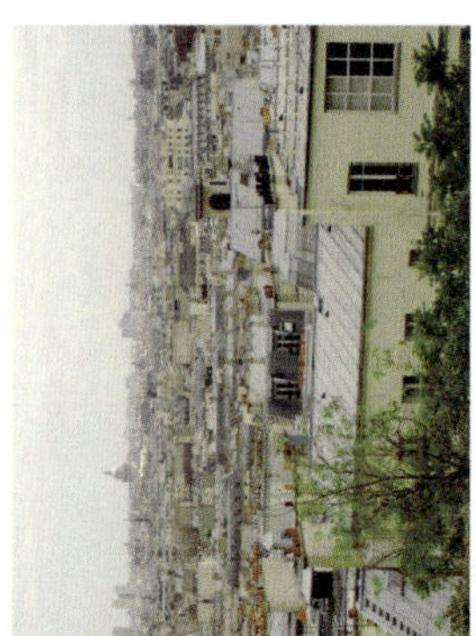
0019G

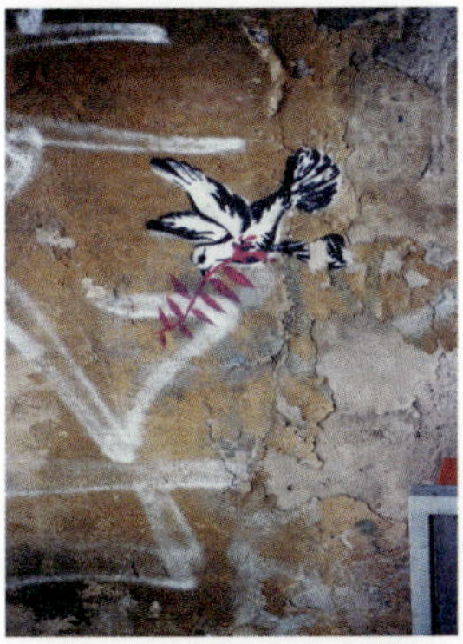
0021P

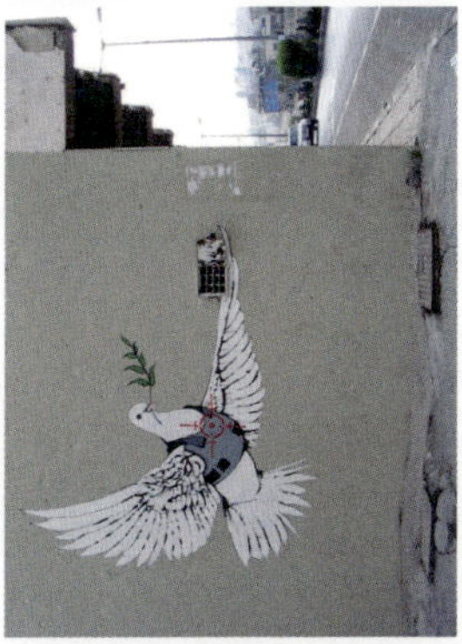
0021G

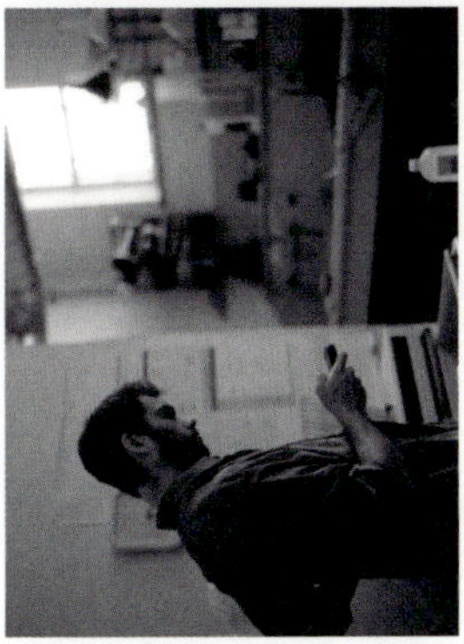
0025P

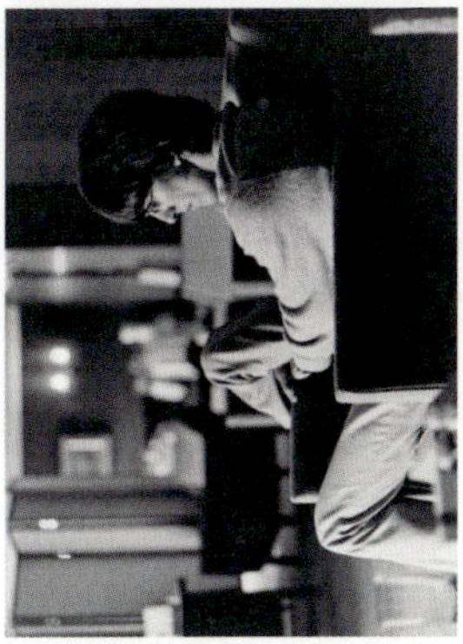
0025G

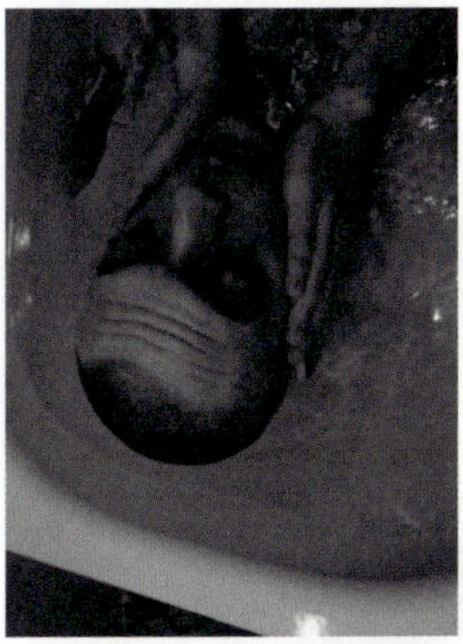
0036P

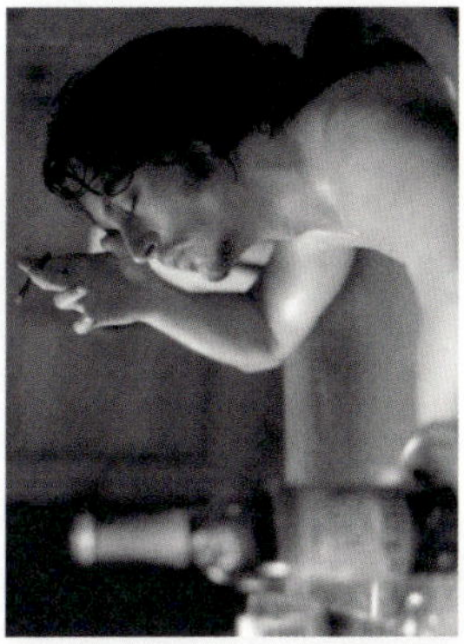
0036G

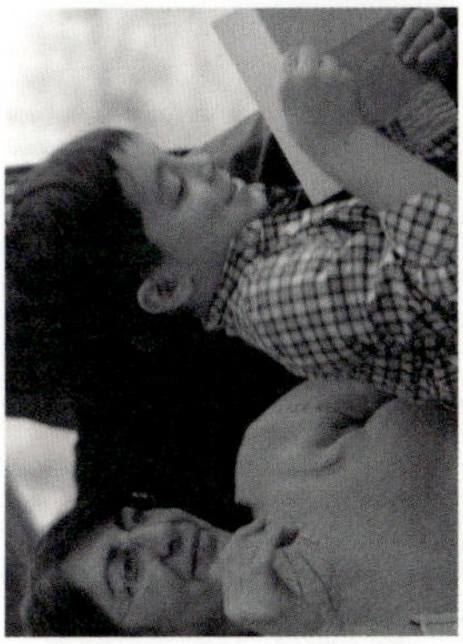
0039P

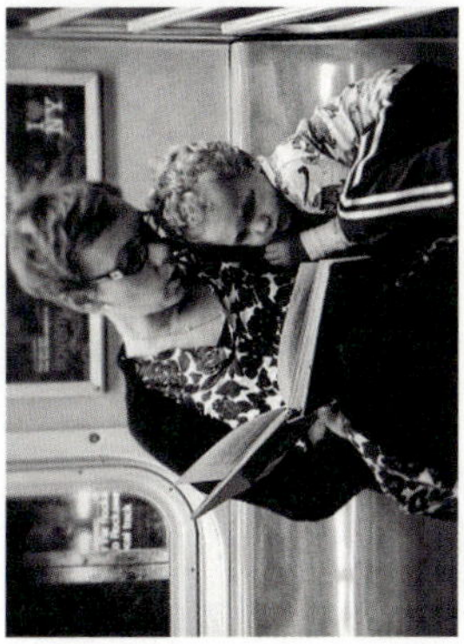
0039G

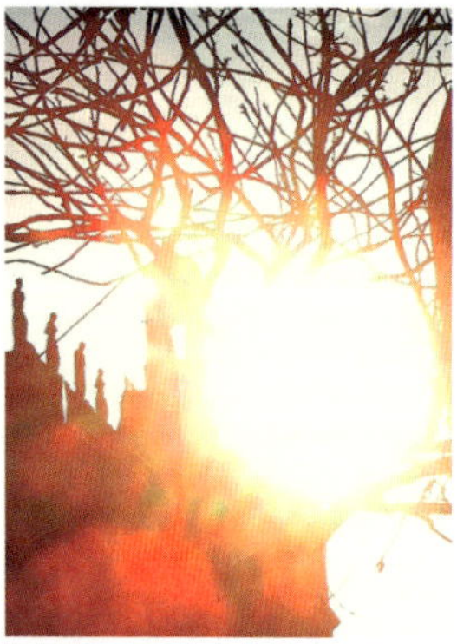
0026P

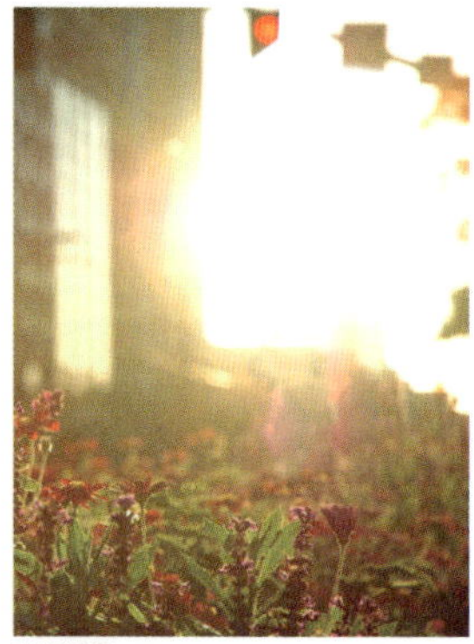
0026G

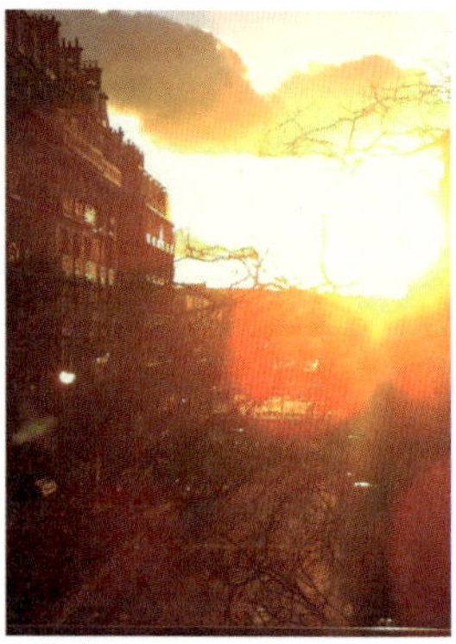
0028P

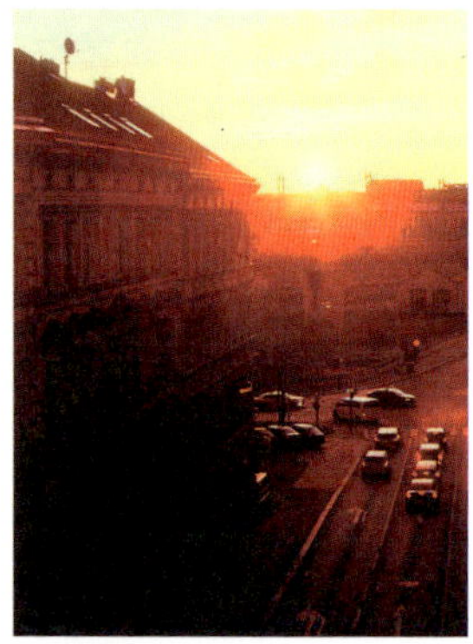
0028G

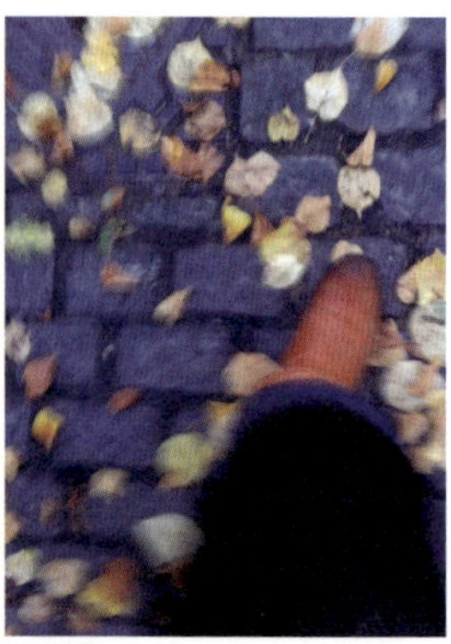
0111P

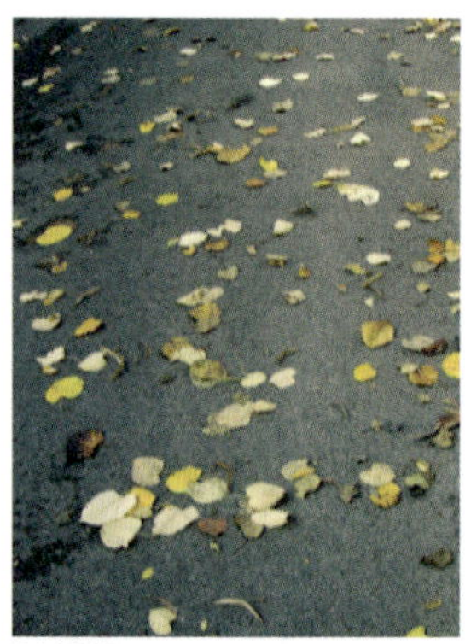
0111G

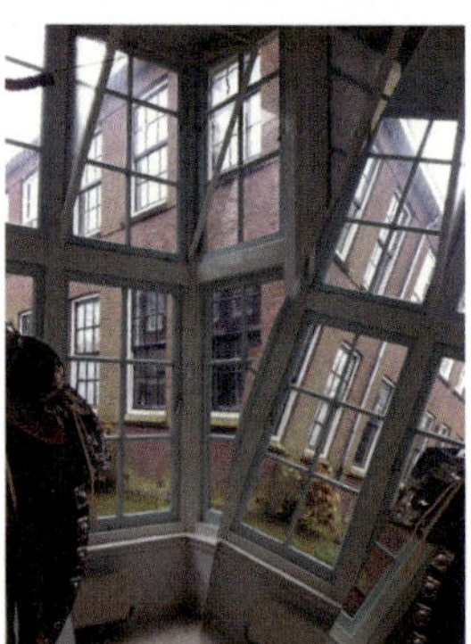
0156P

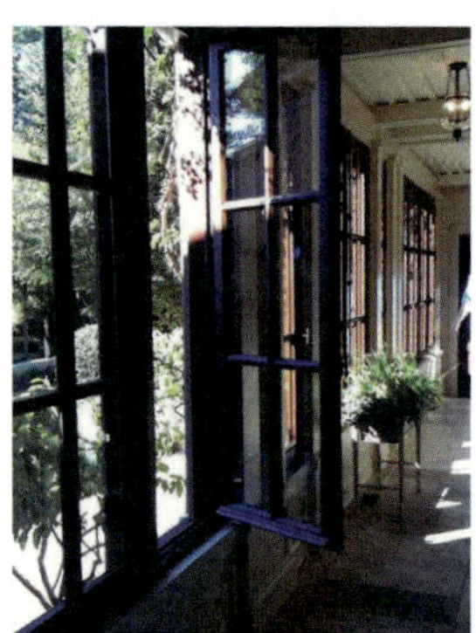
0156G

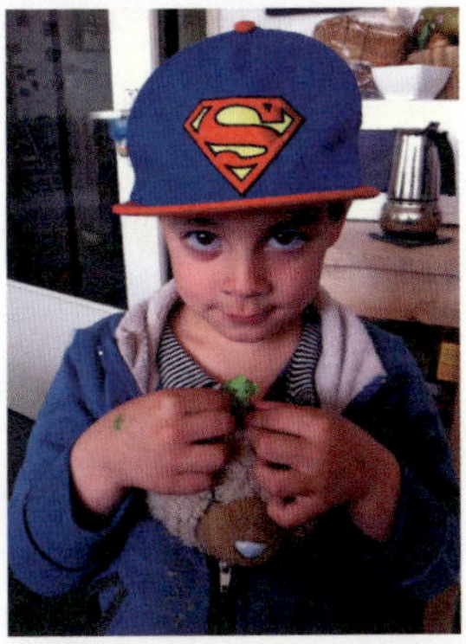
0199P

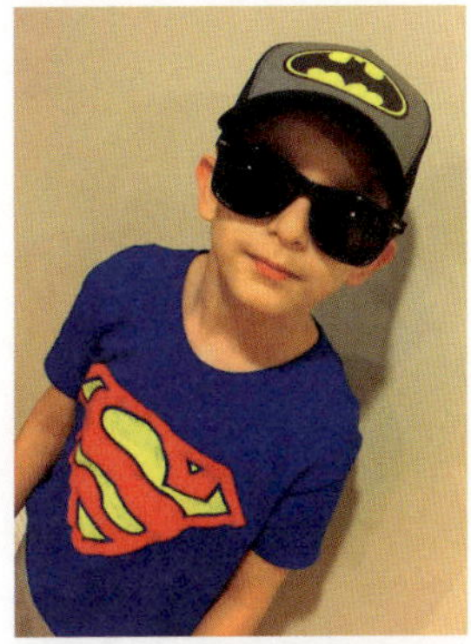
0199G

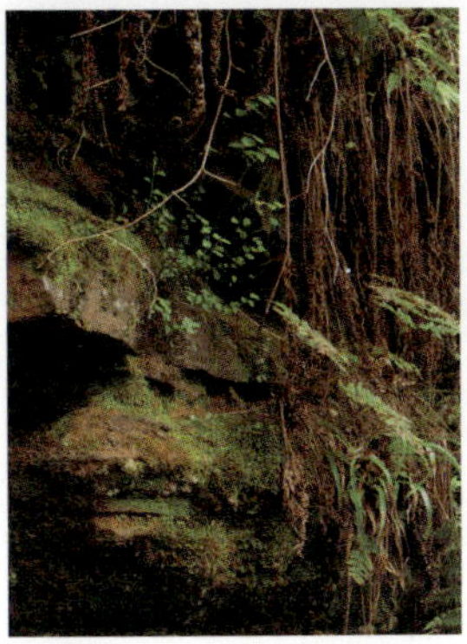
0232P

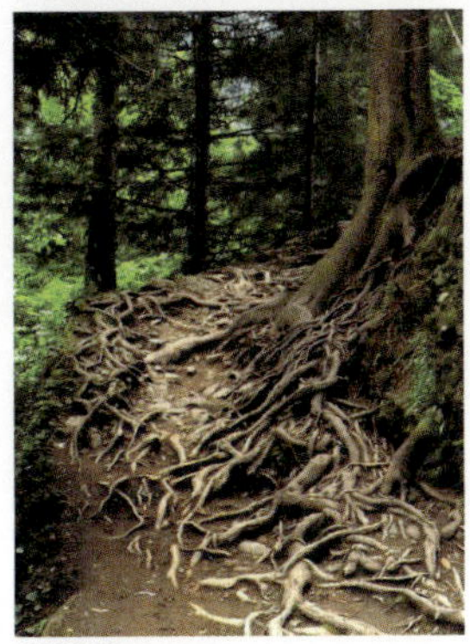
0232G

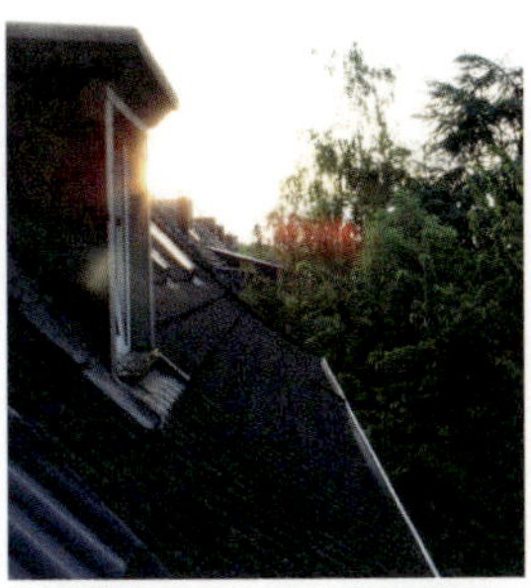
0240P

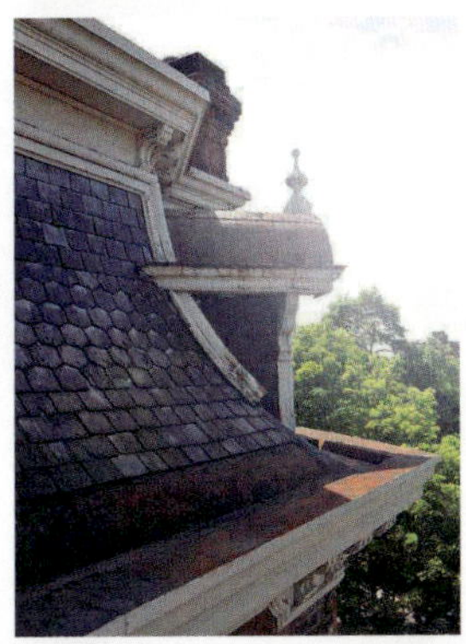
0240G

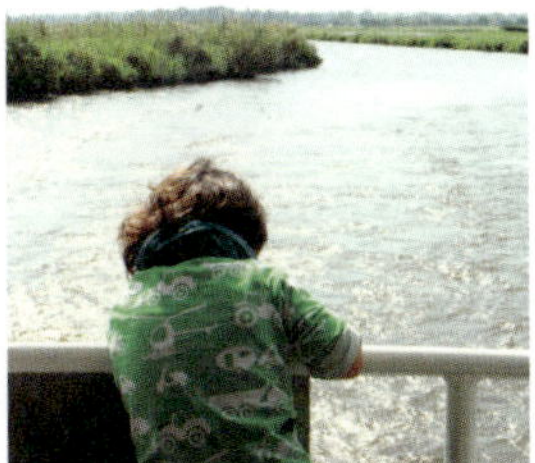
0278P

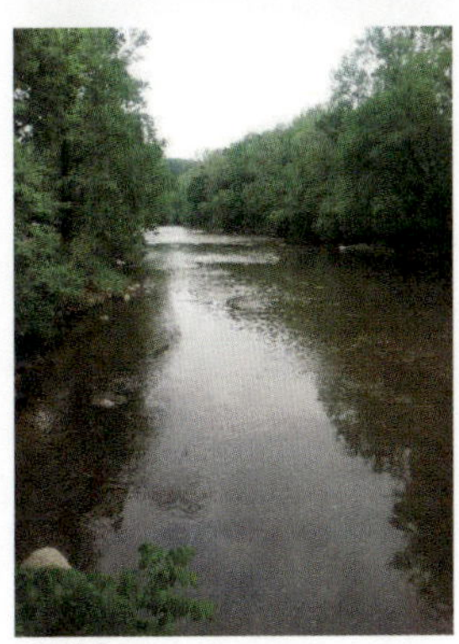
0278G

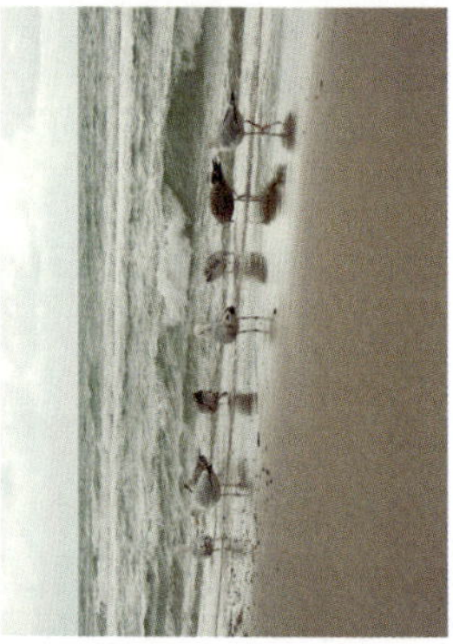
0048P

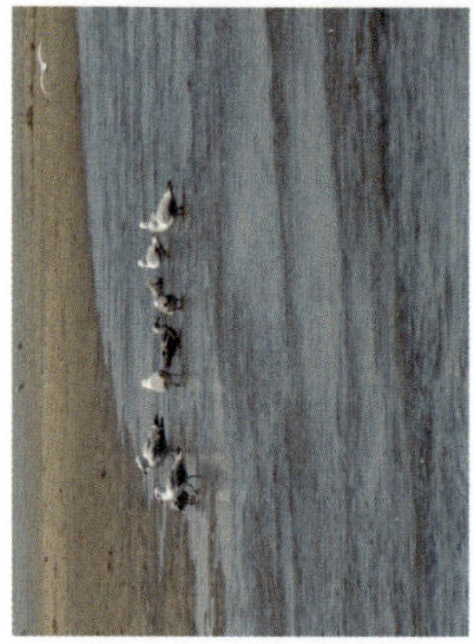
0048G

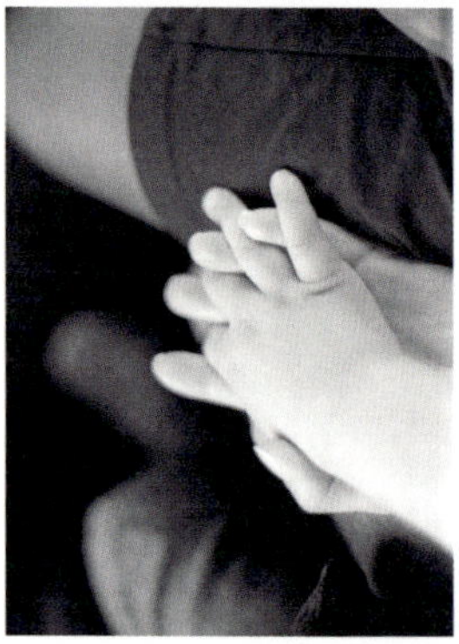
0058P

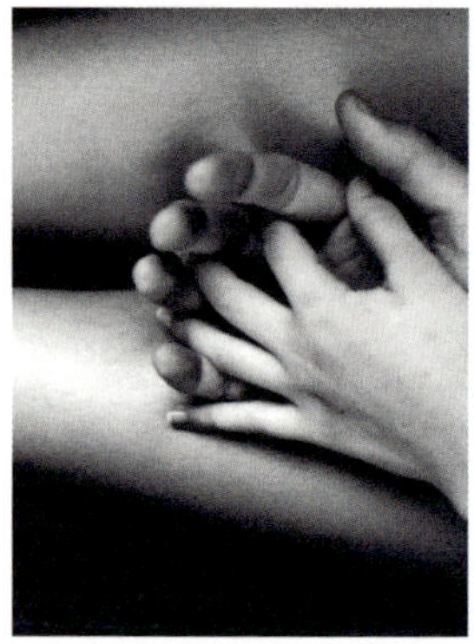
0058G

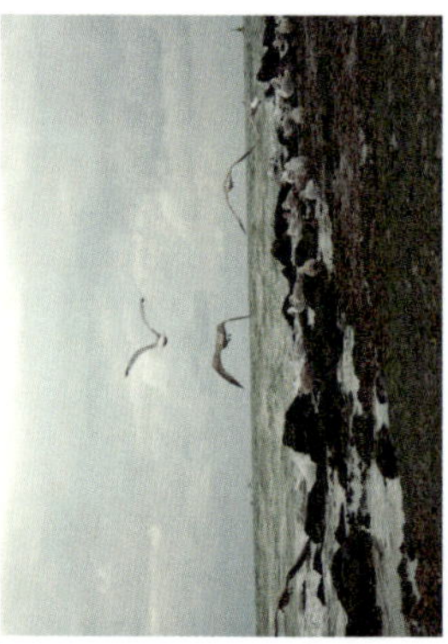
0068P

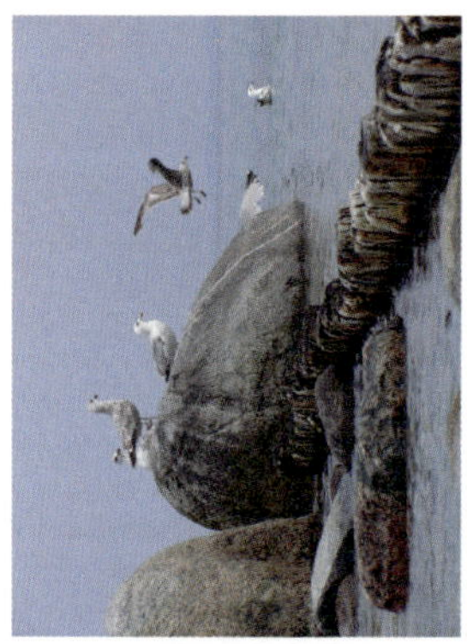
0068G

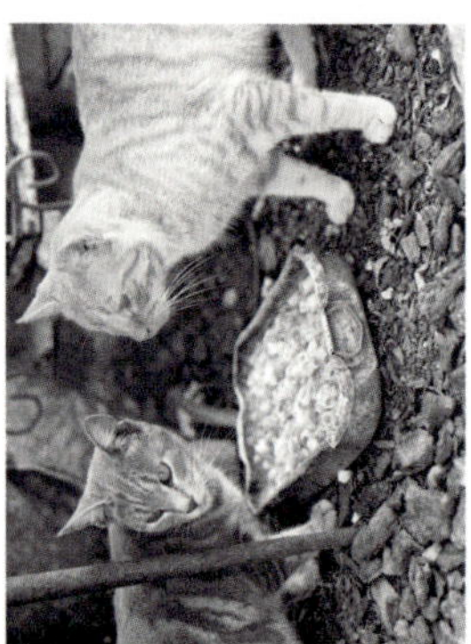
0114P

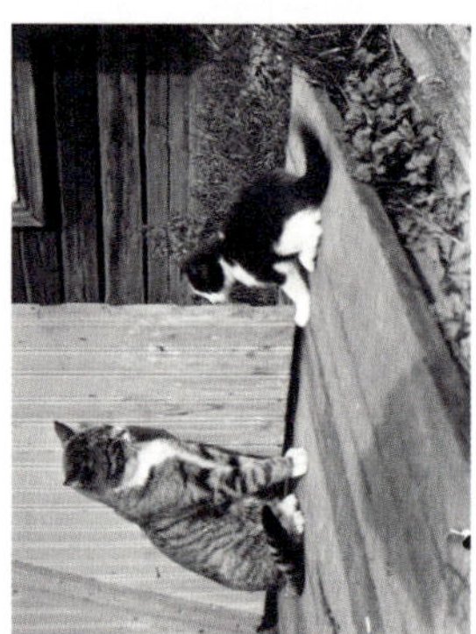
0114G

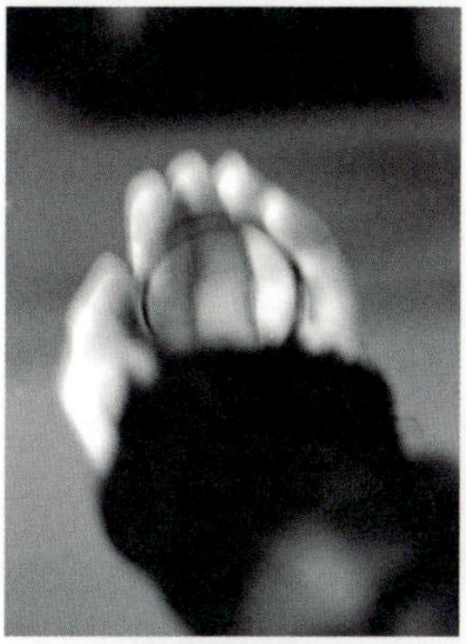
0116P

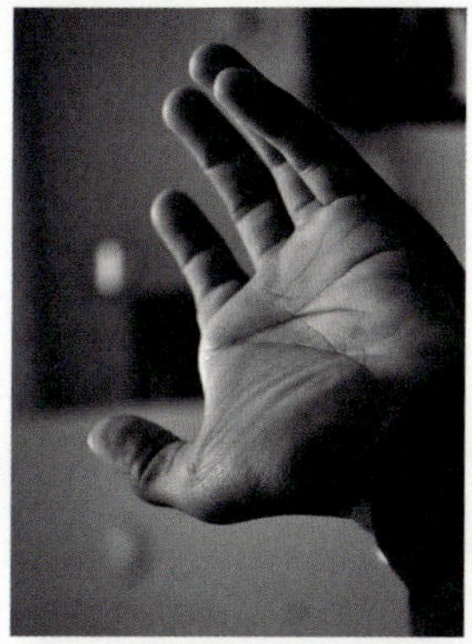
0116G

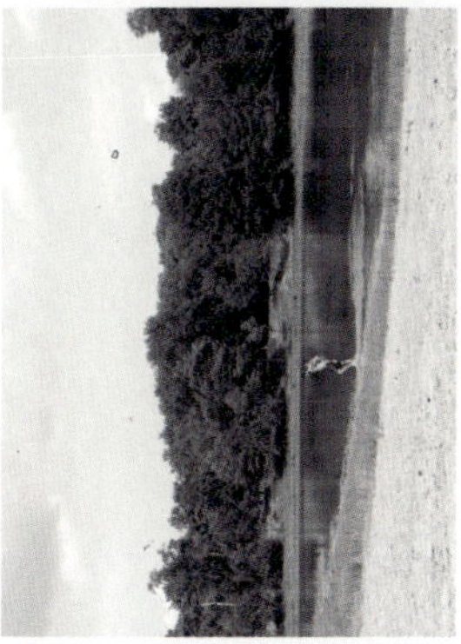
0117P

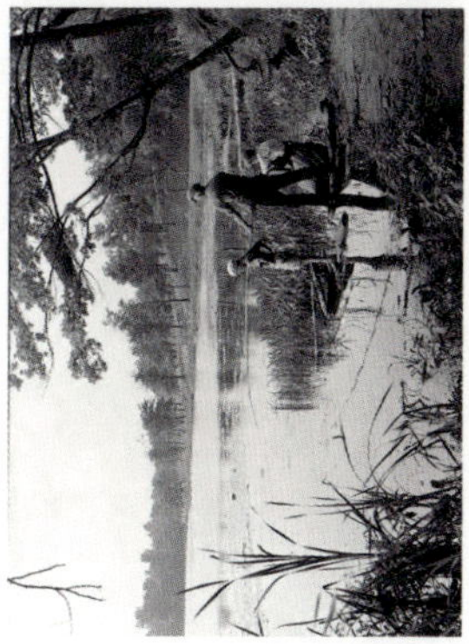
0117G

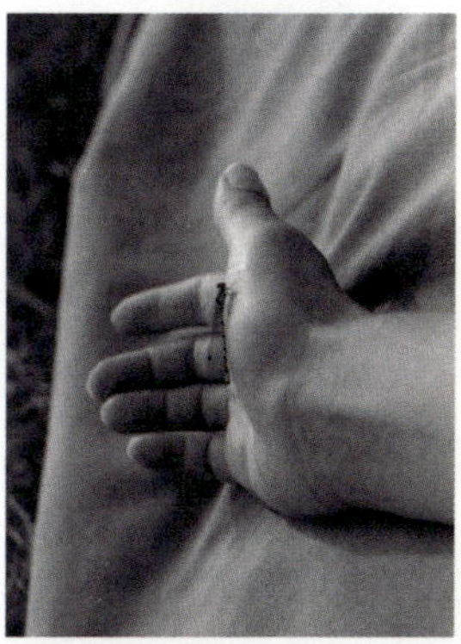
0179P

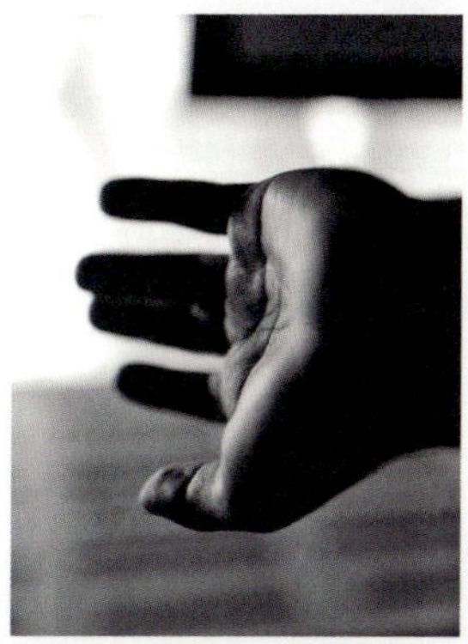
0179G

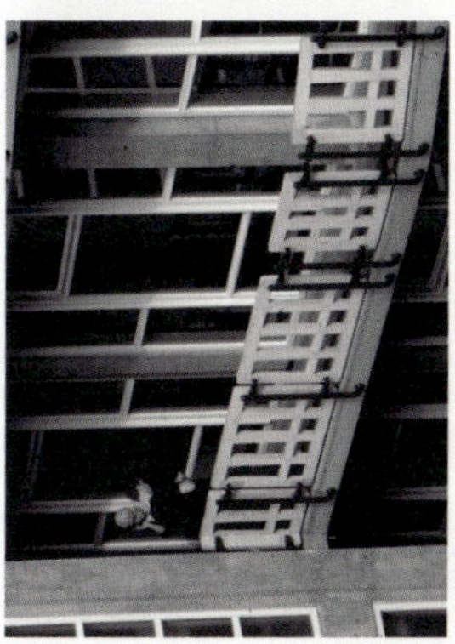
0180P

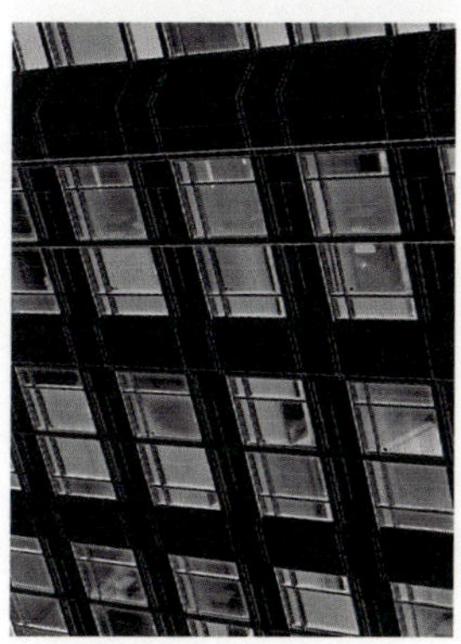
0180G

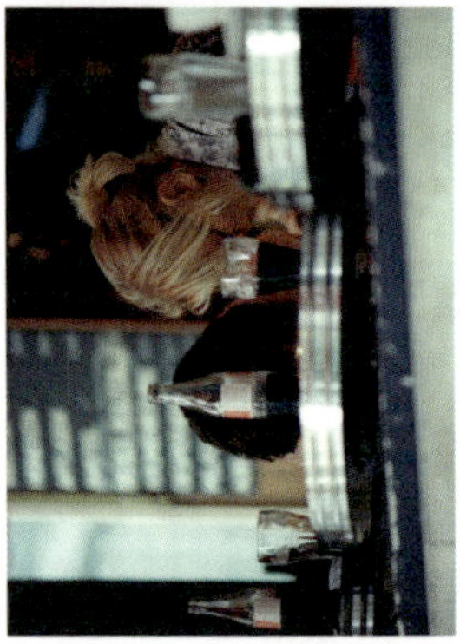

0220P

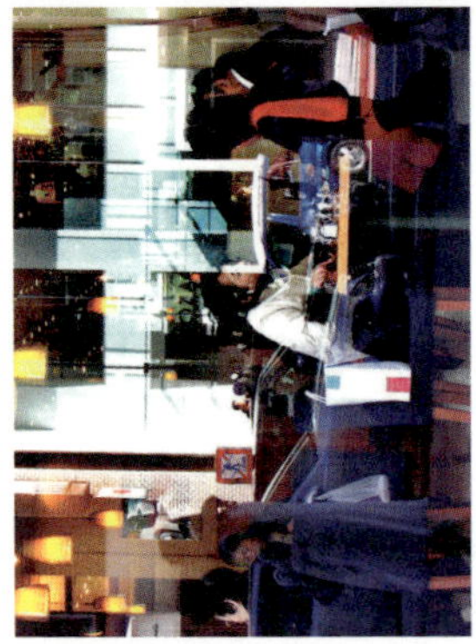

0220G

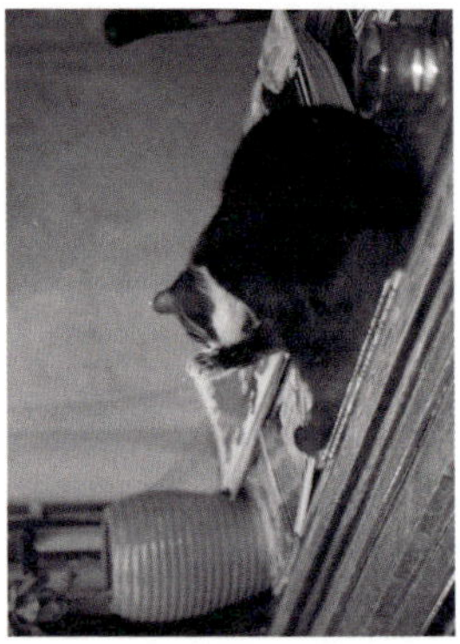

0241P

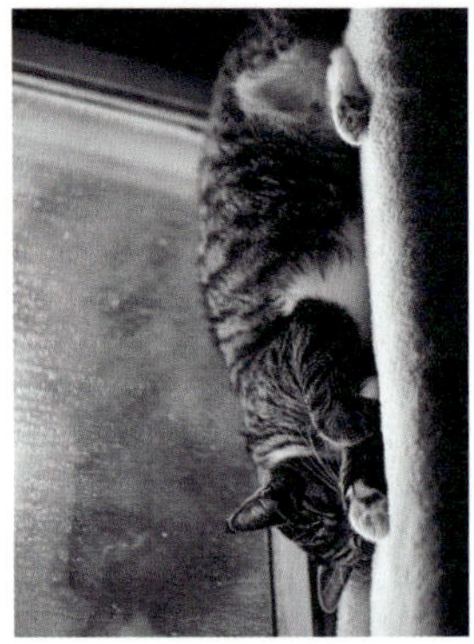

0241G

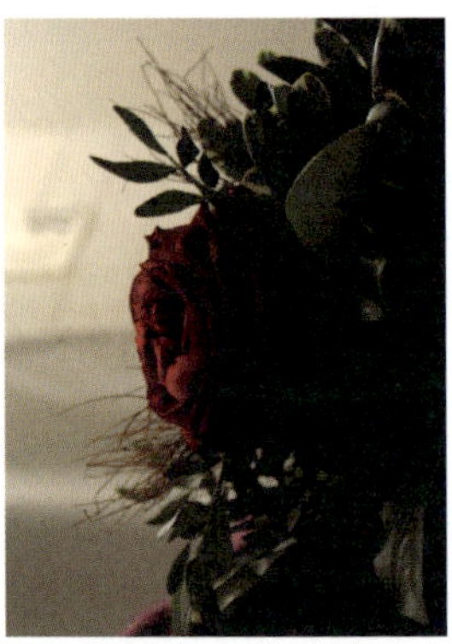

0248P

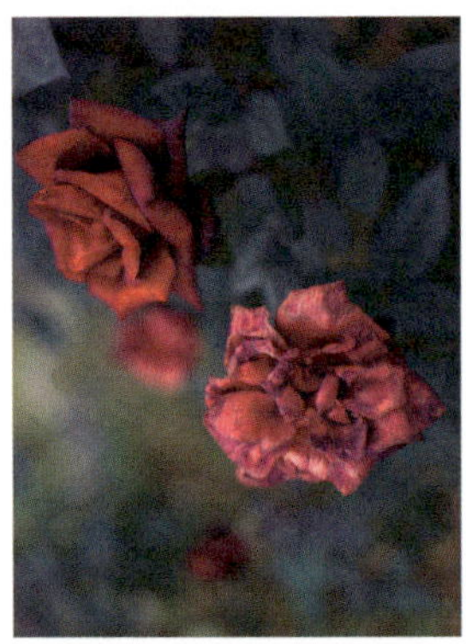

0248G

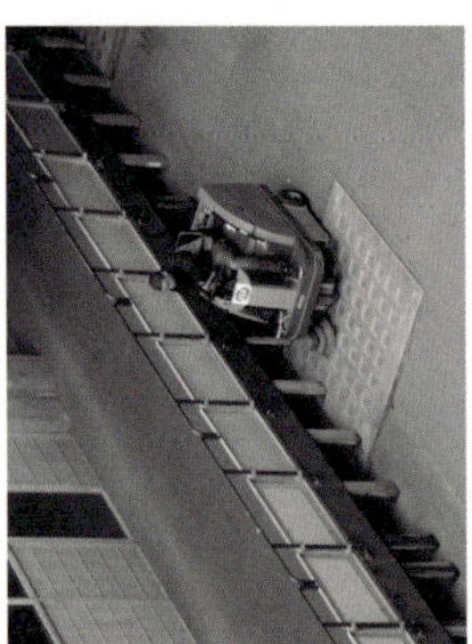

0260P

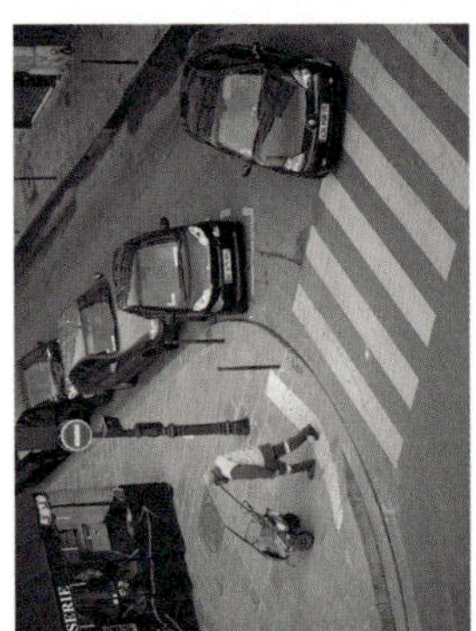

0260G

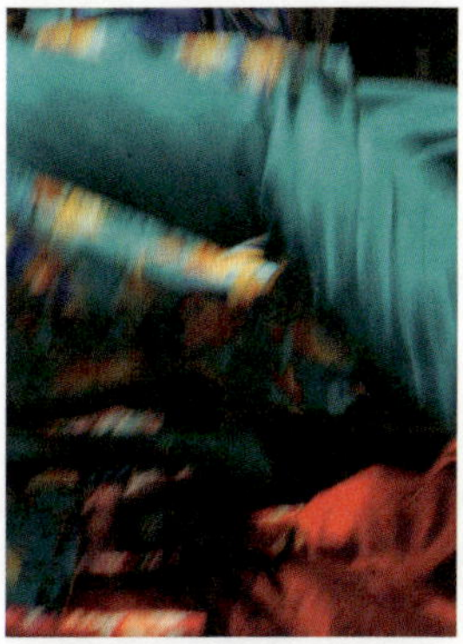

0290P

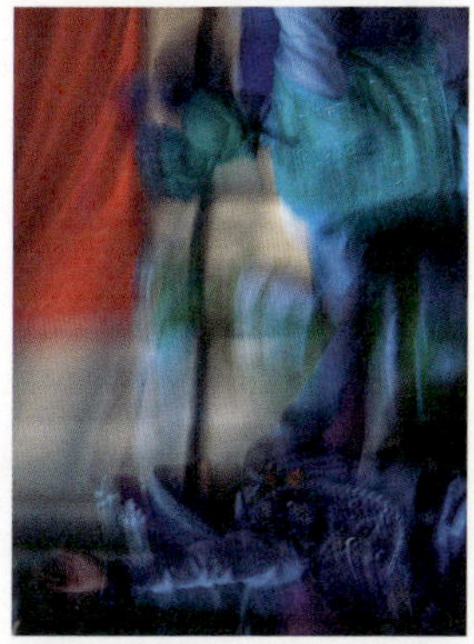

0290G

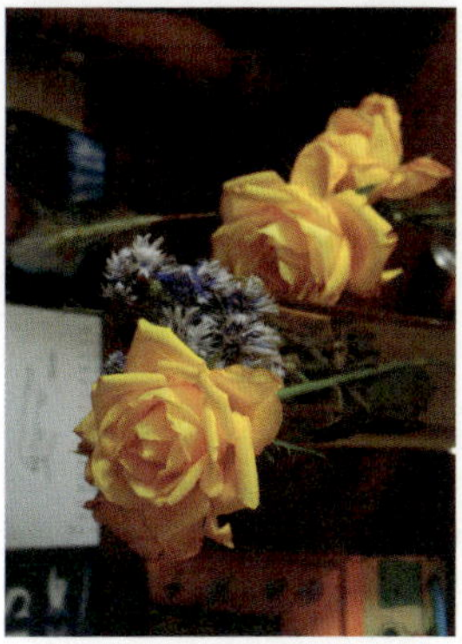

0305P

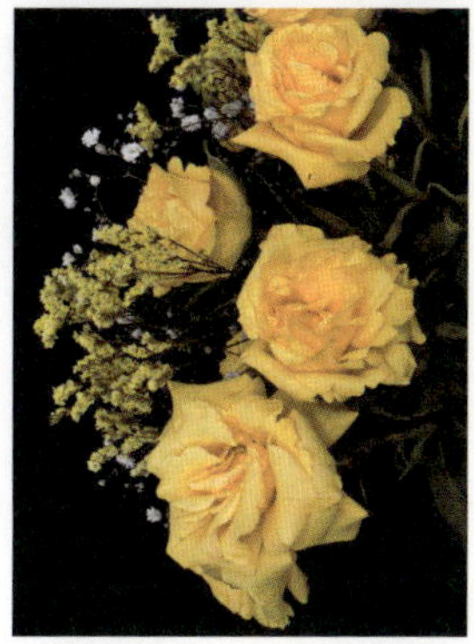

0305G

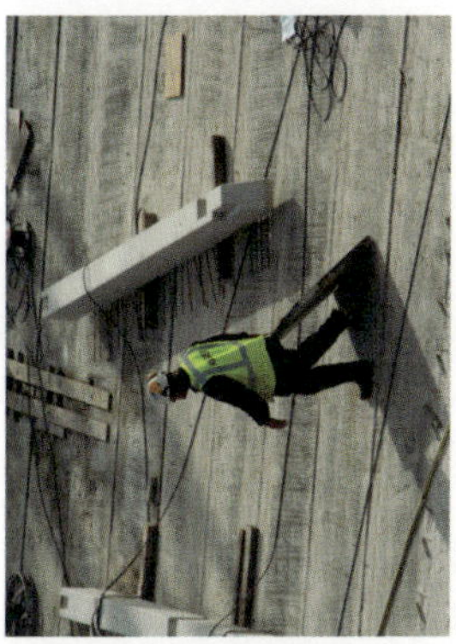

0311P

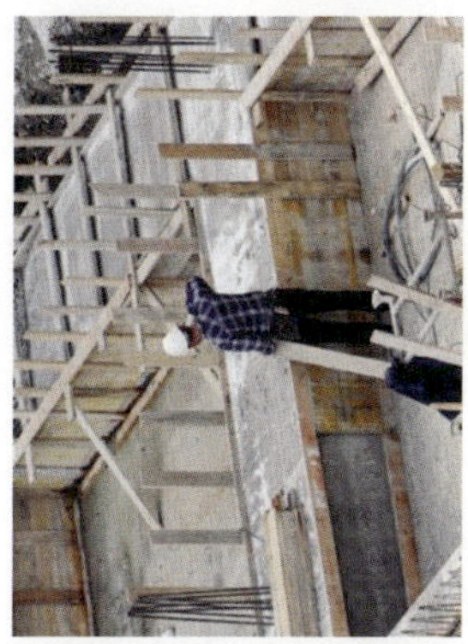

0311G

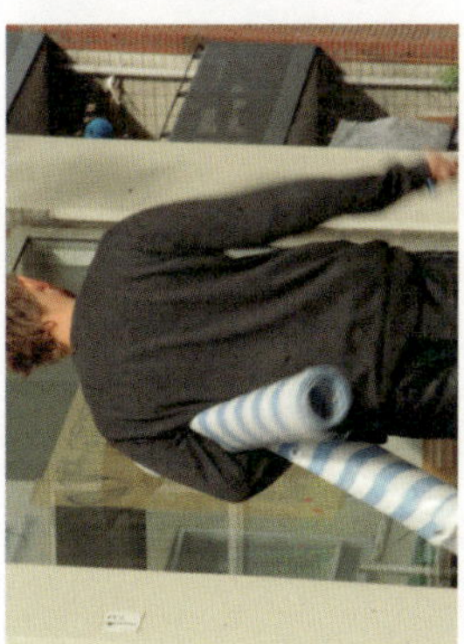

0316P

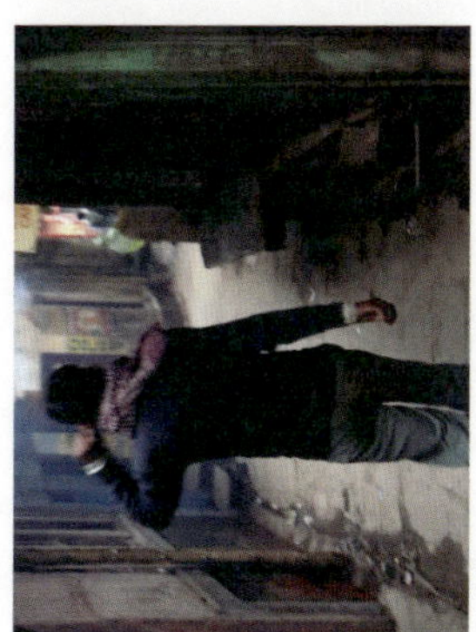

0316G

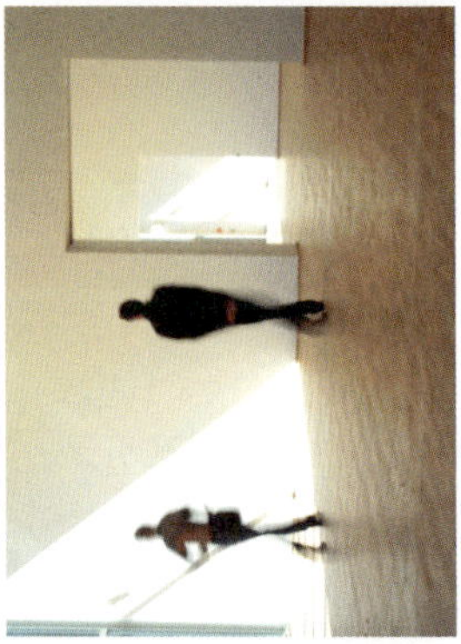
0334P

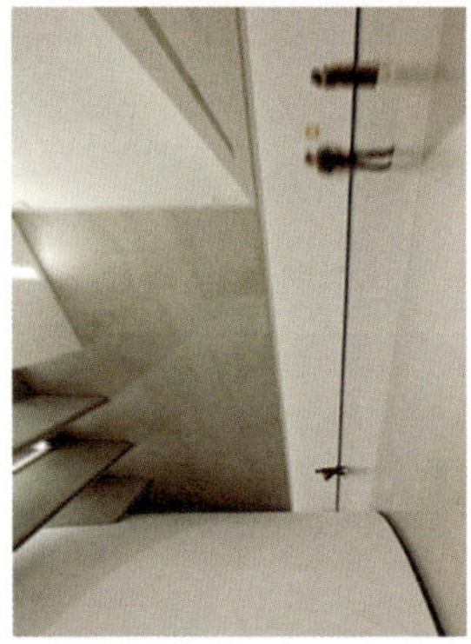
0334G

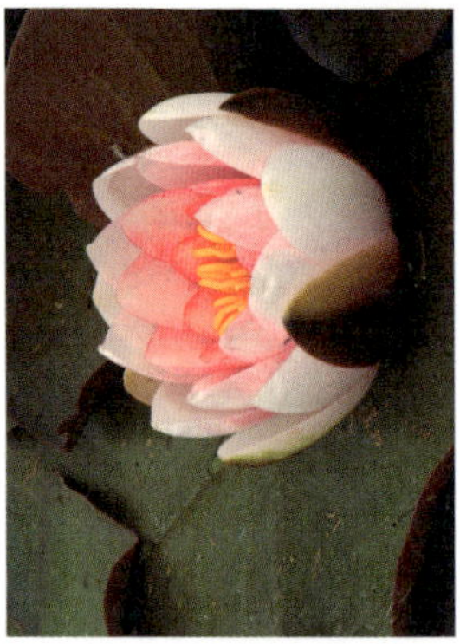
0335P

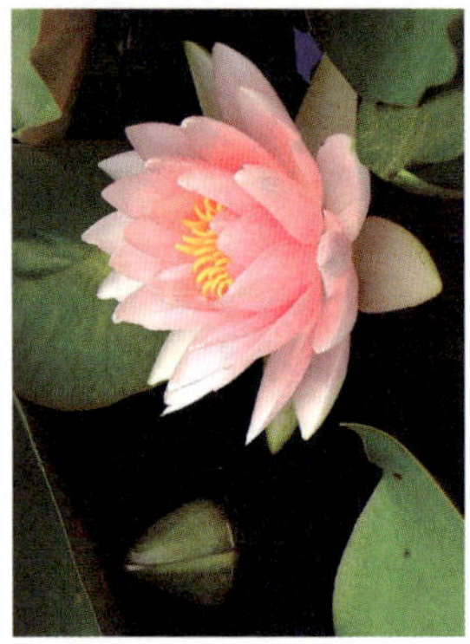
0335G

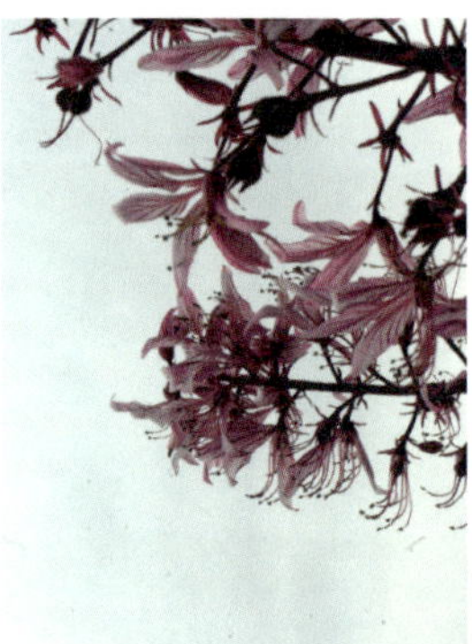
0337P

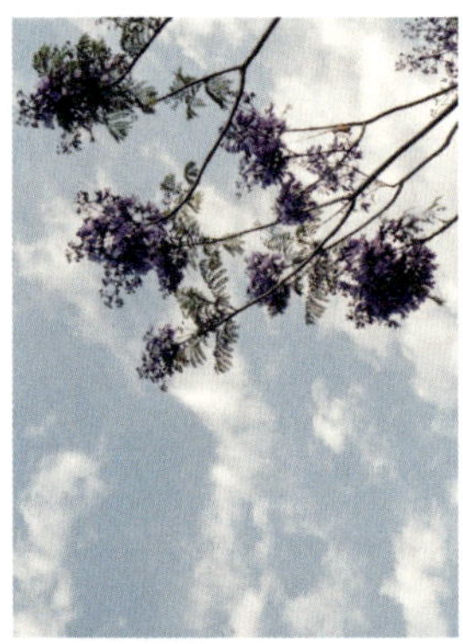
0337G

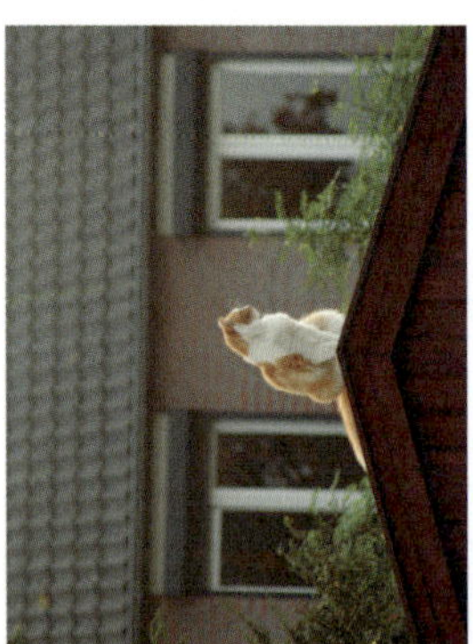
0342P

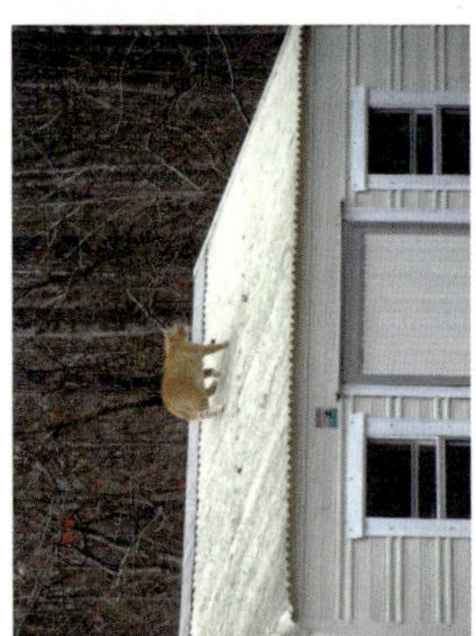
0342G

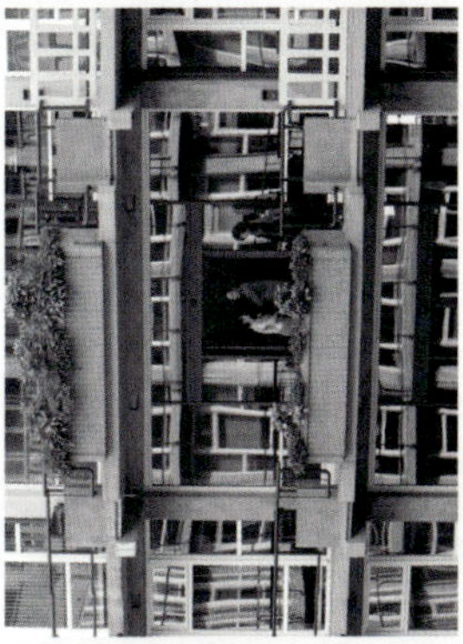

0364P

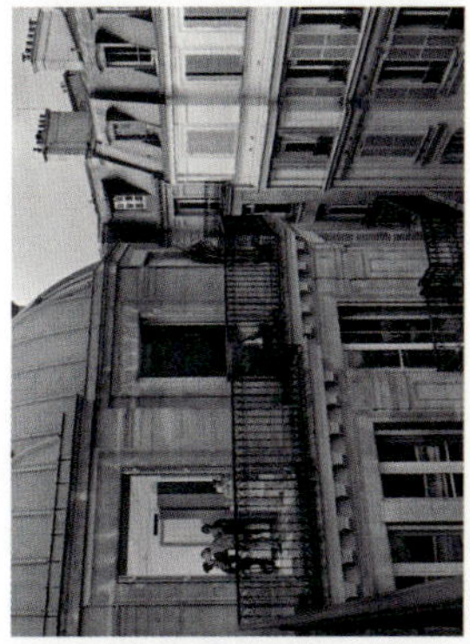

0364G

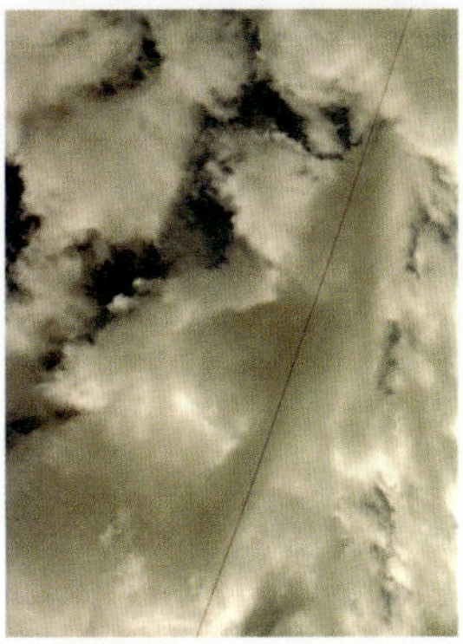

0365P

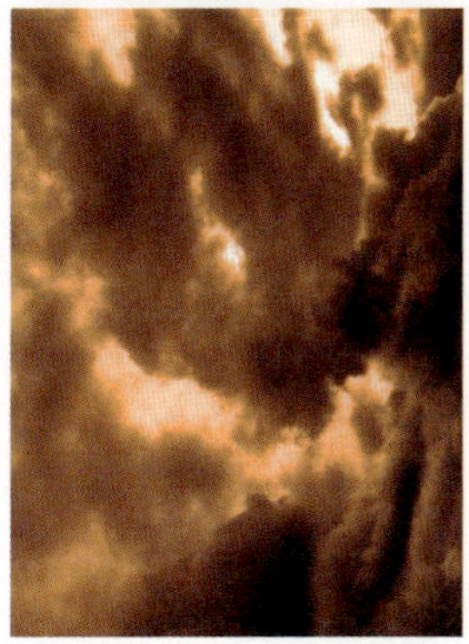

0365G

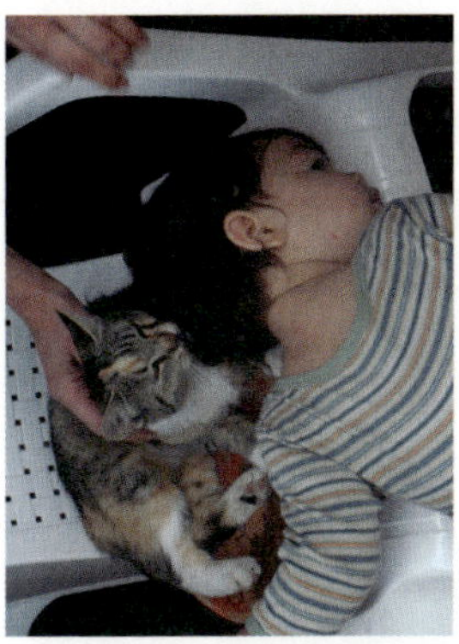

0368P

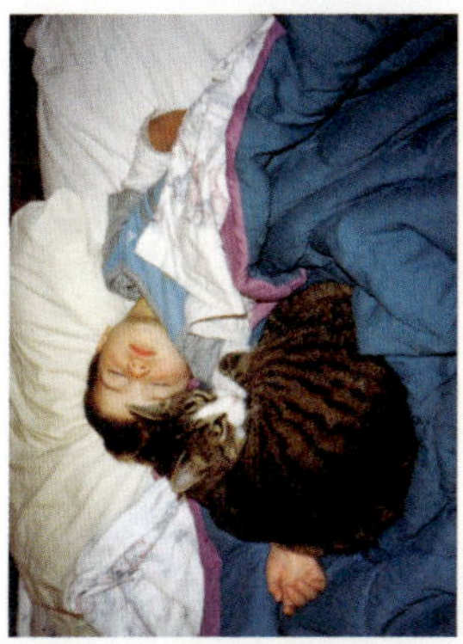

0368G

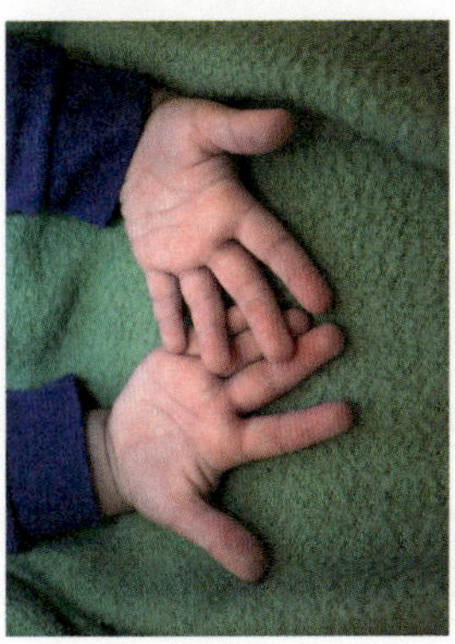

0380P

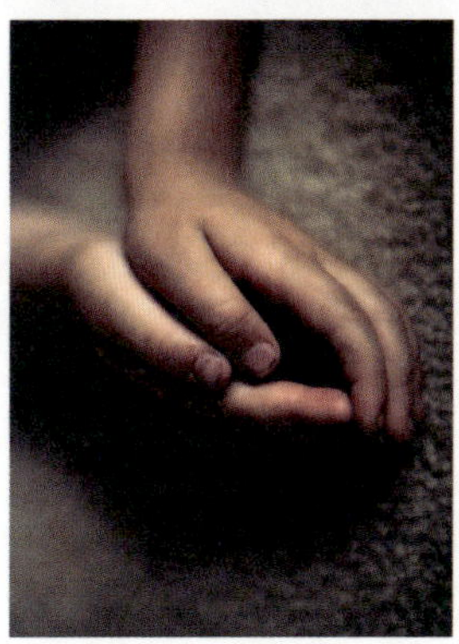

0380G

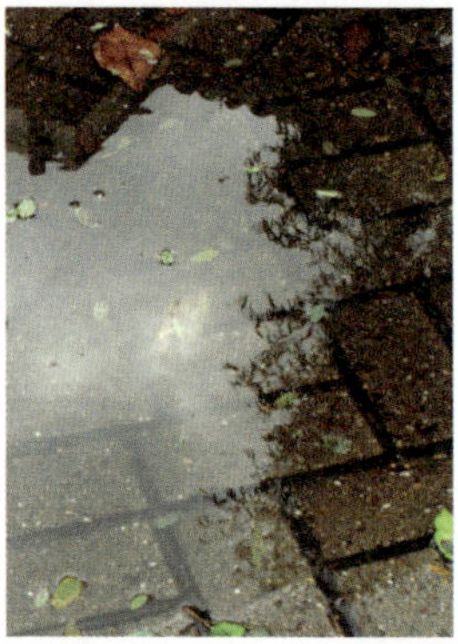
0339P

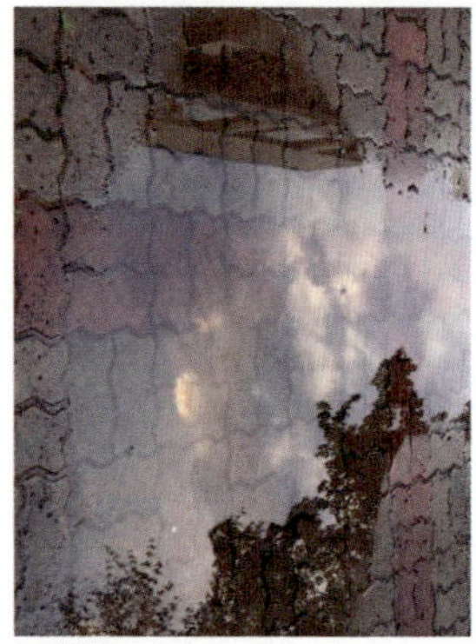
0339G

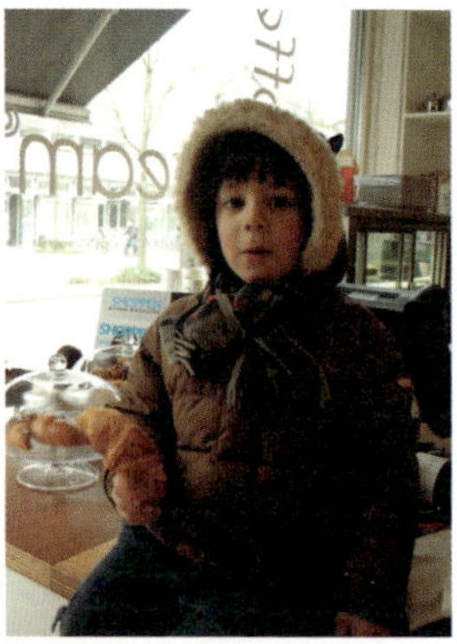
0353P

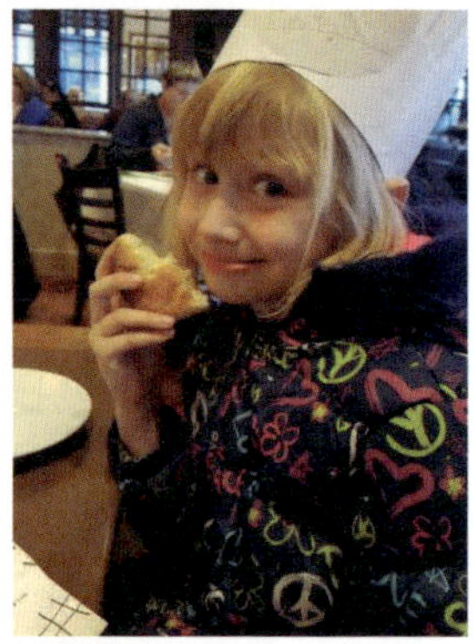
0353G

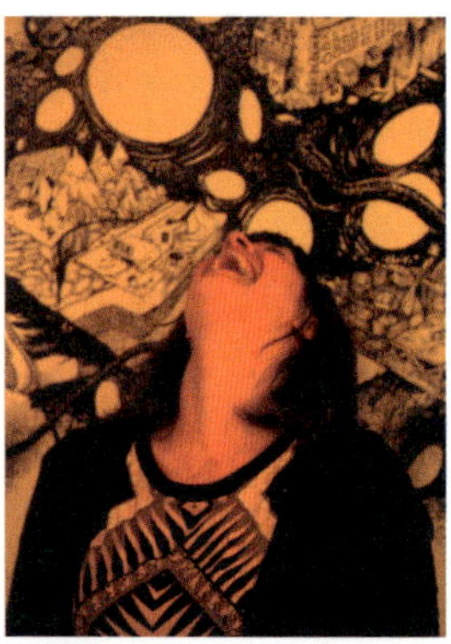
0356P

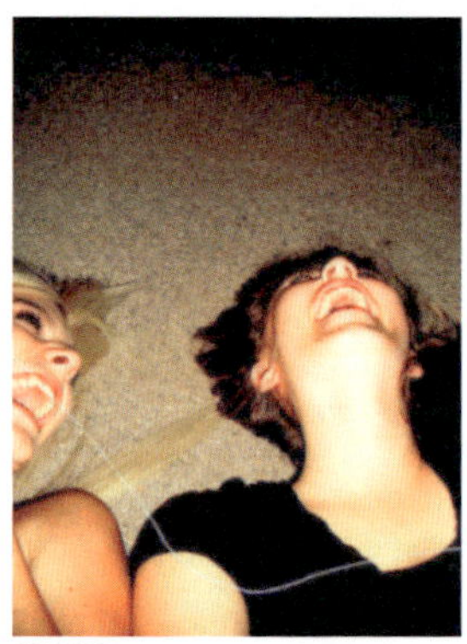
0356G

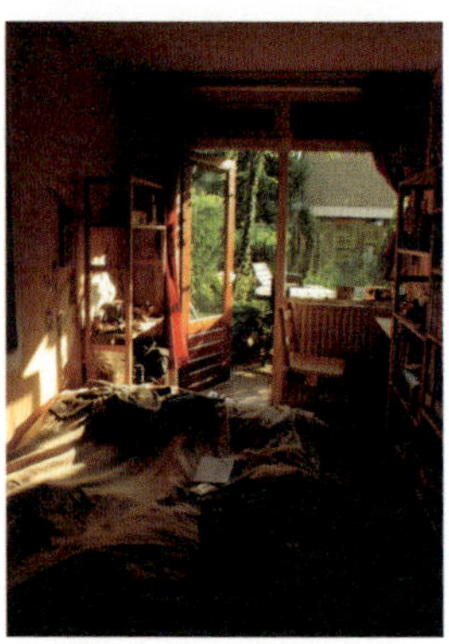
0366P

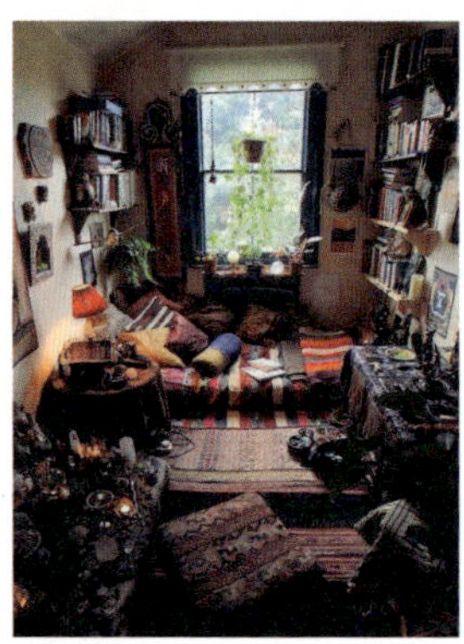
0366G

0398P

0398G

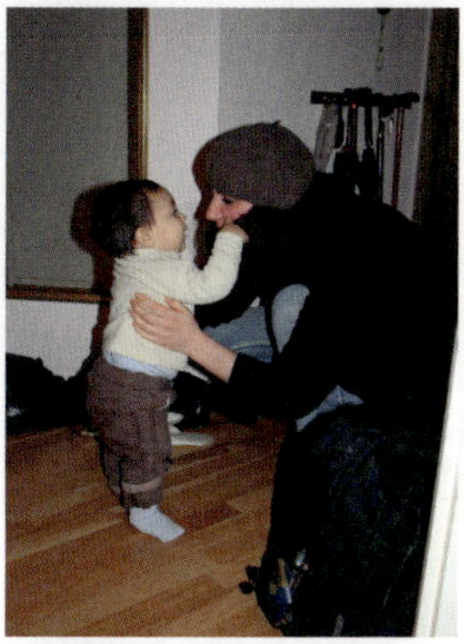

0403P

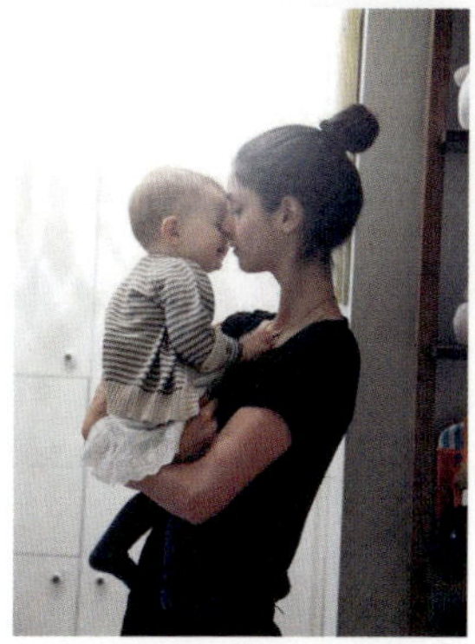

0403G

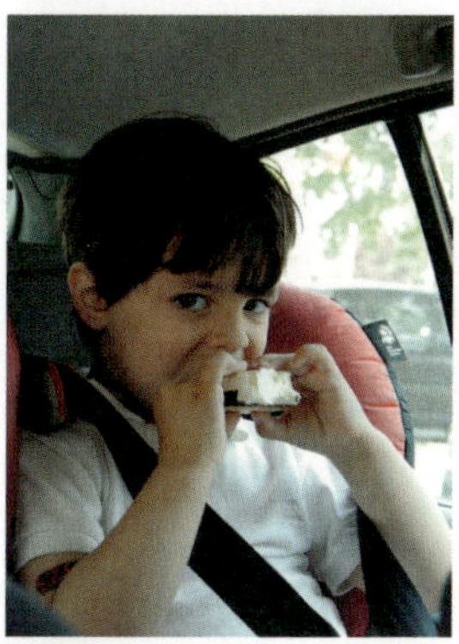

0439P

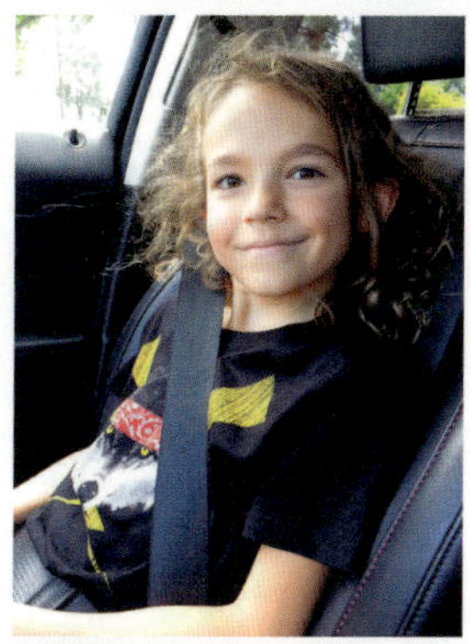

0439G

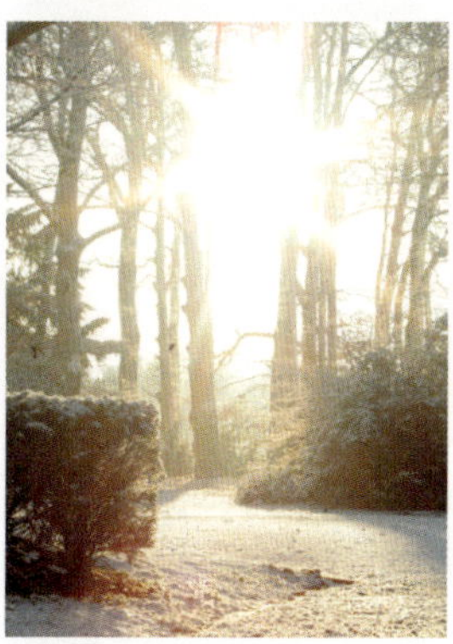

0472P

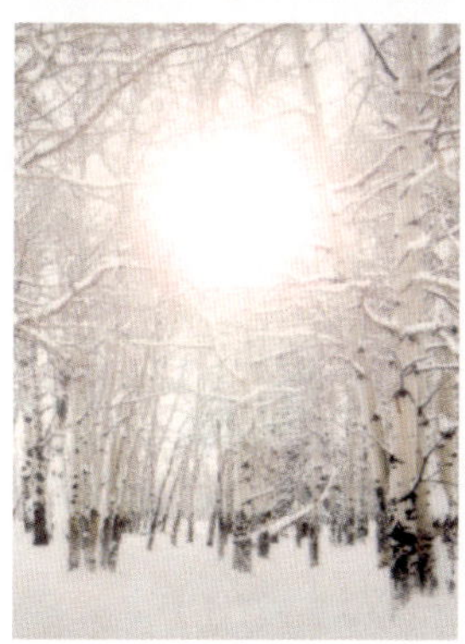

0472G

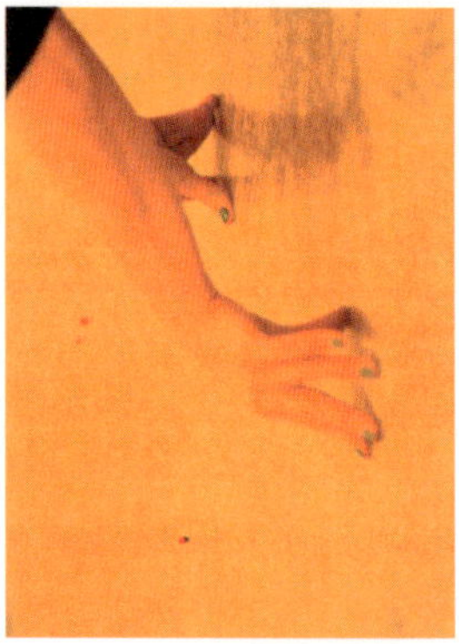
0399P

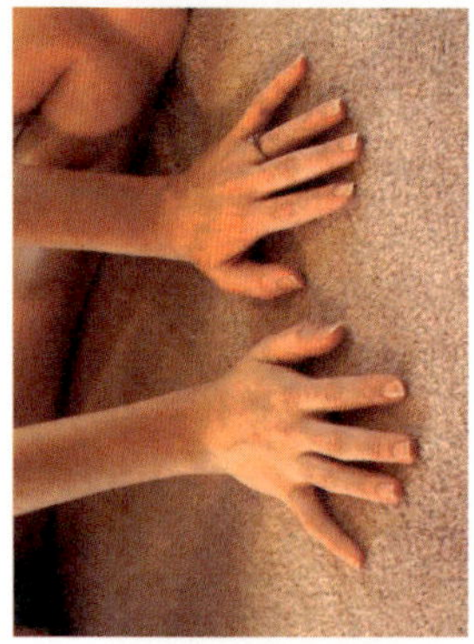
0399G

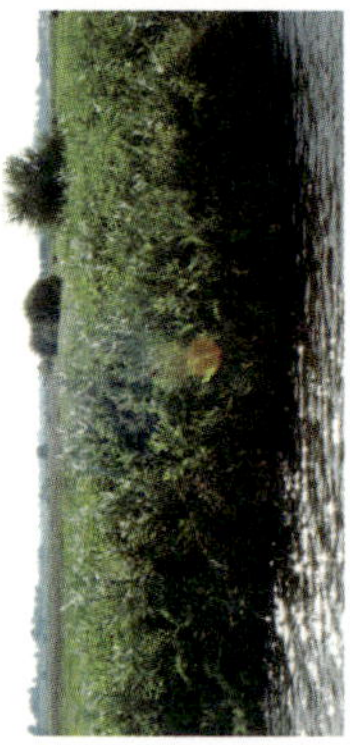
0410P

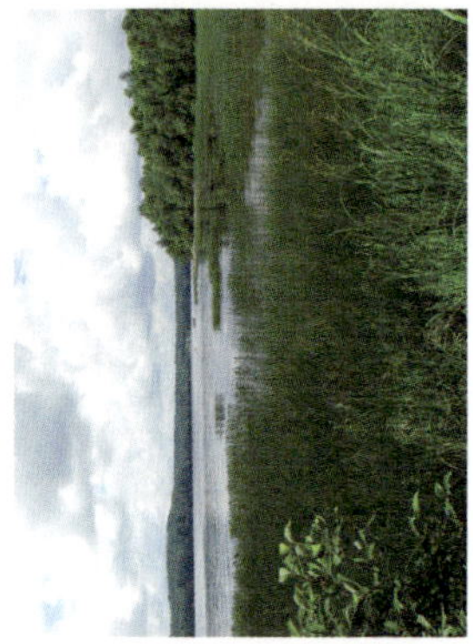
0410G

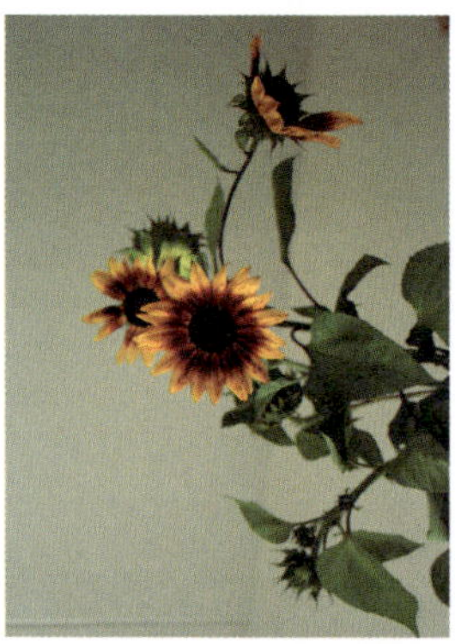
0430P

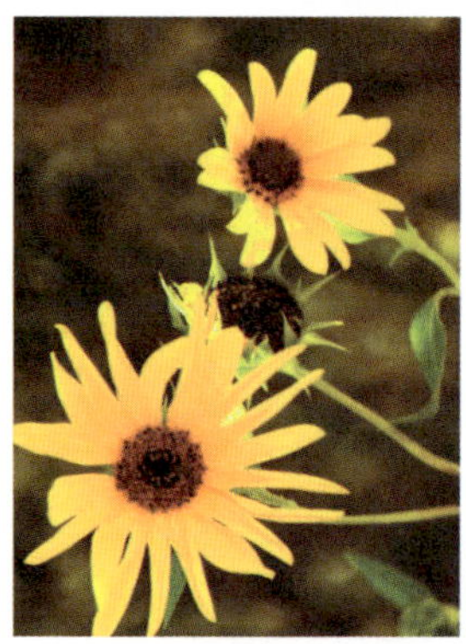
0430G

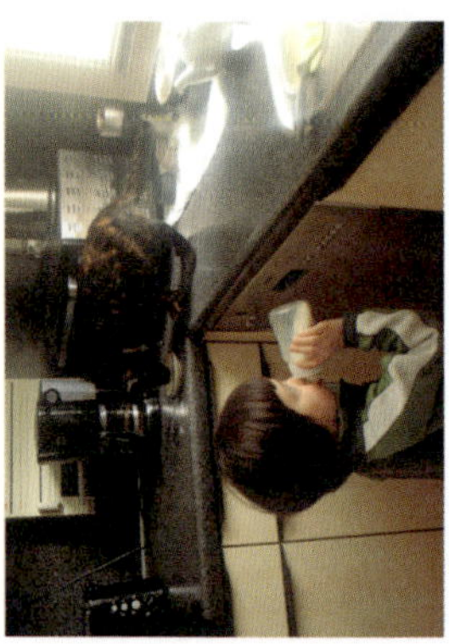
0431P

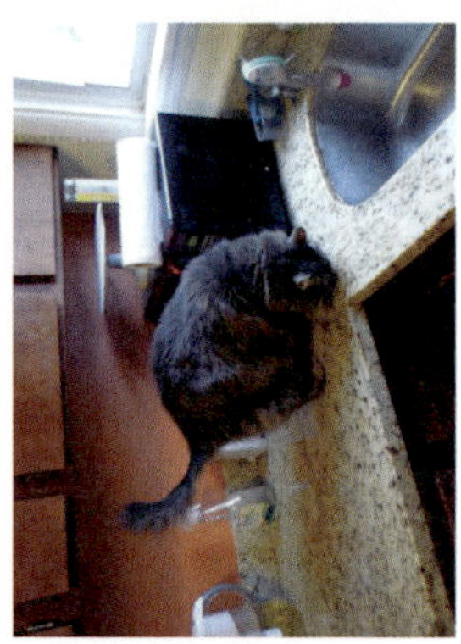
0431G

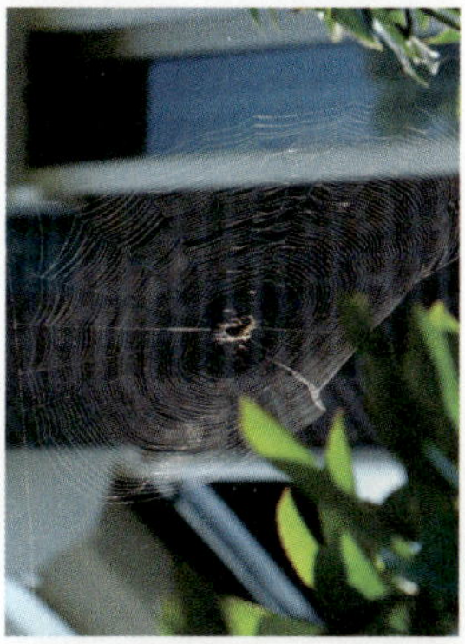
0437P

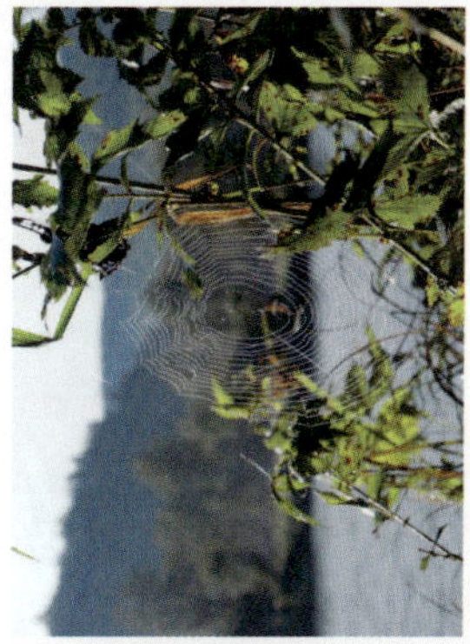
0437G

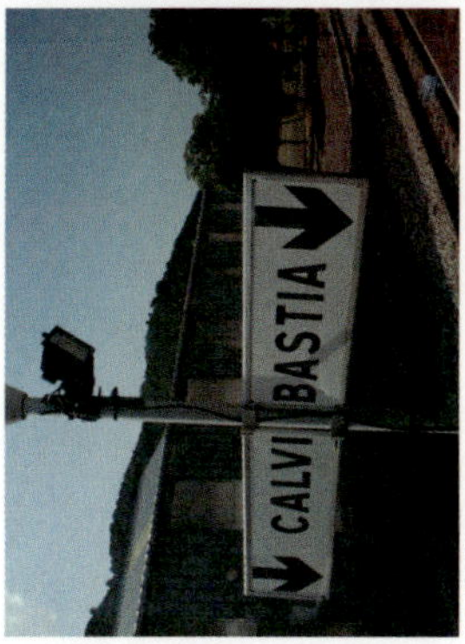

0445P

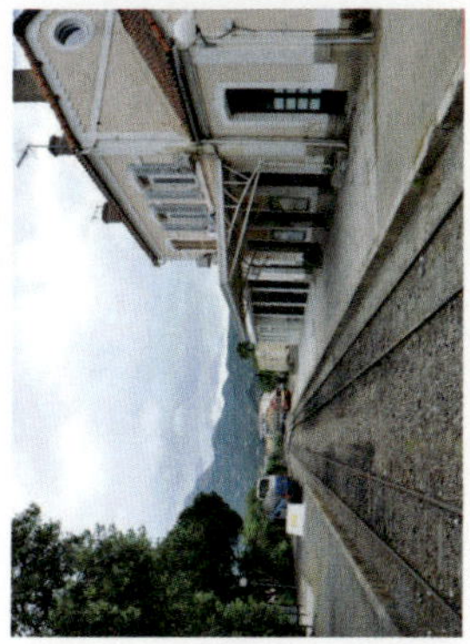
0445G

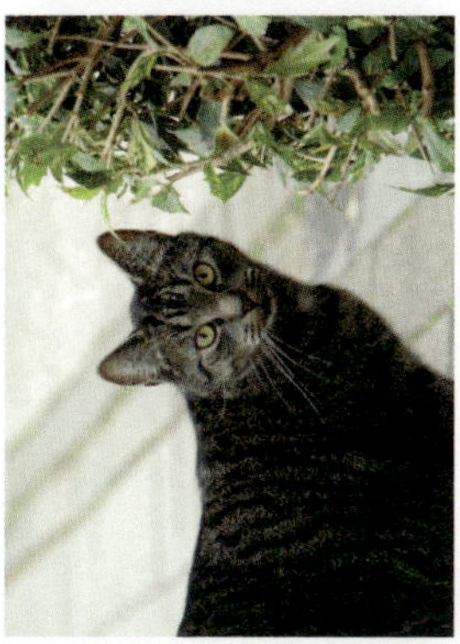
0453P

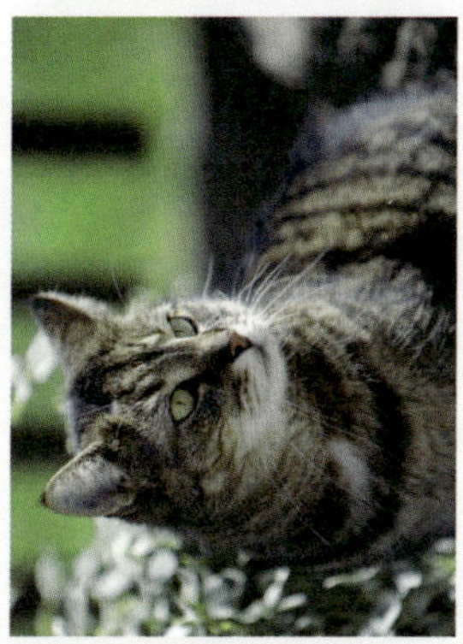
0453G

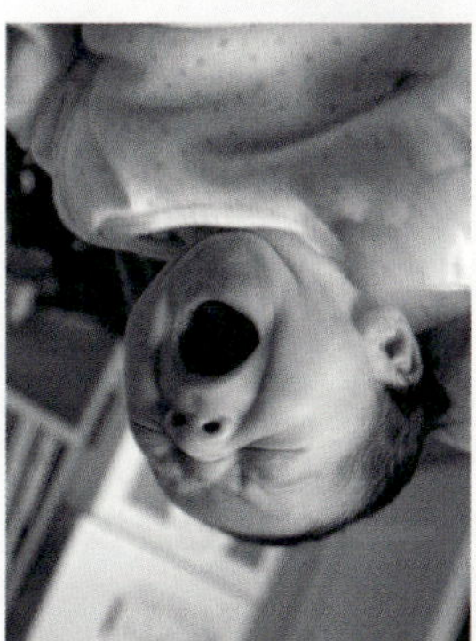
0459P

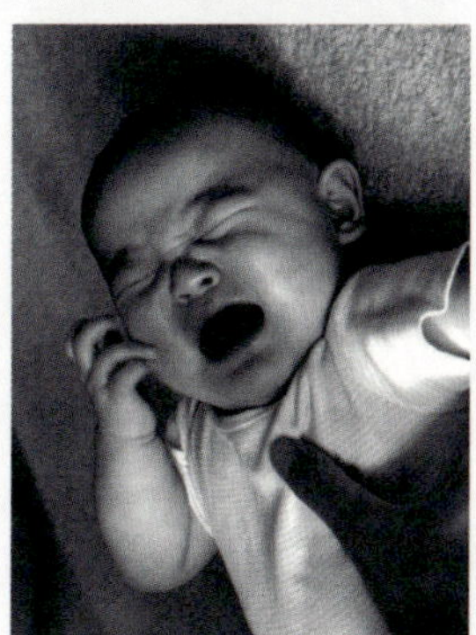
0459G

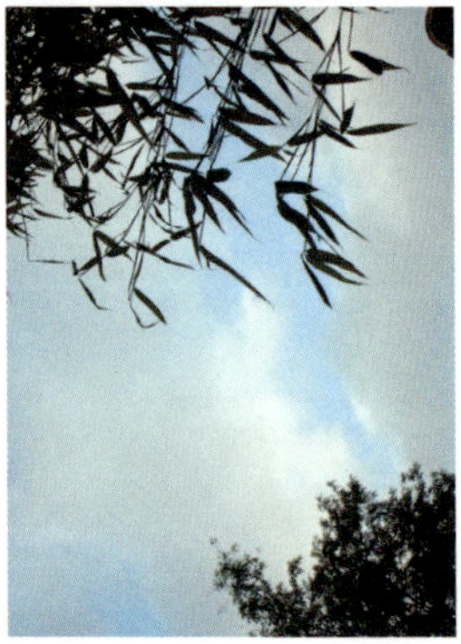
0466P

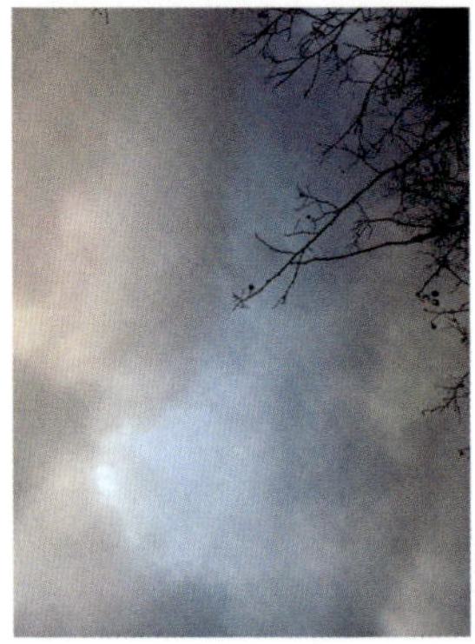
0466G

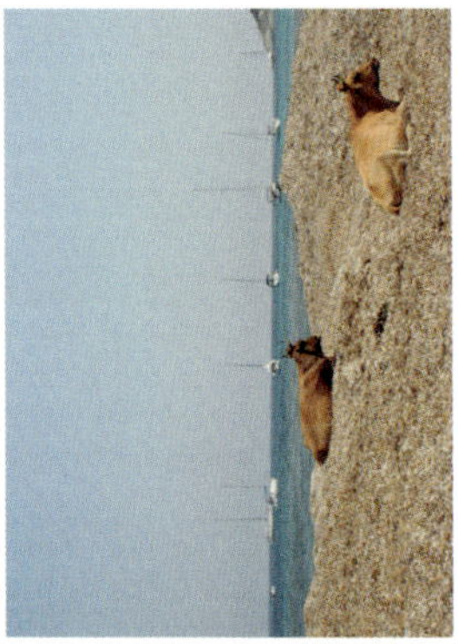
0485P

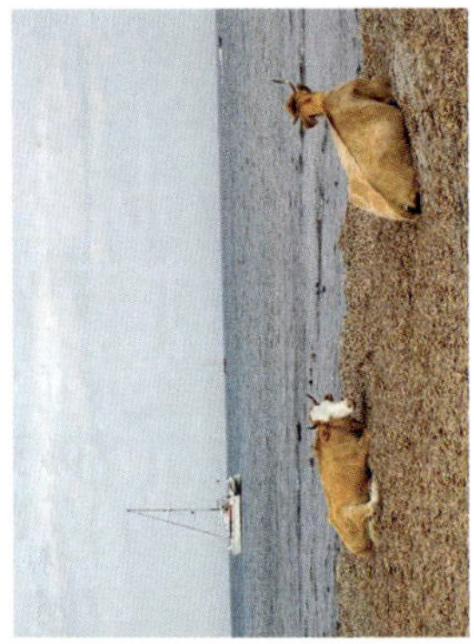
0485G

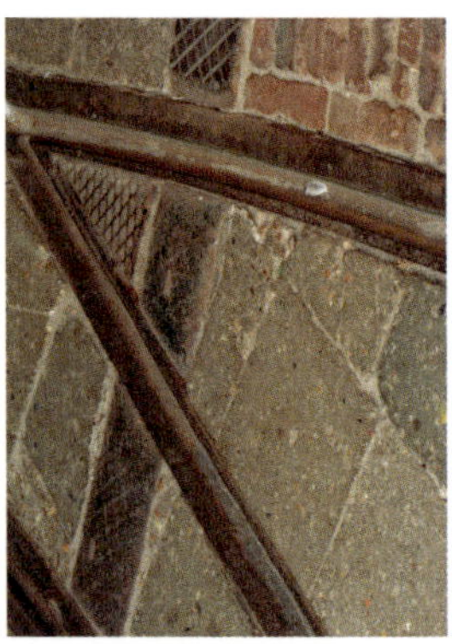
0486P

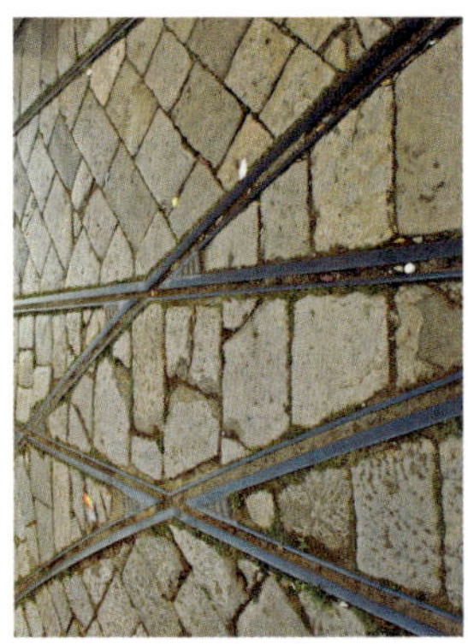
0486G

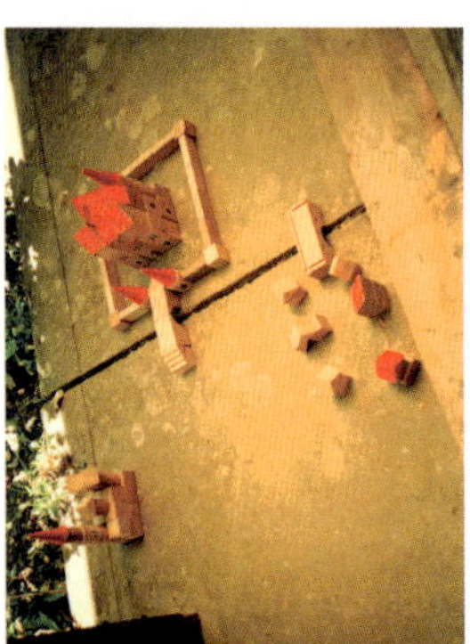
0487P

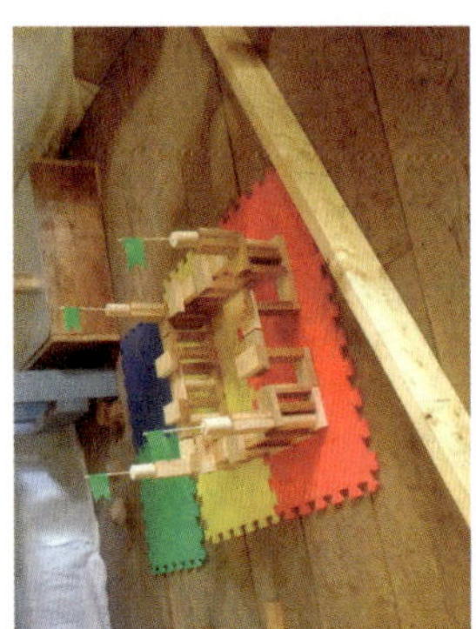
0487G

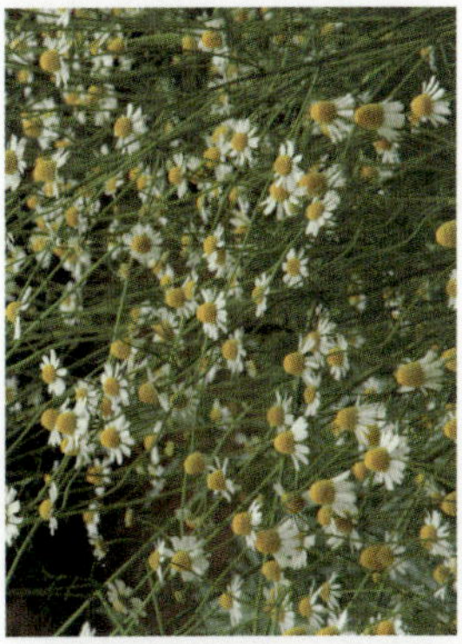
0496P

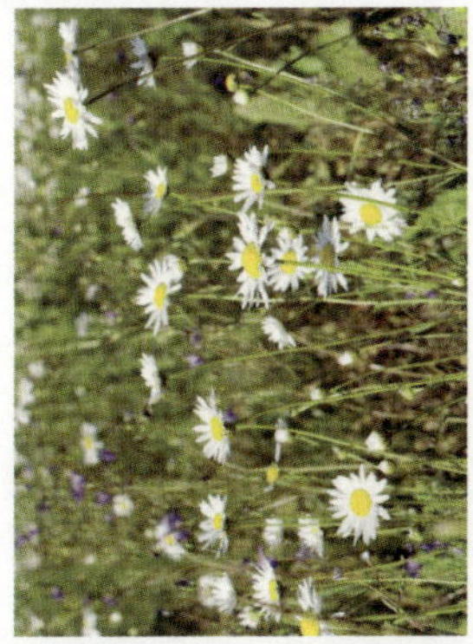
0496G

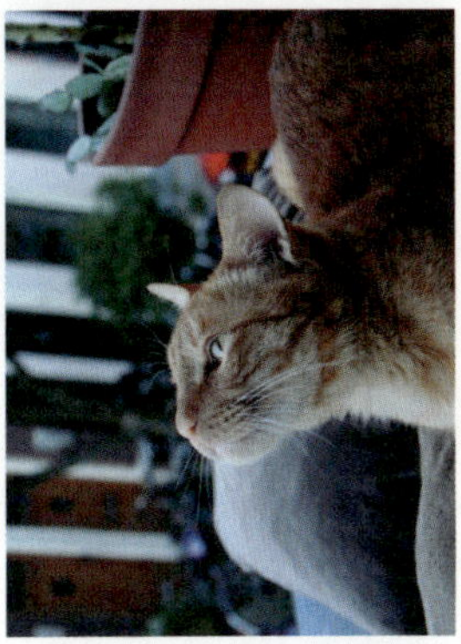
0533P

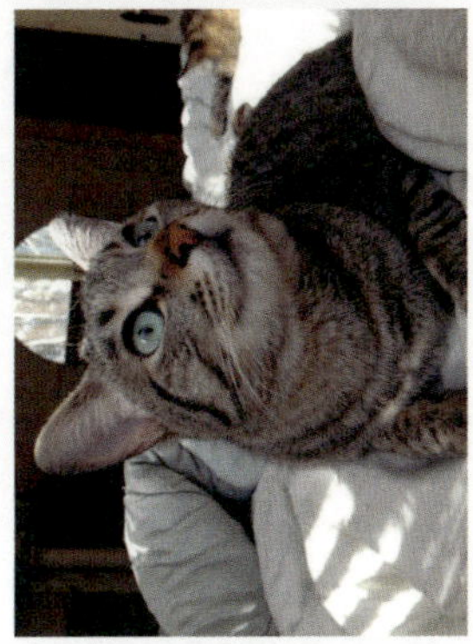
0533G

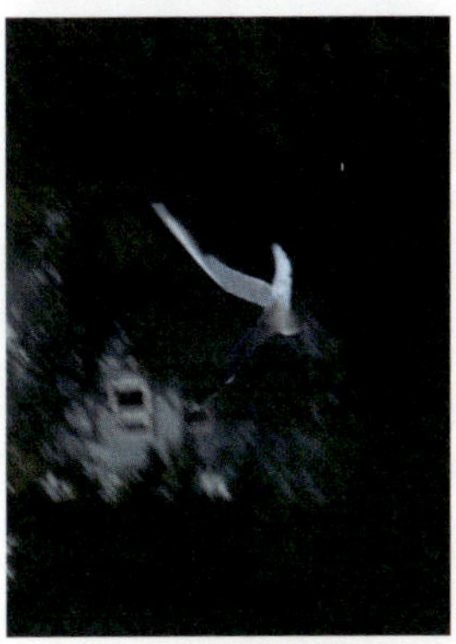
0534P

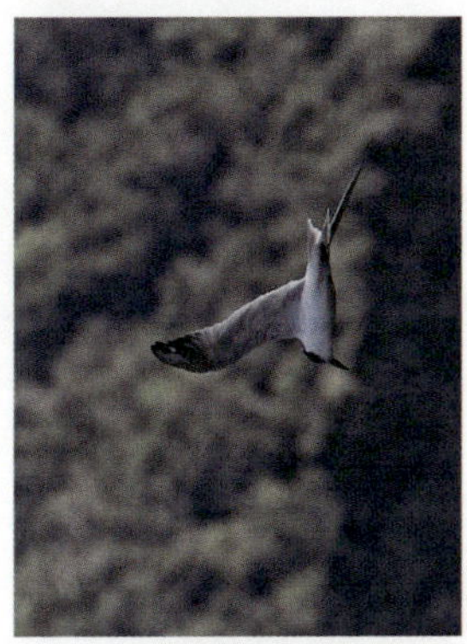
0534G

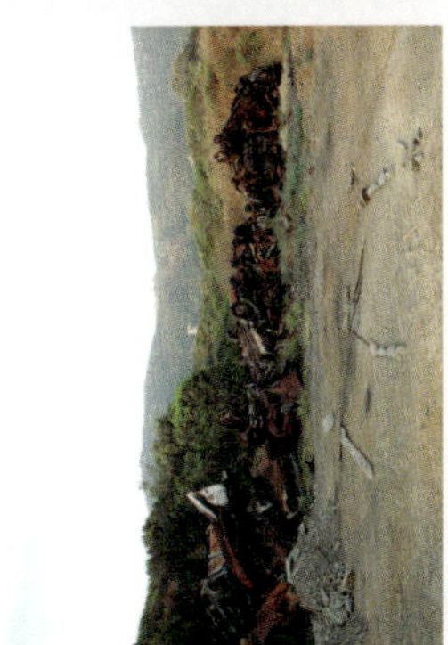
0542P

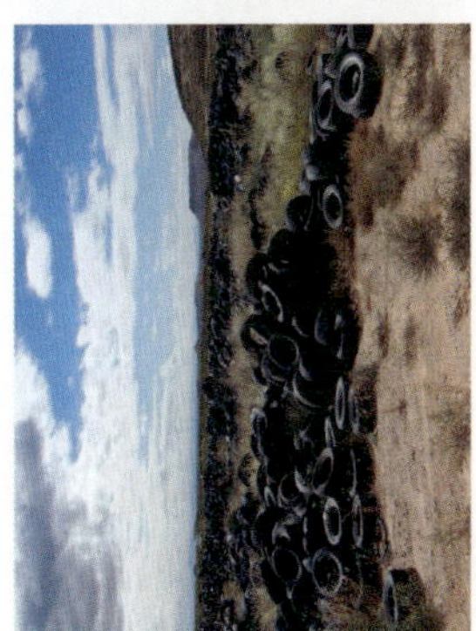
0542G

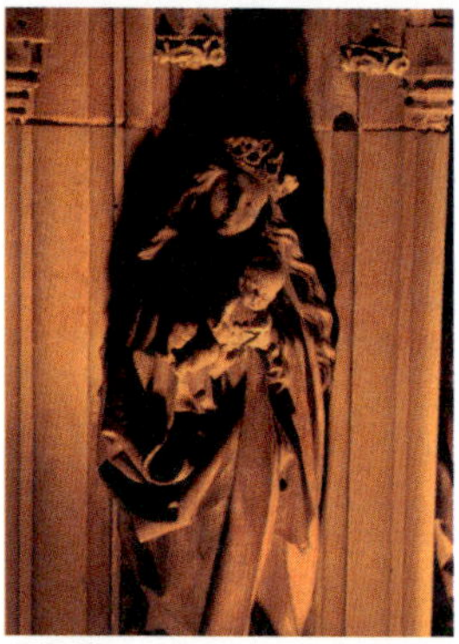

0474P

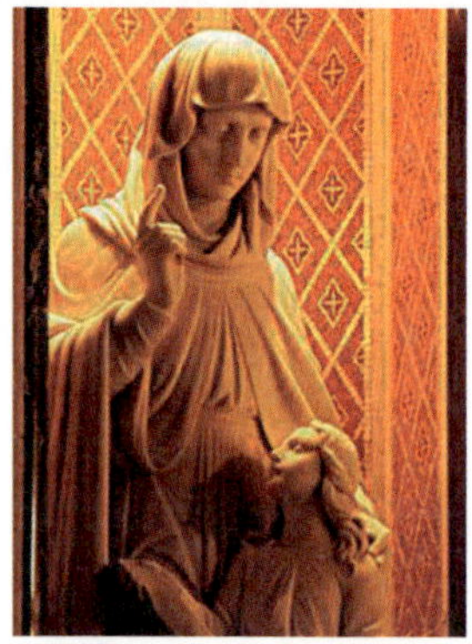

0474G

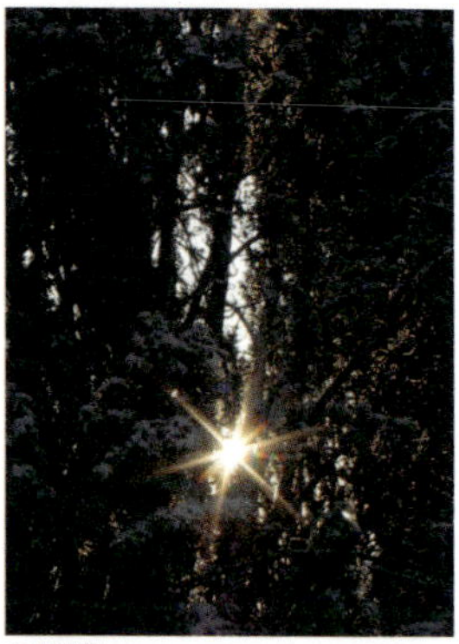

0482P

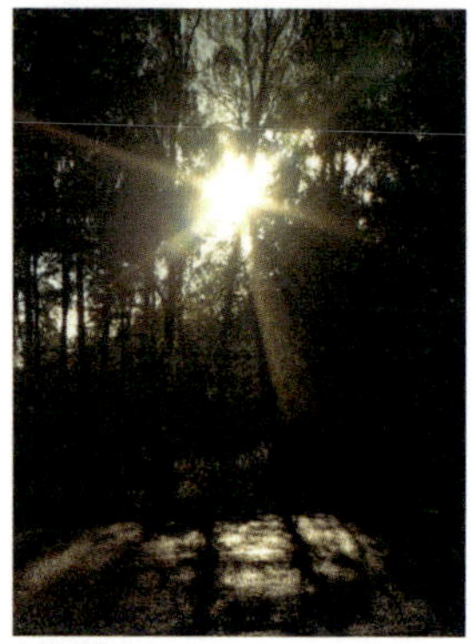

0482G

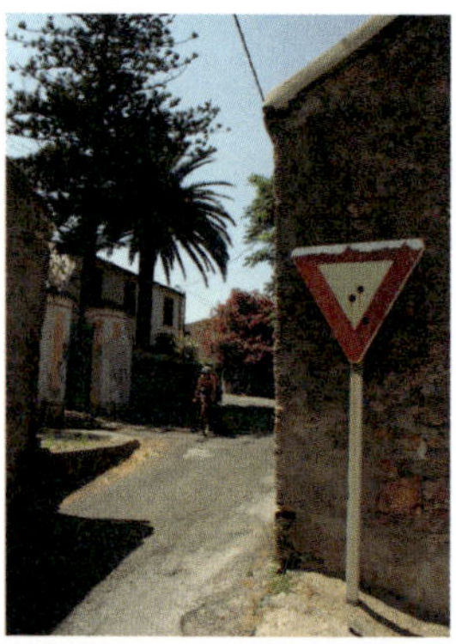

0505P

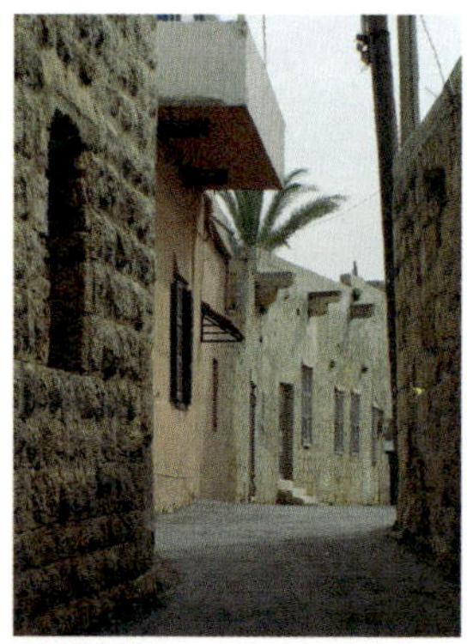

0505G

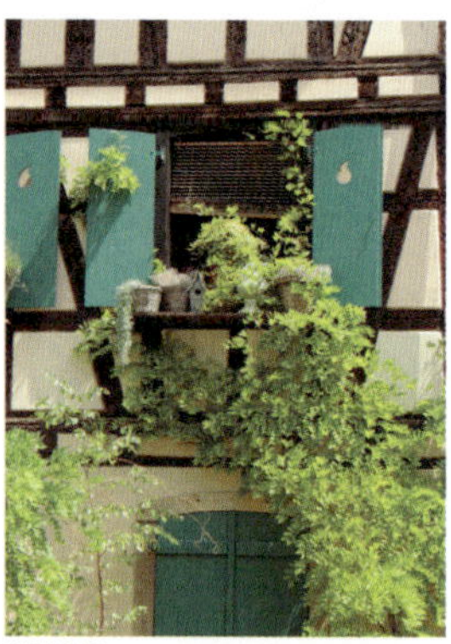

0530P

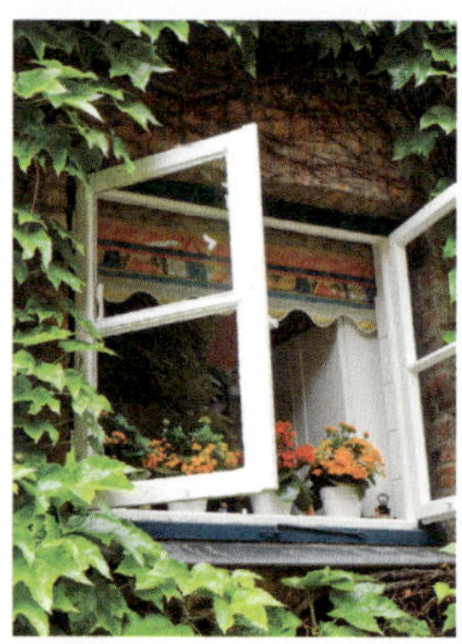

0530G

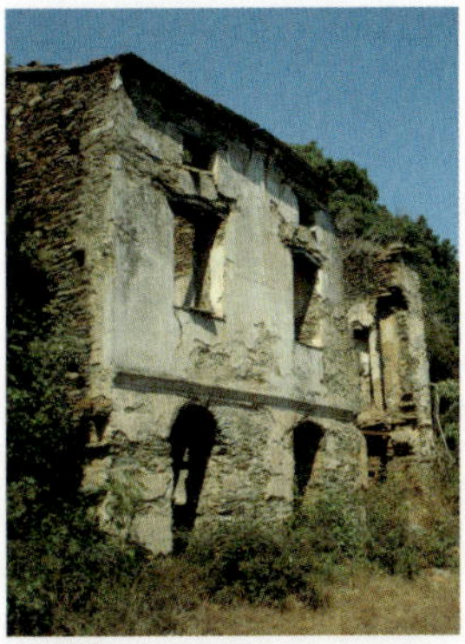

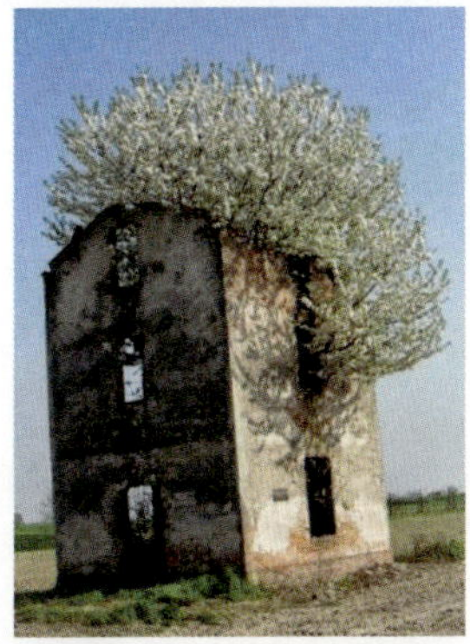

0540P

0540G

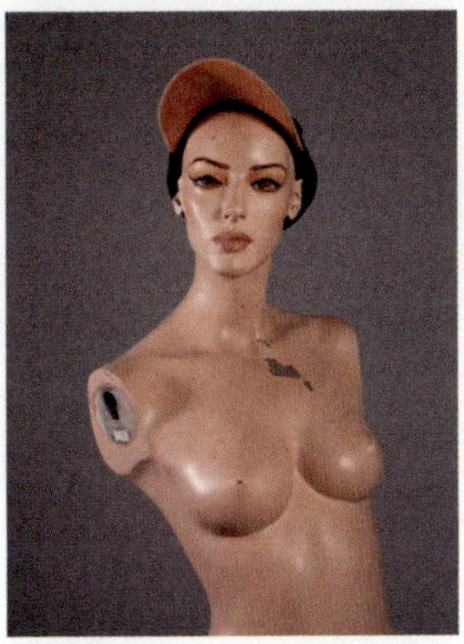

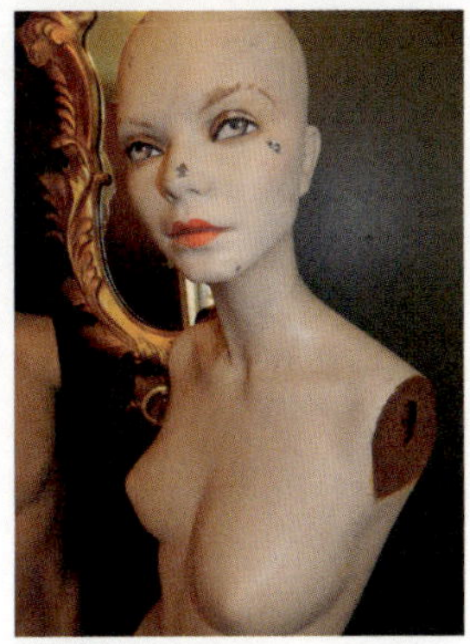

0541P

0541G

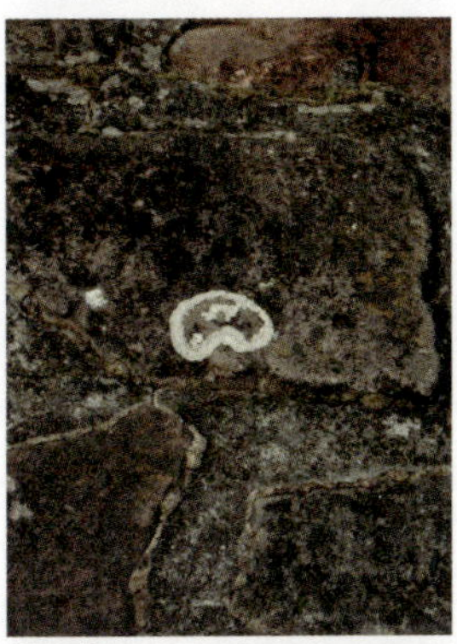

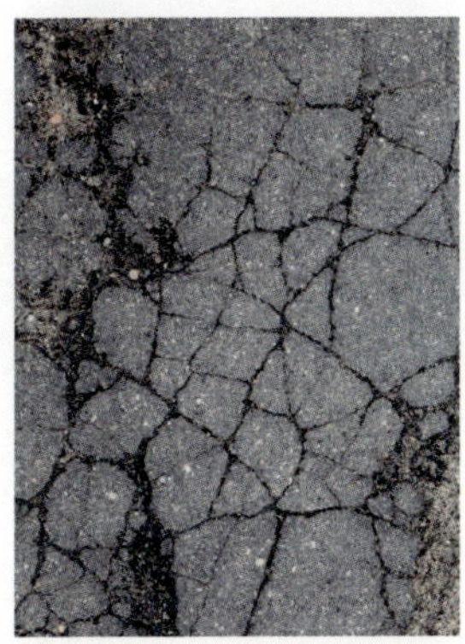

0547P

0547G

0564P

0564G

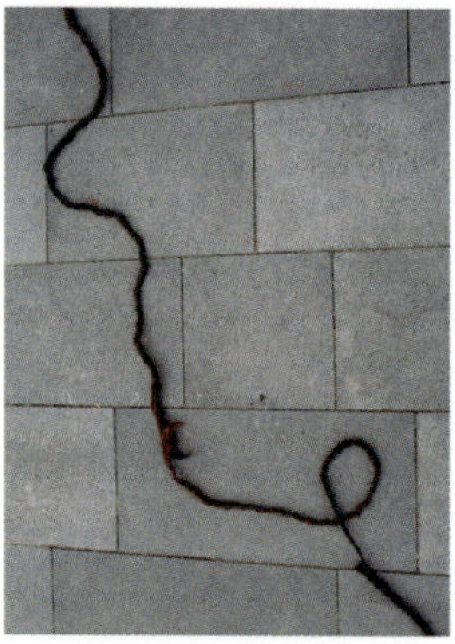
0548P

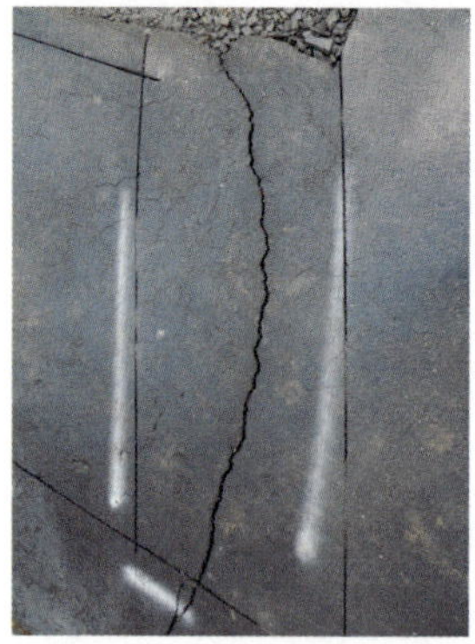
0548G

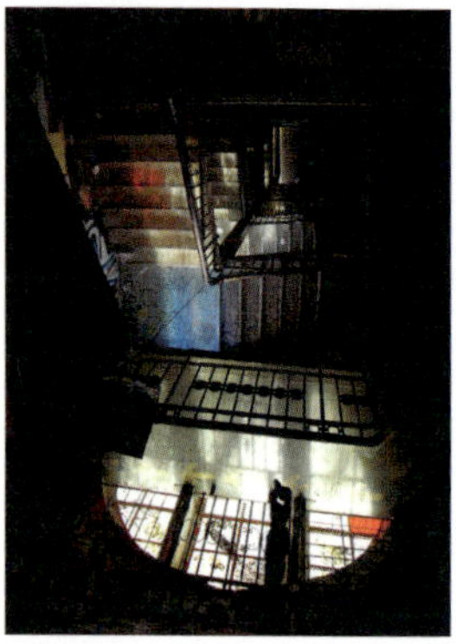
0552P

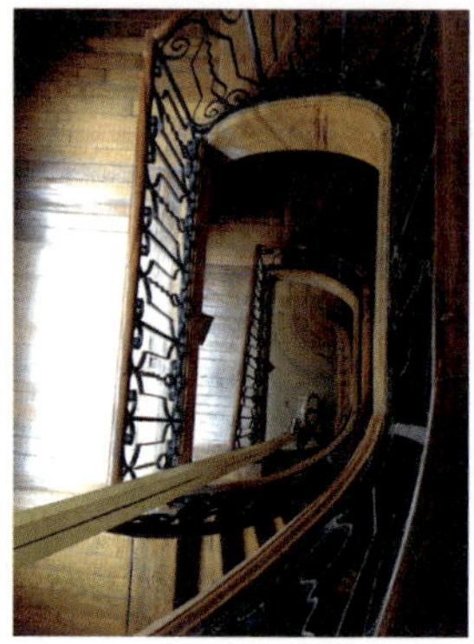
0552G

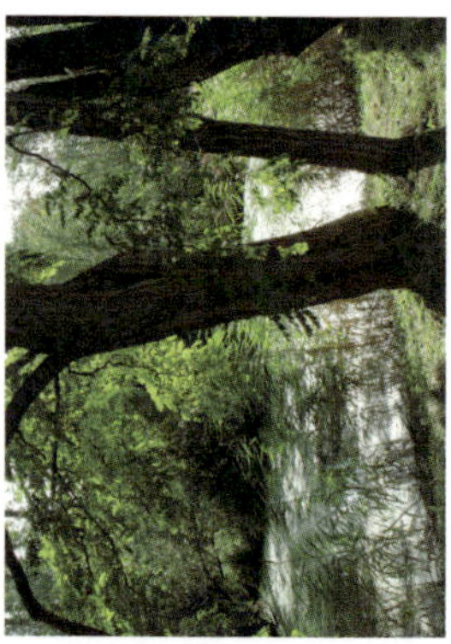
0557P
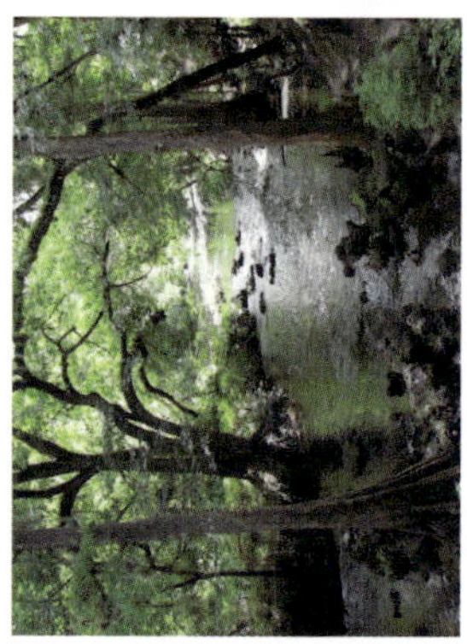
0557G

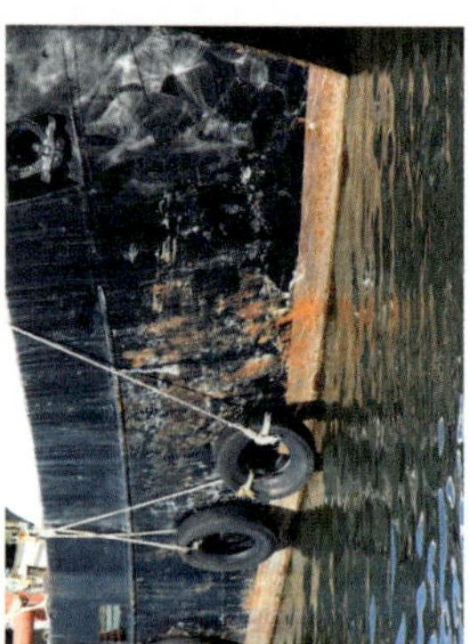
0562P
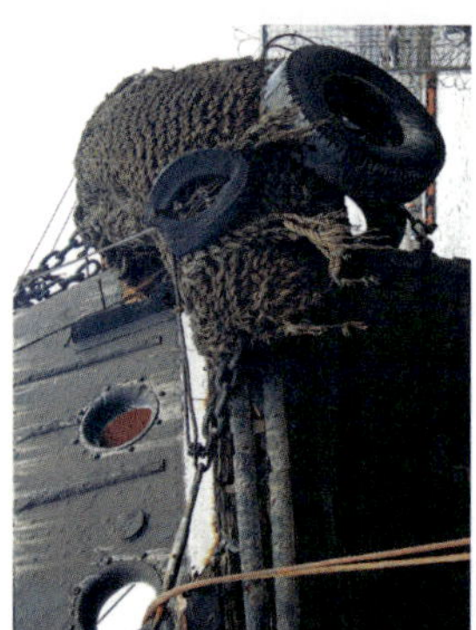
0562G

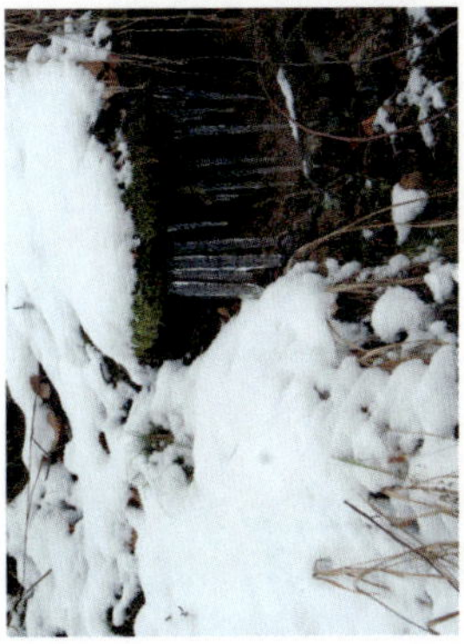
0596P

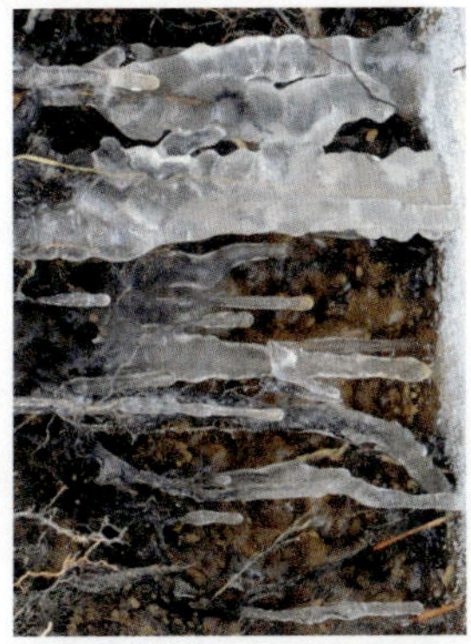
0596G

0624P

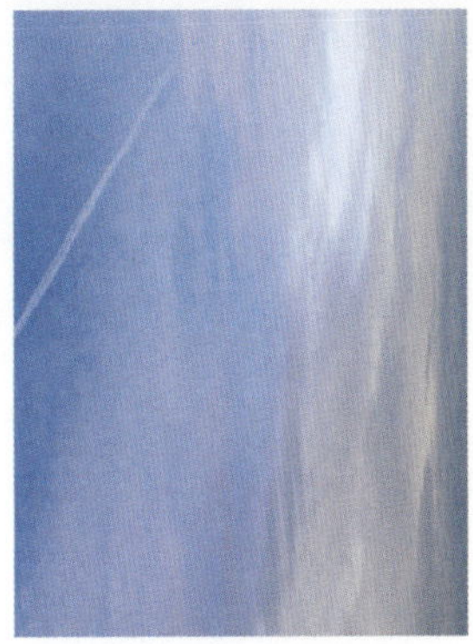
0624G

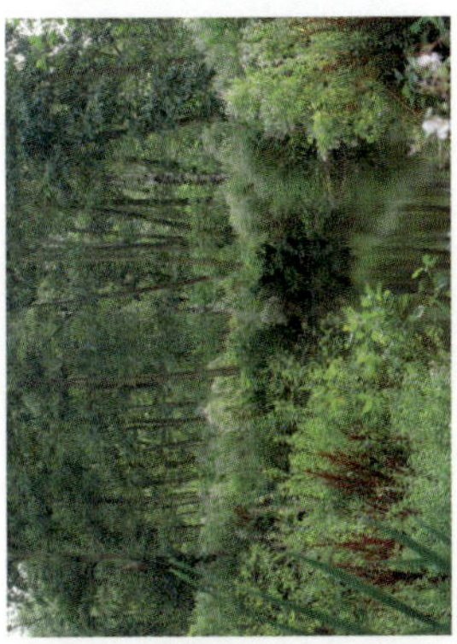
0629P

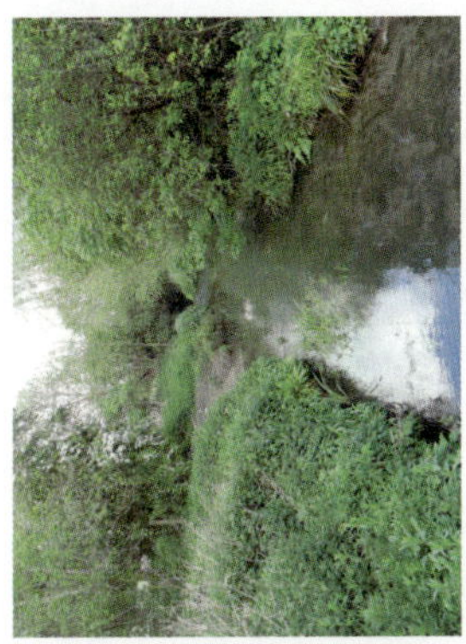
0629G

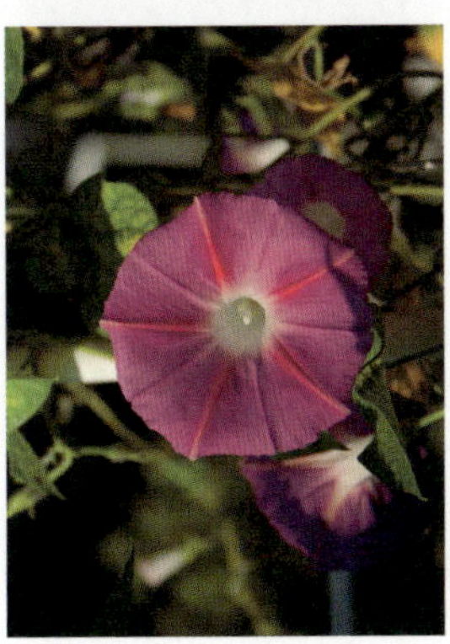
0630P

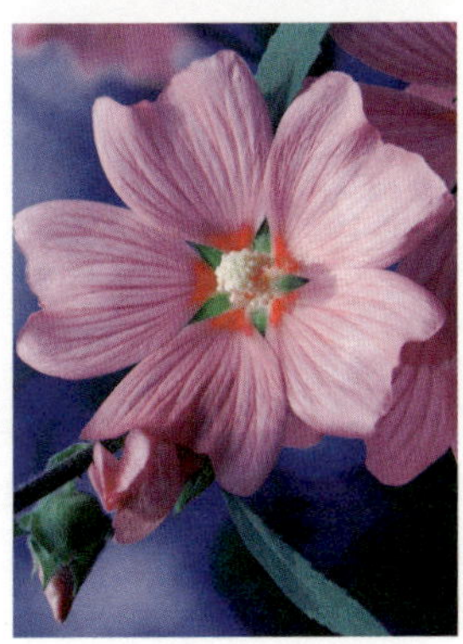
0630G

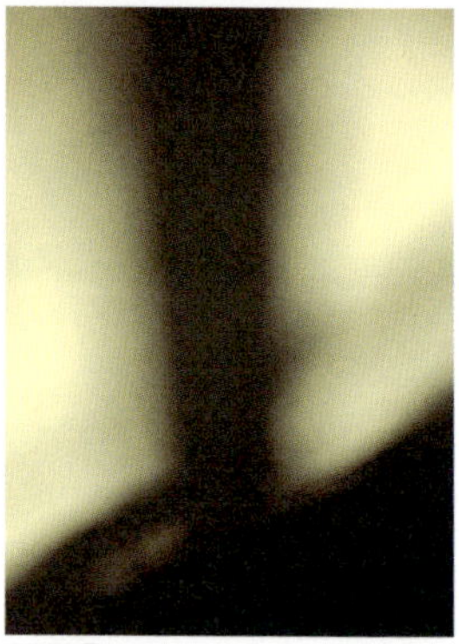

0565P

0565G

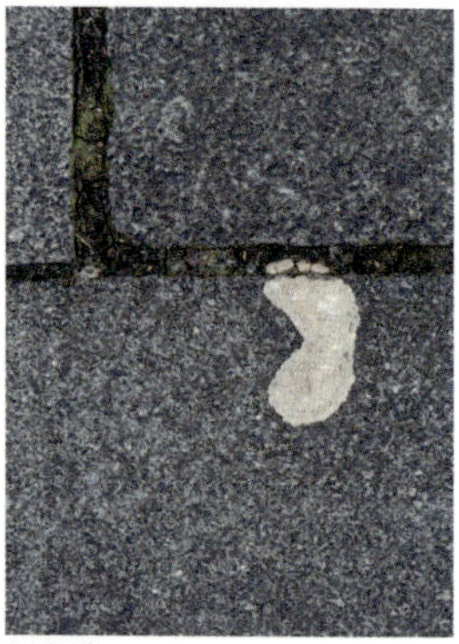

0604P

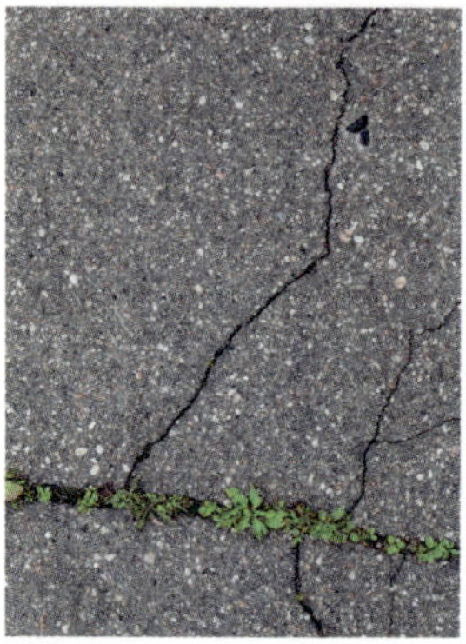

0604G

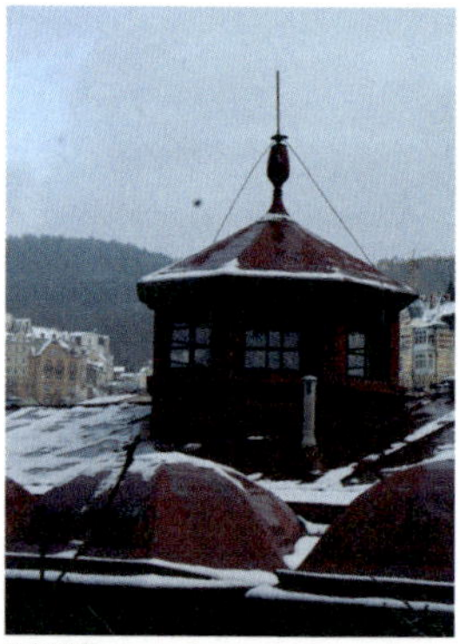

0613P

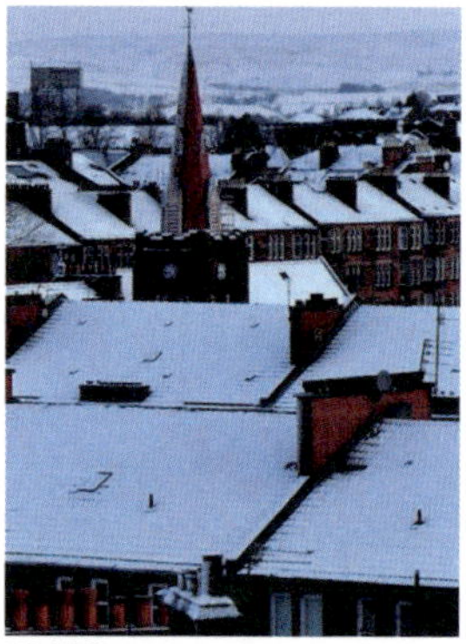

0613G

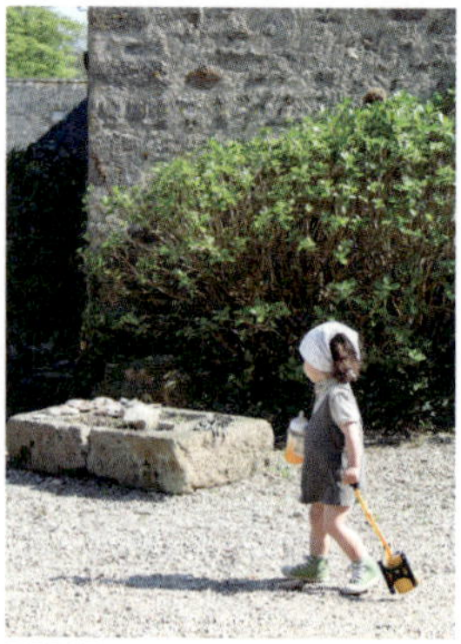

0632P

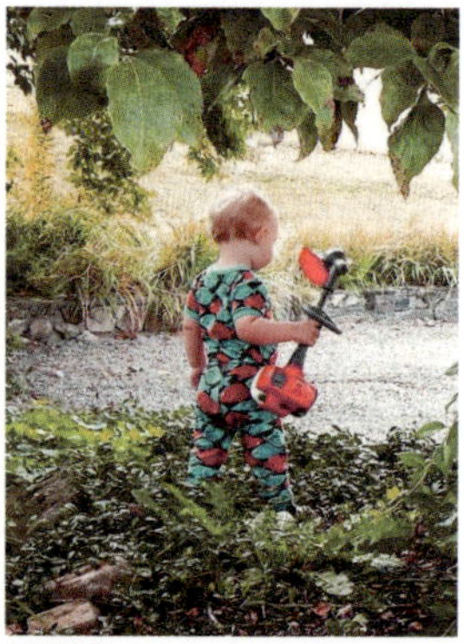

0632G

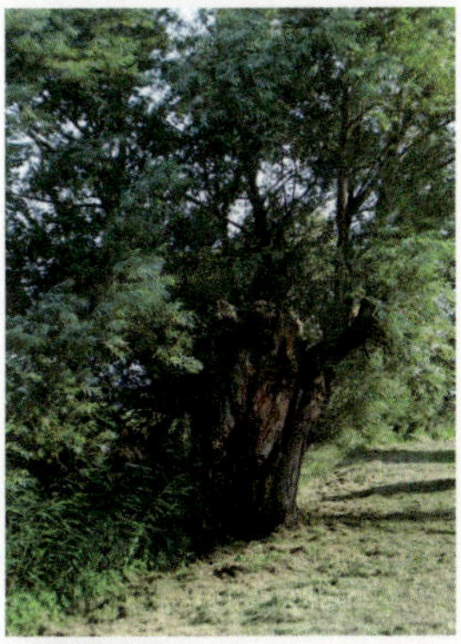
0635P

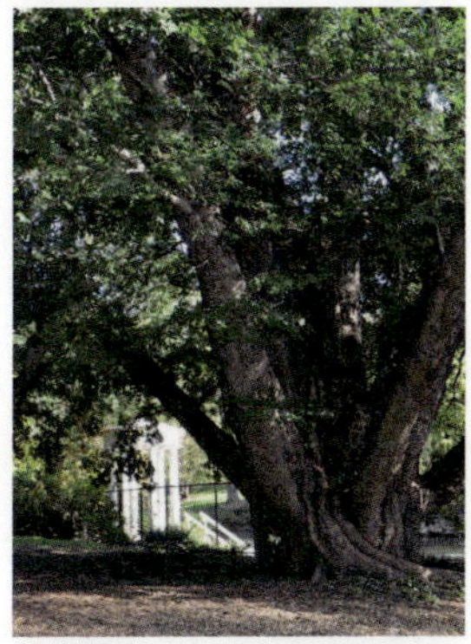
0635G

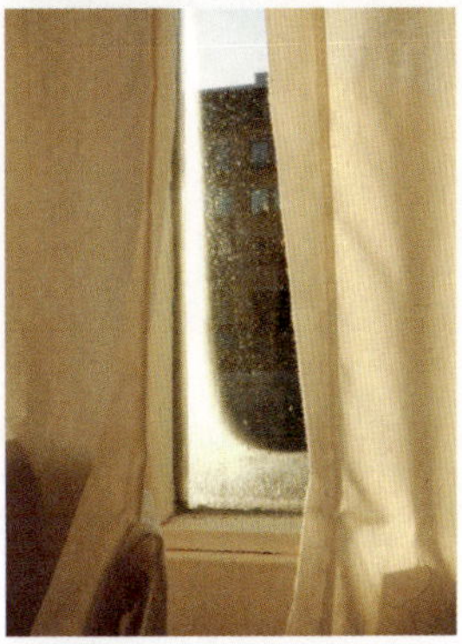
0718P

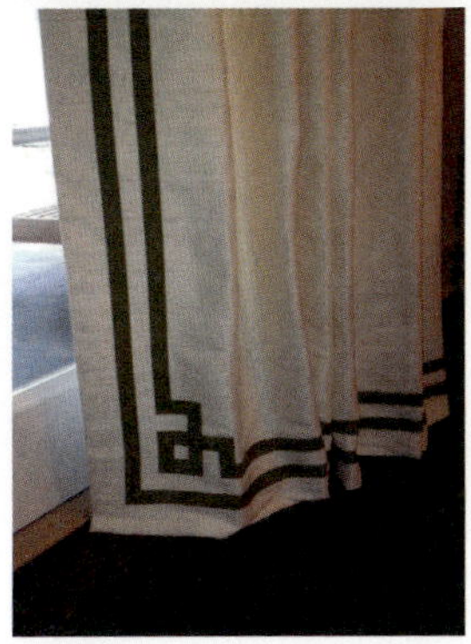
0718G

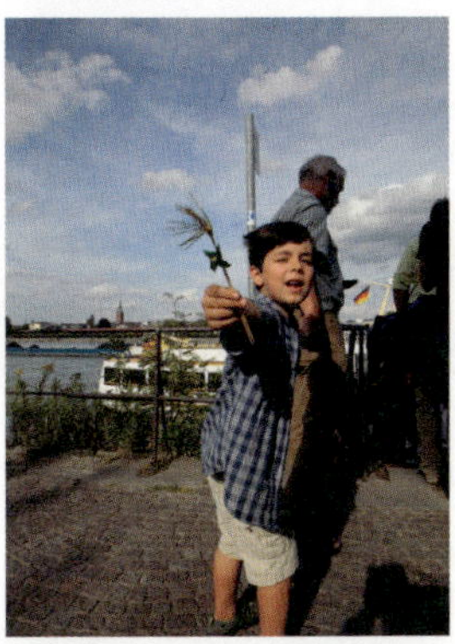
0723P

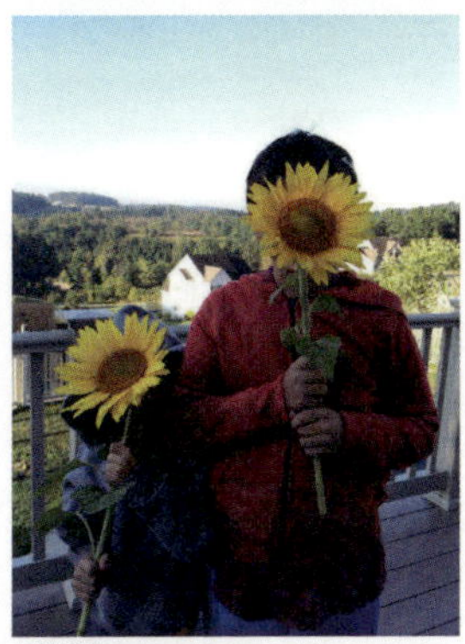
0723G

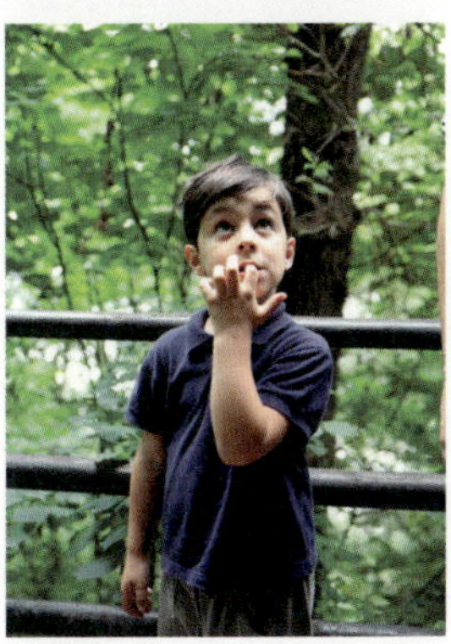
0732P

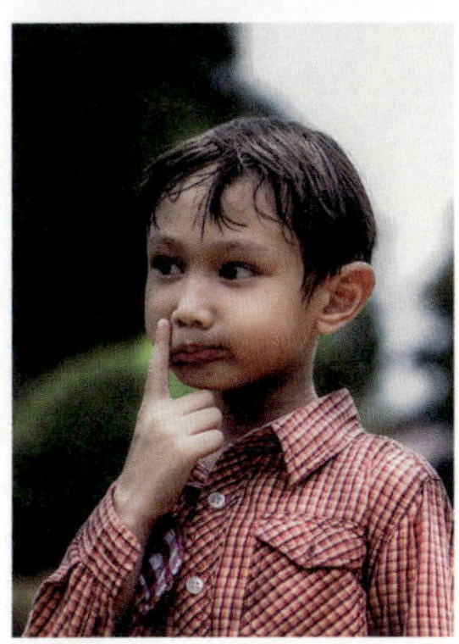
0732G

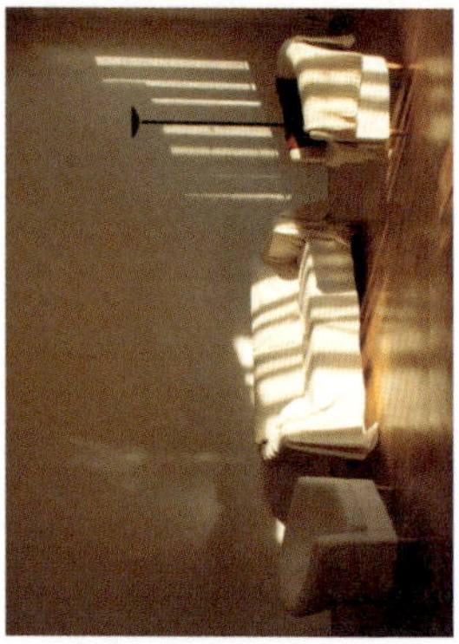

0636P

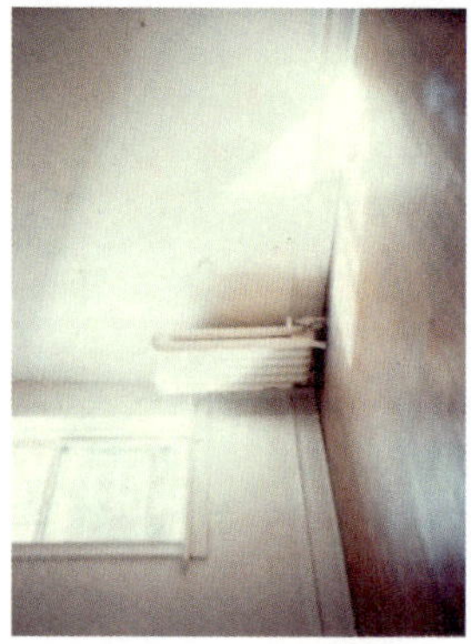

0636G

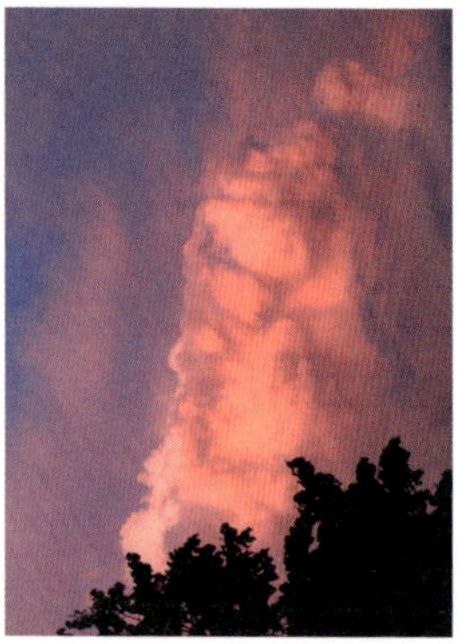

0654P

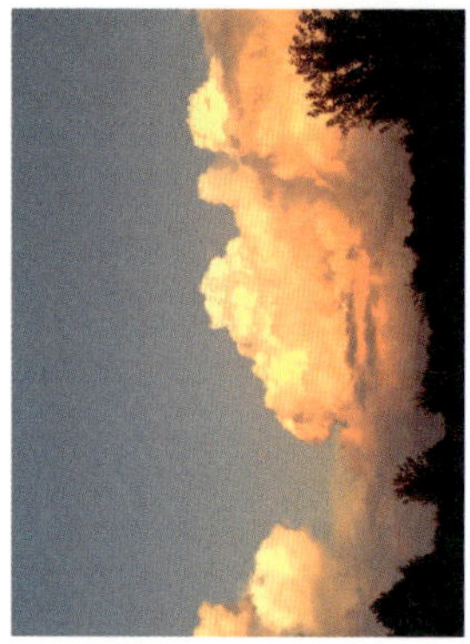

0654G

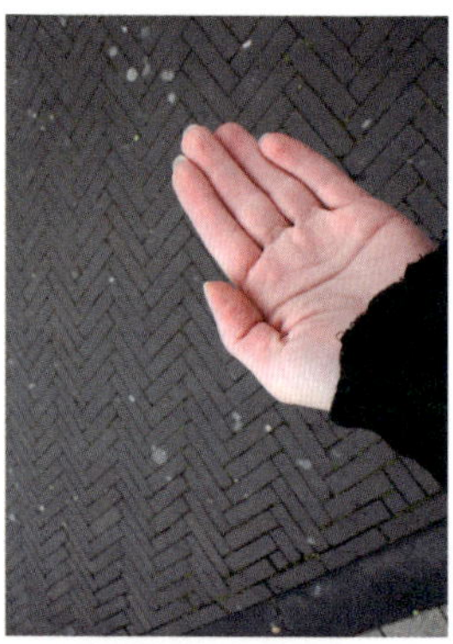

0671P

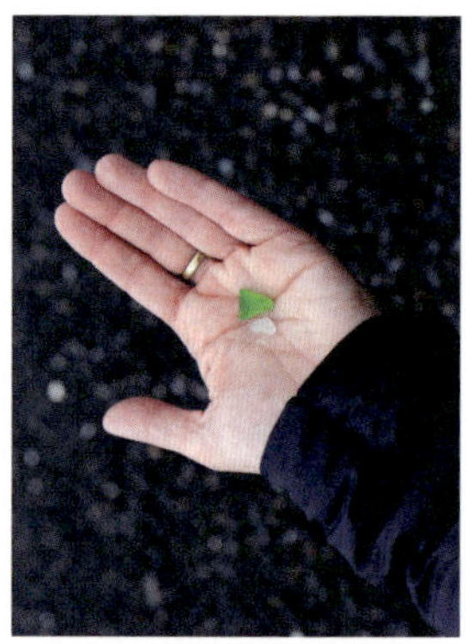

0671G

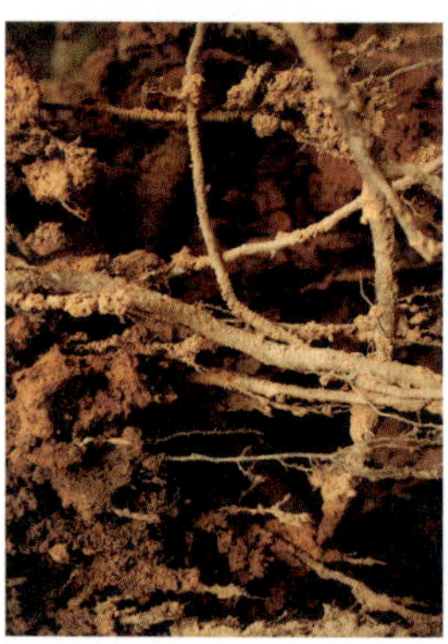

0674P

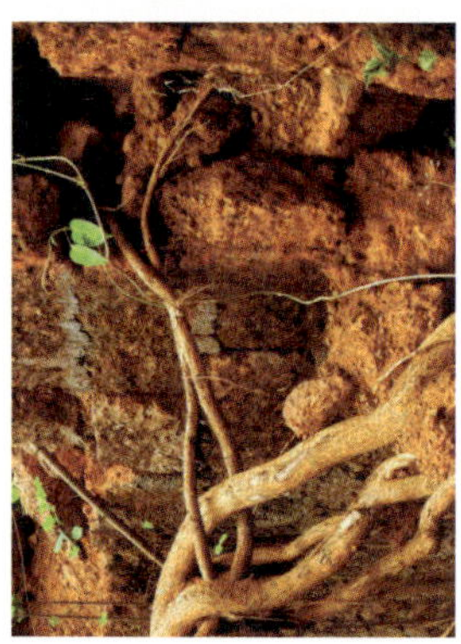

0674G

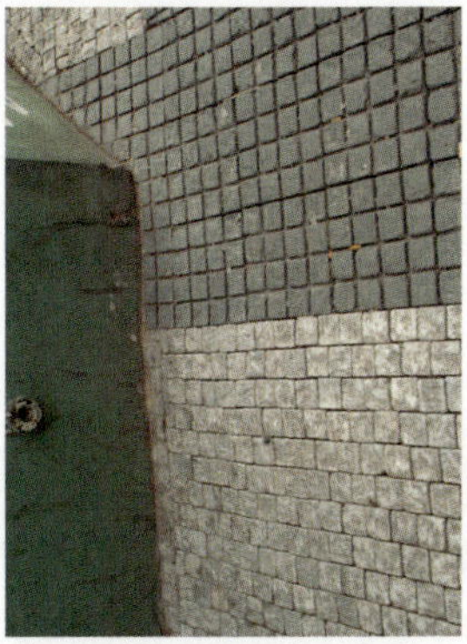

0681P

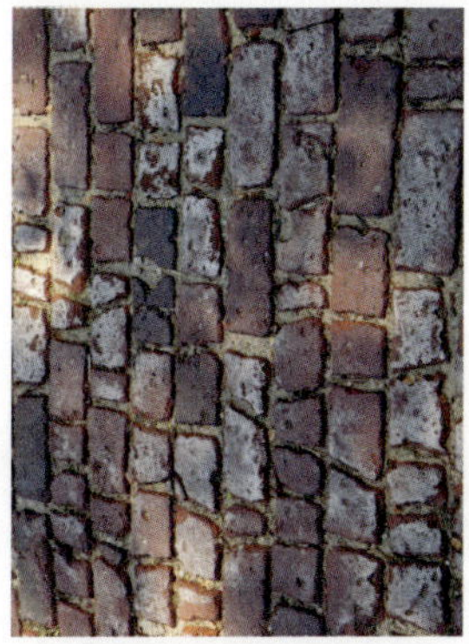

0681G

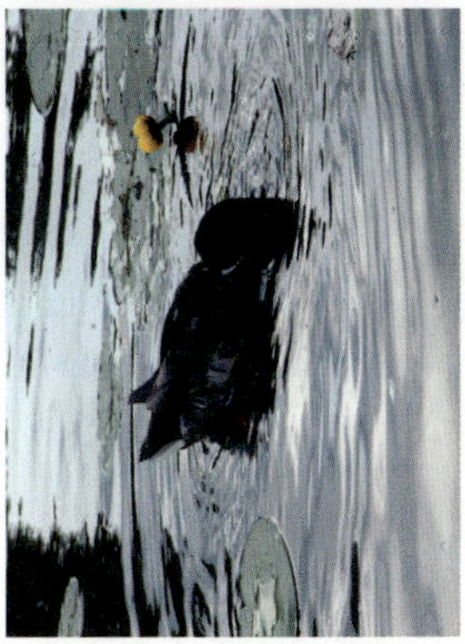

0689P

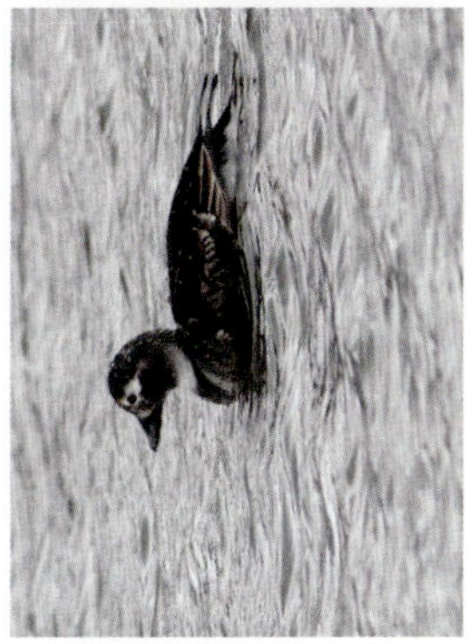

0689G

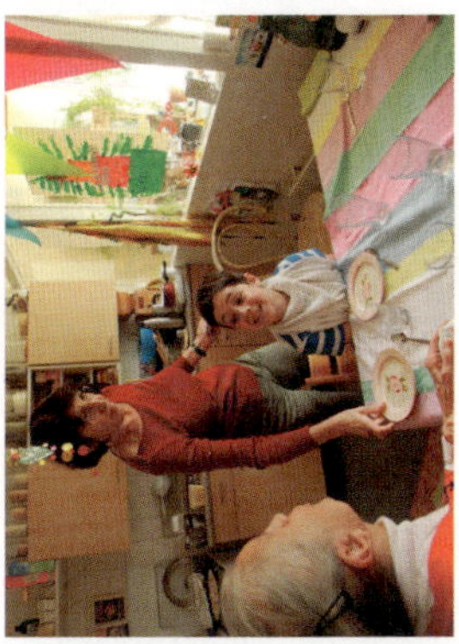

0713P

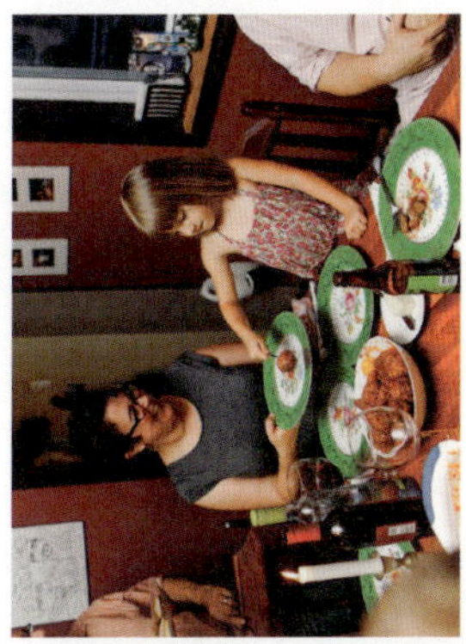

0713G

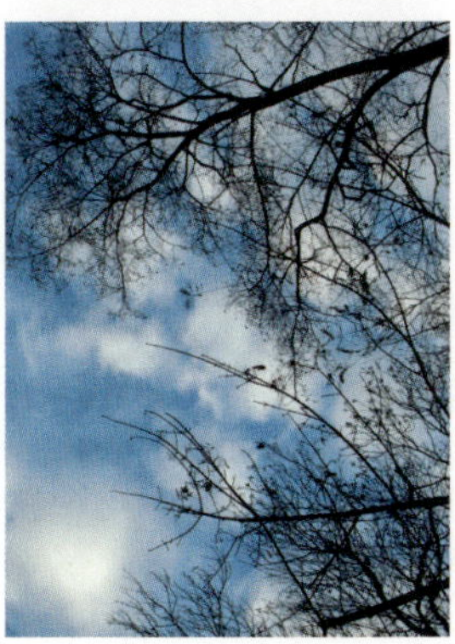

0729P

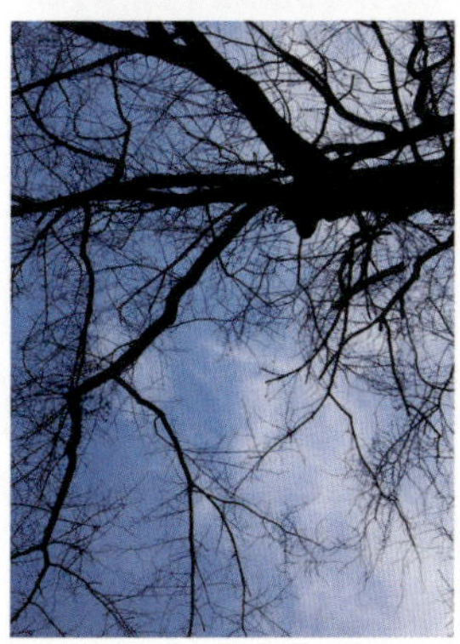

0729G

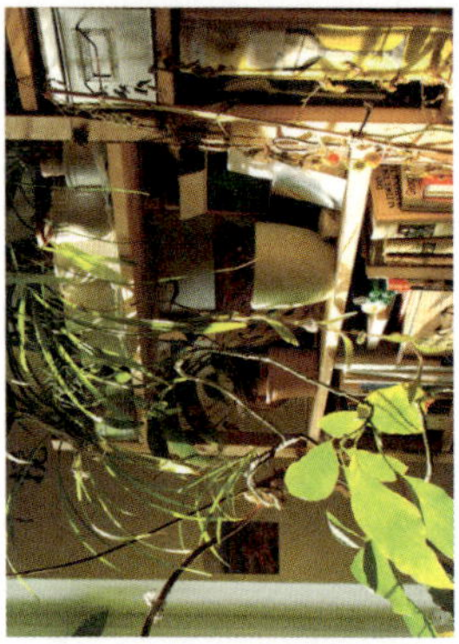

0740P

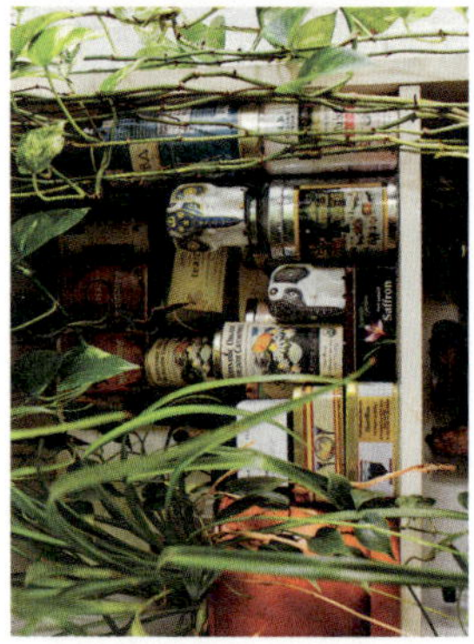

0740G

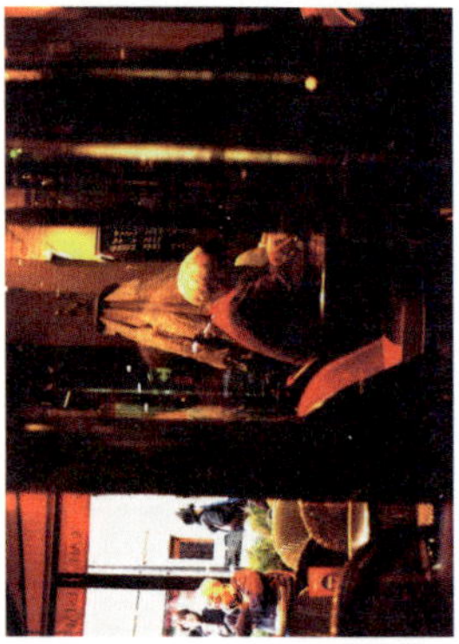

0754P

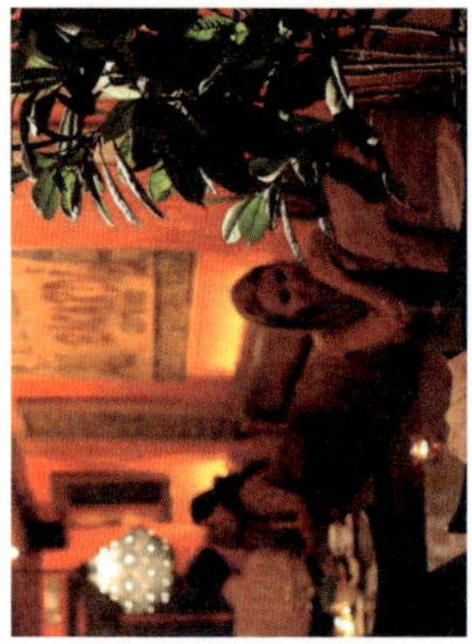

0754G

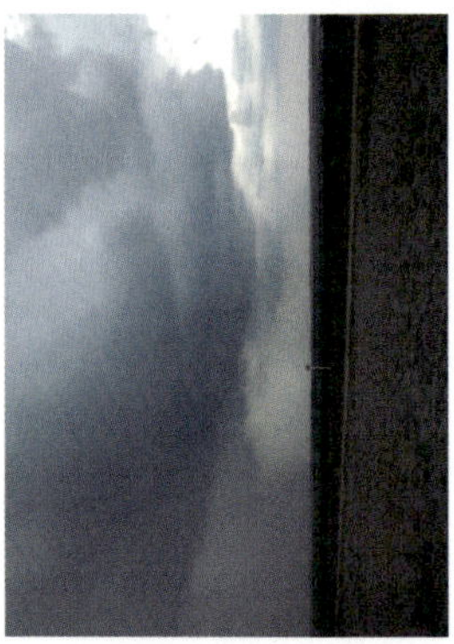

0764P

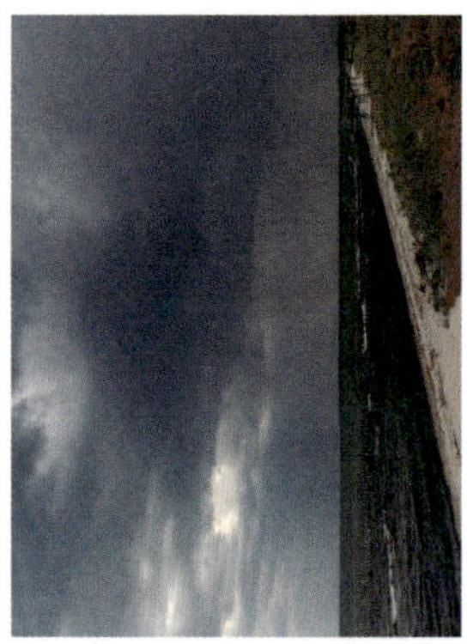

0764G

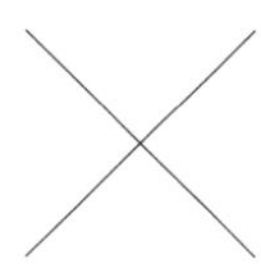

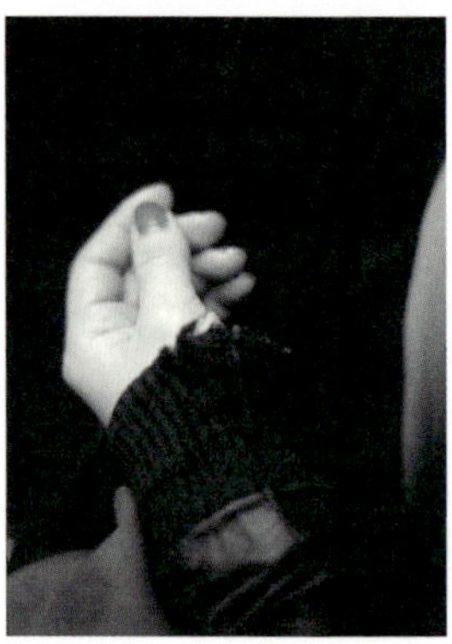

0767P

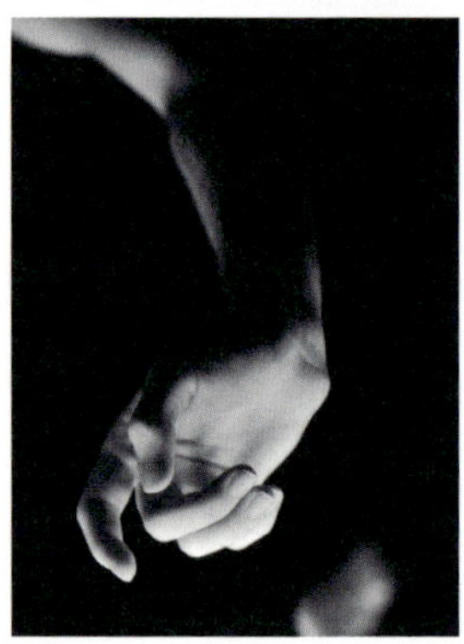

0767G

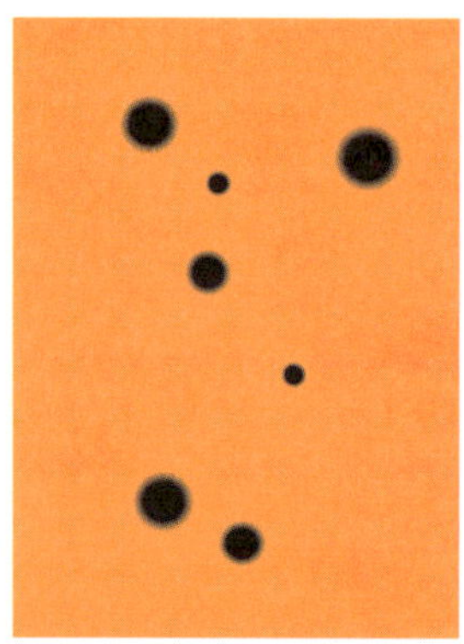

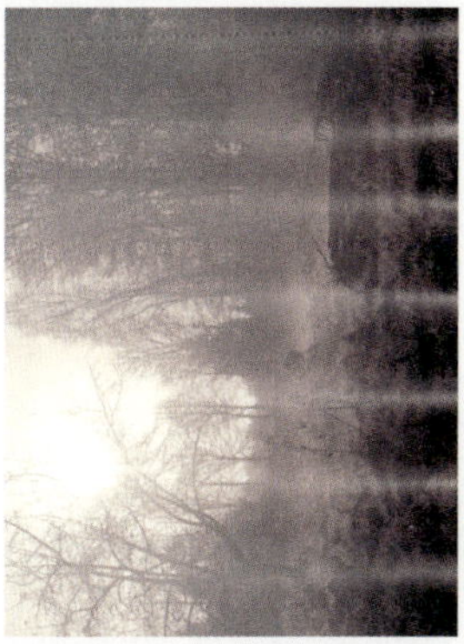

0769P

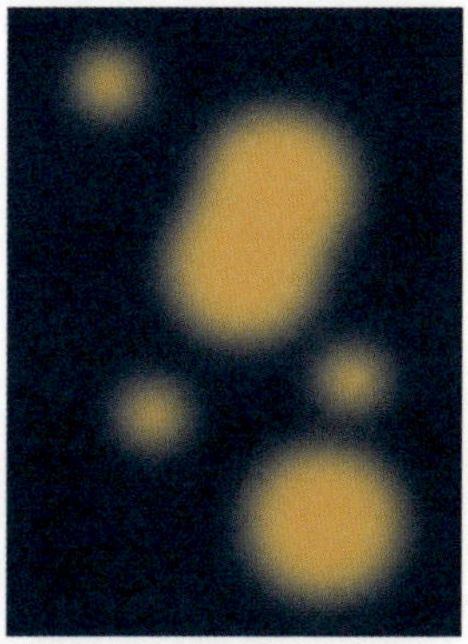

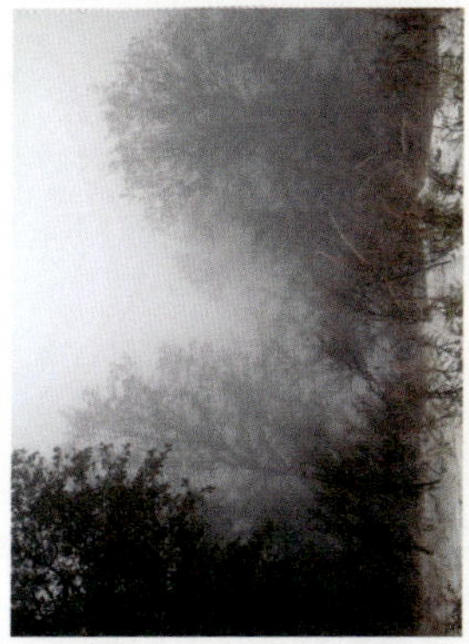

0769G

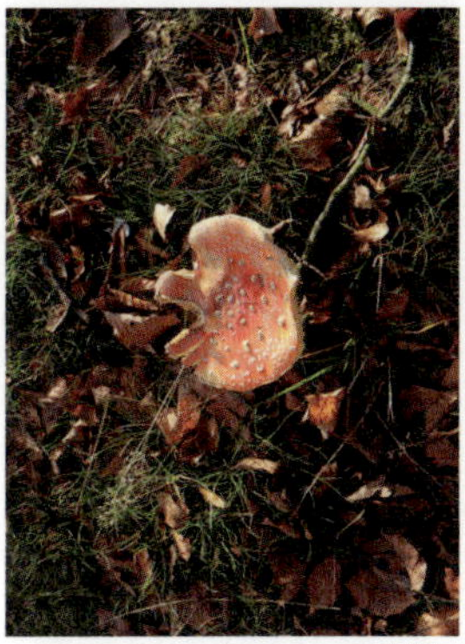

0771P

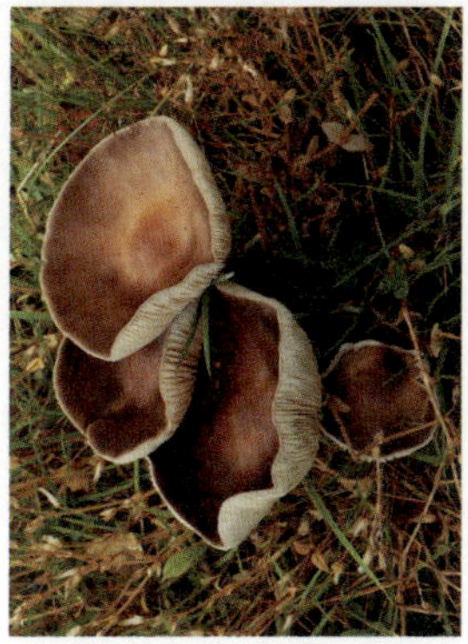

0771G

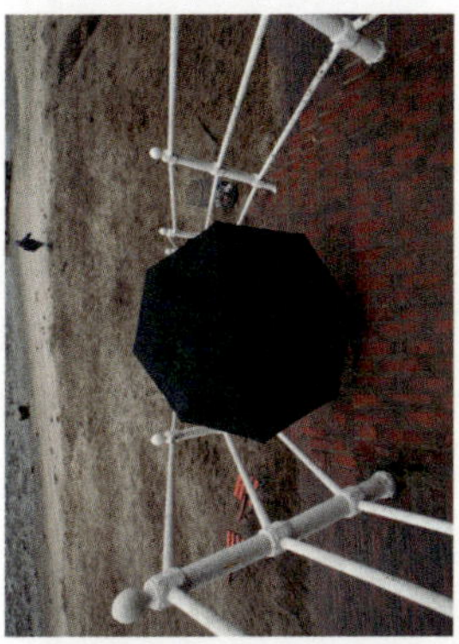

0775P

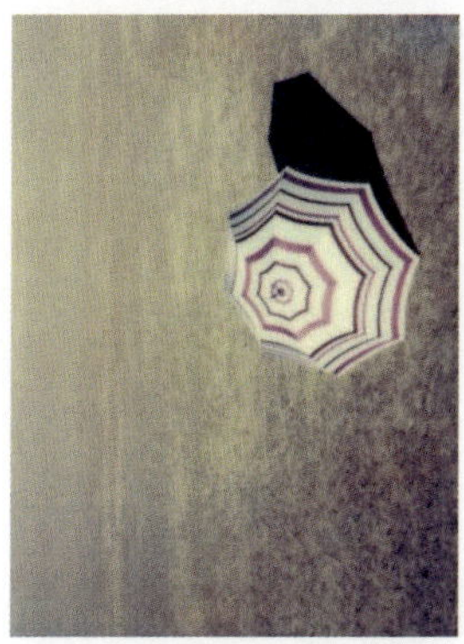

0775G

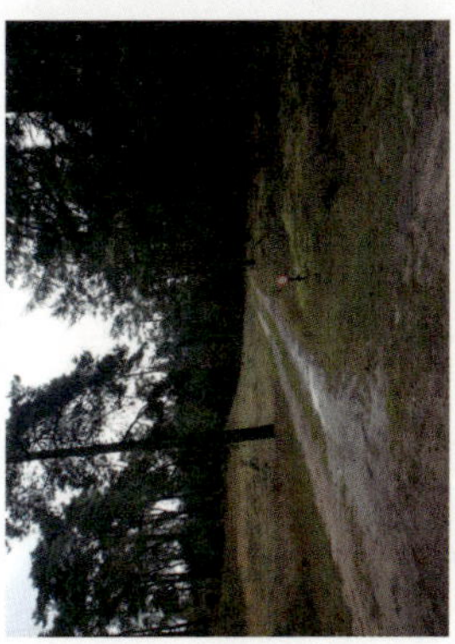

0797P

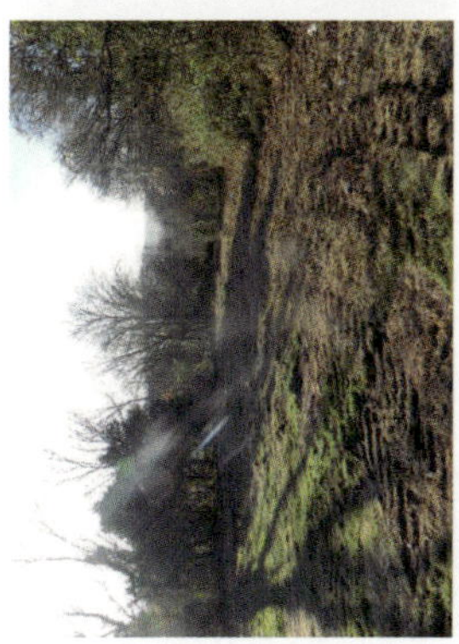

0797G

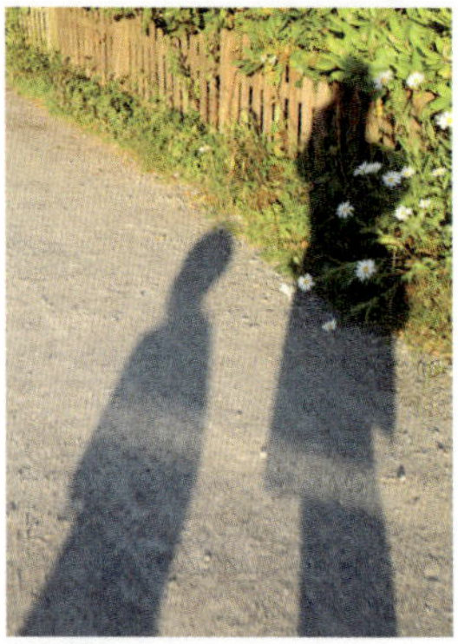
0768P

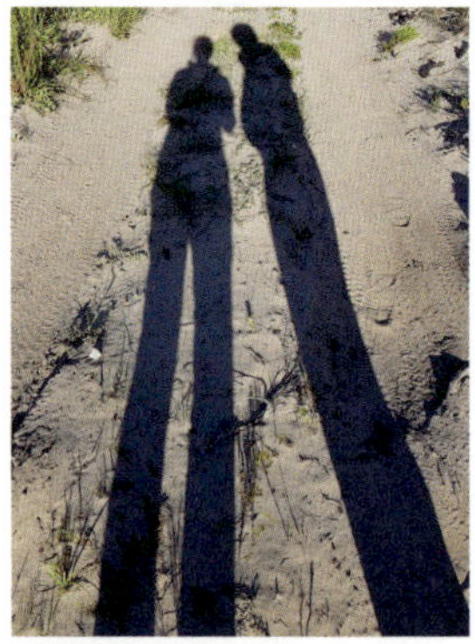
0768G

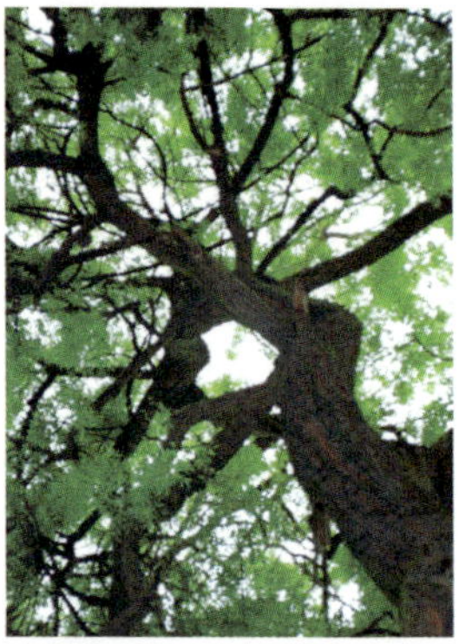
0820P

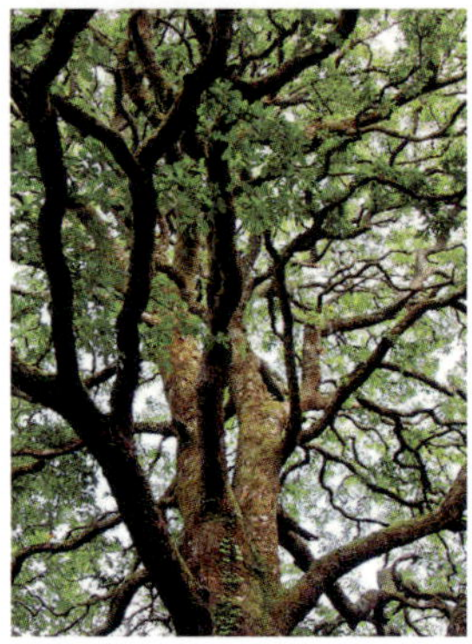
0820G

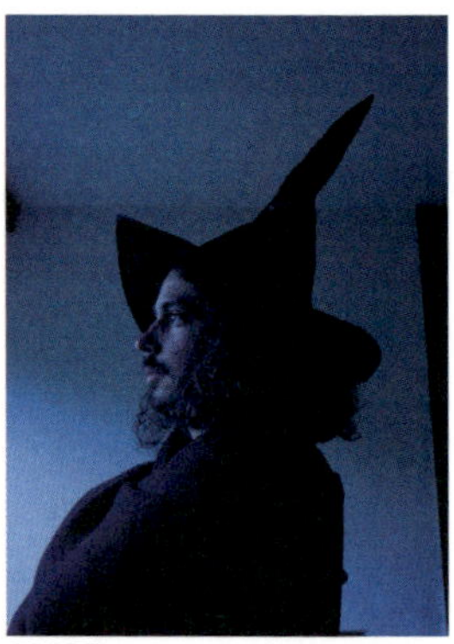
0868P

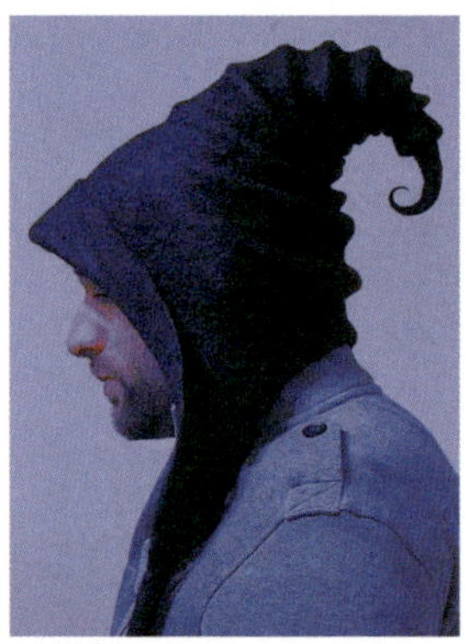
0868G

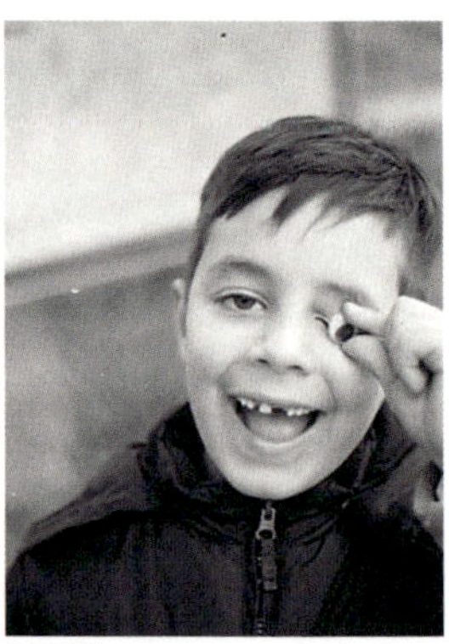
0897P

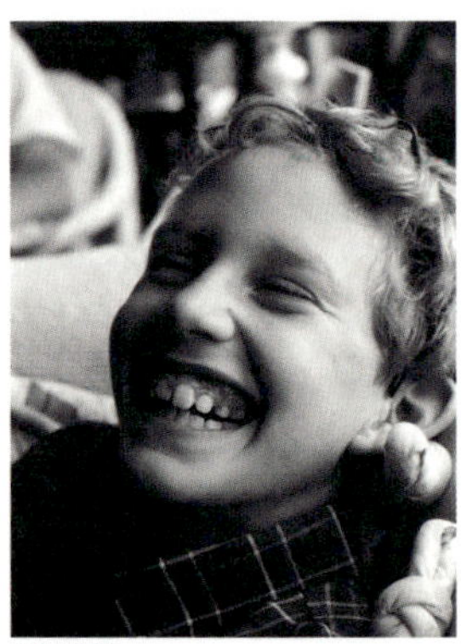
0897G

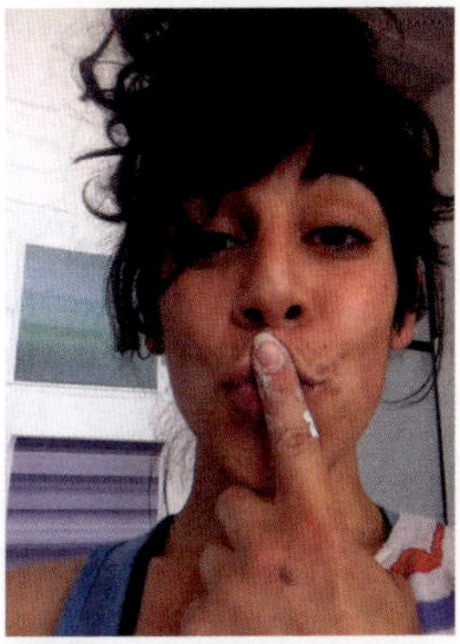

0903P

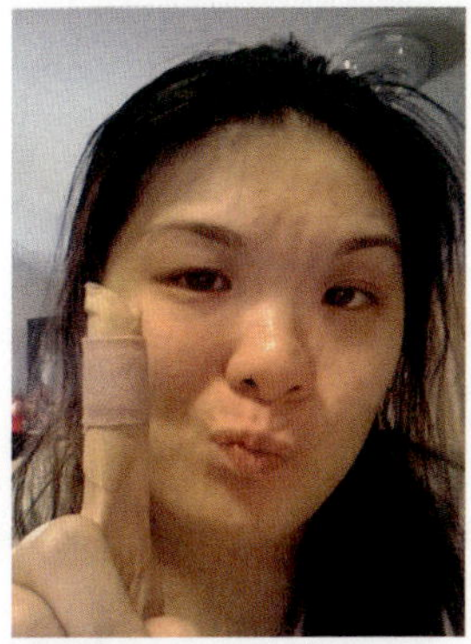

0903G

0917P

0917G

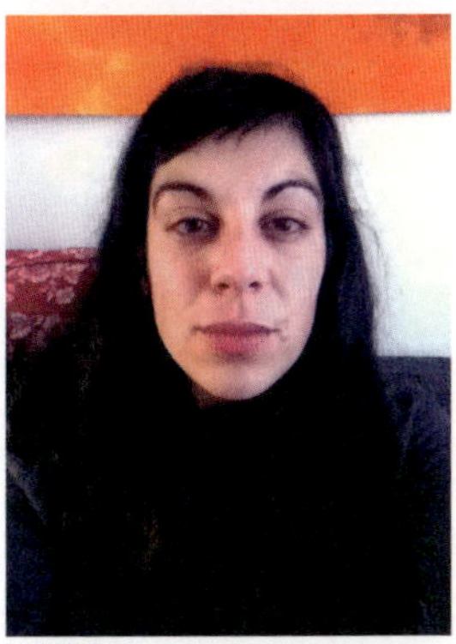

0920P

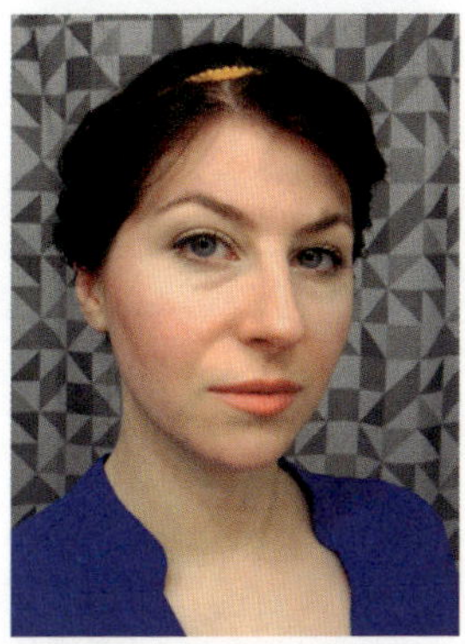

0920G

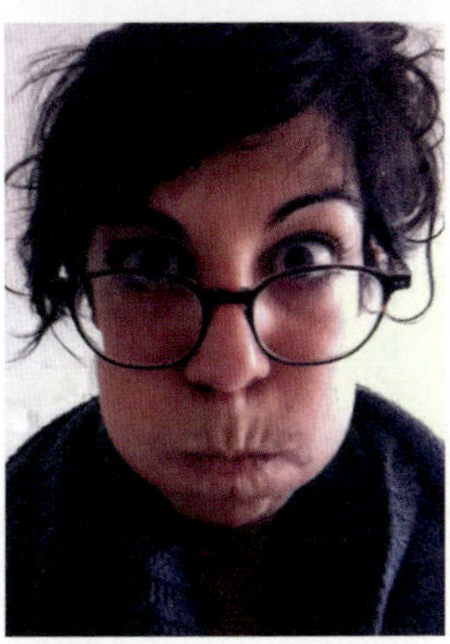

0922P

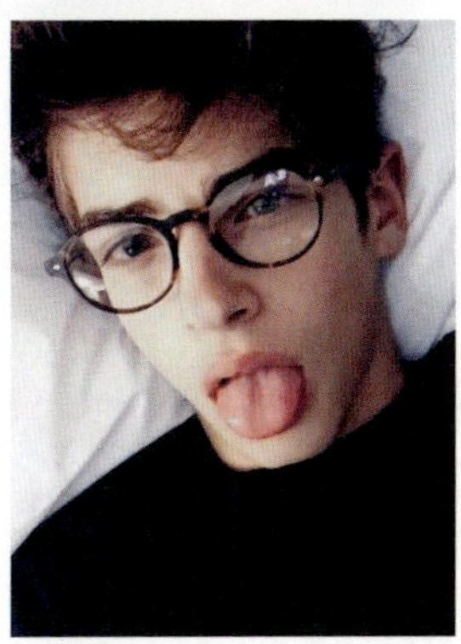

0922G

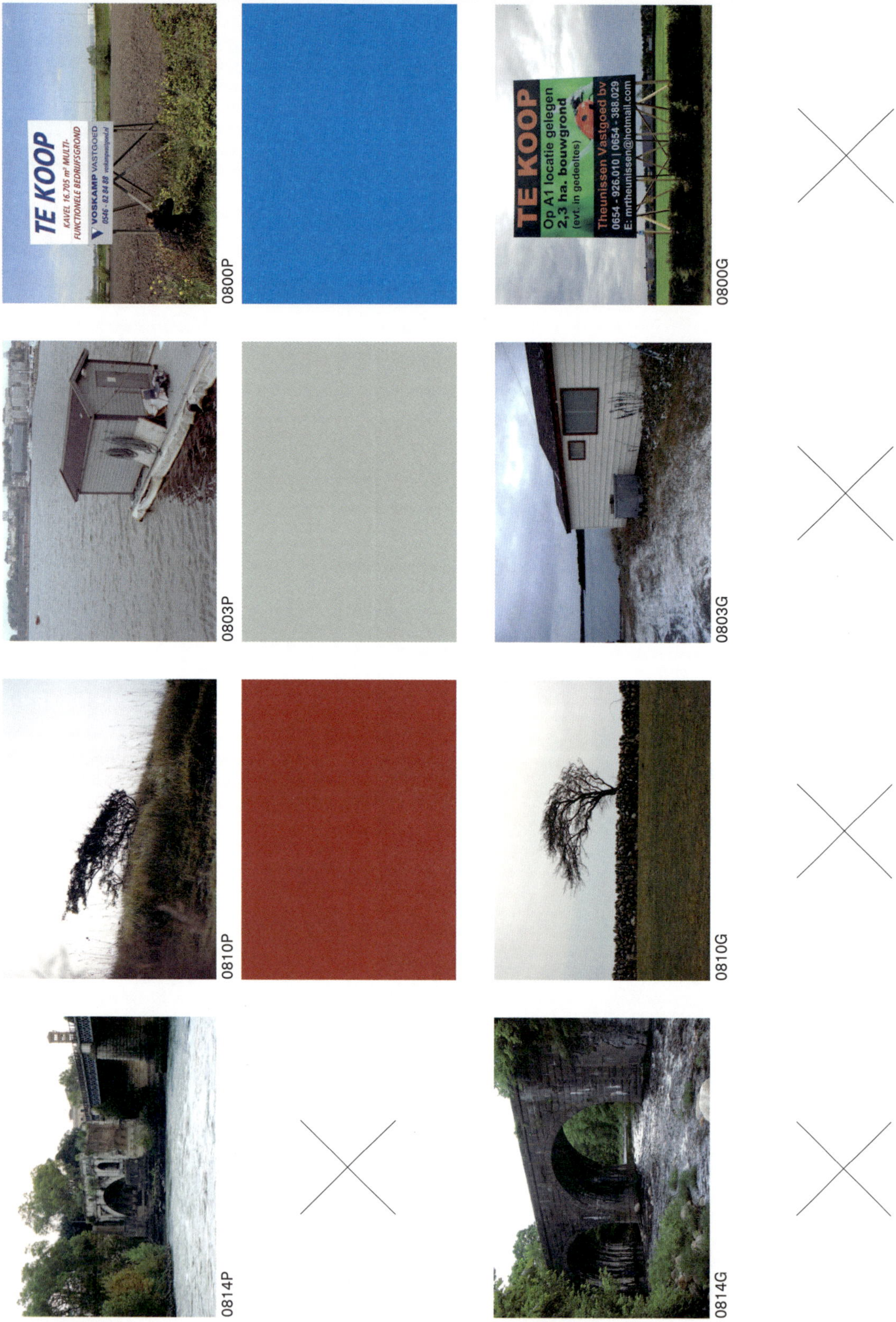
0800P
TE KOOP
KAVEL 16.705 m² MULTI-
FUNCTIONELE BEDRIJFSGROND
VOSKAMP VASTGOED
0546 - 82 84 88
0803P
0810P
0814P
0800G
TE KOOP
Op A1 locatie gelegen
2,3 ha. bouwgrond
(evt. in gedeeltes)
Theunissen Vastgoed bv
0654 - 926.010 | 0654 - 388.029
E: mrtheunissen@hotmail.com
0803G
0810G
0814G

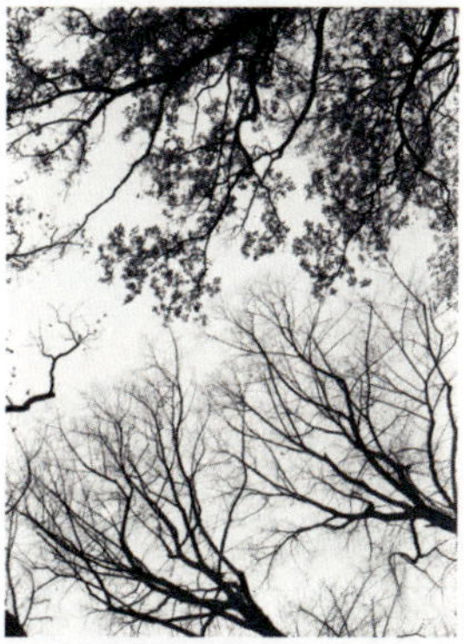
0816P

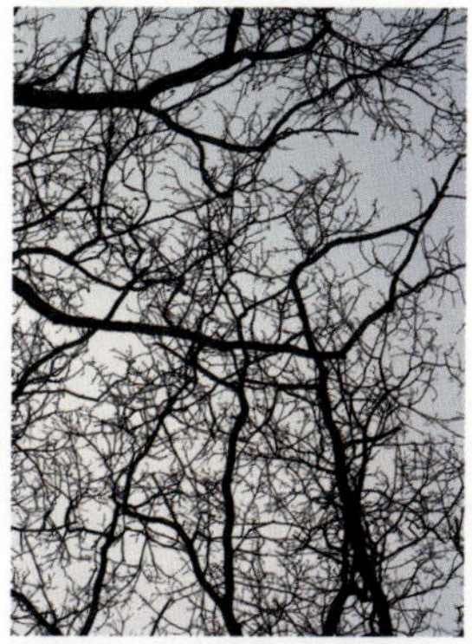
0816G

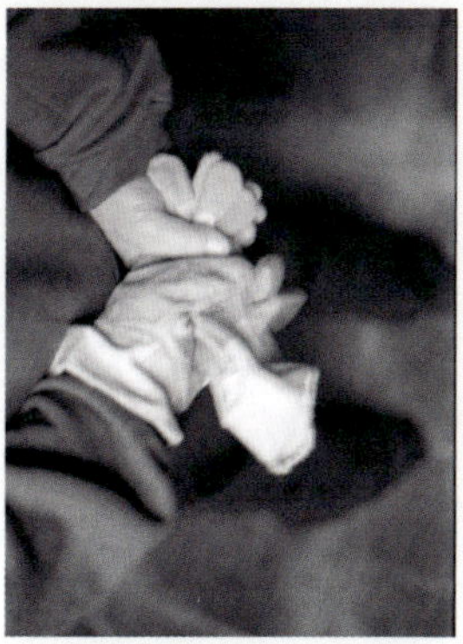
0823P

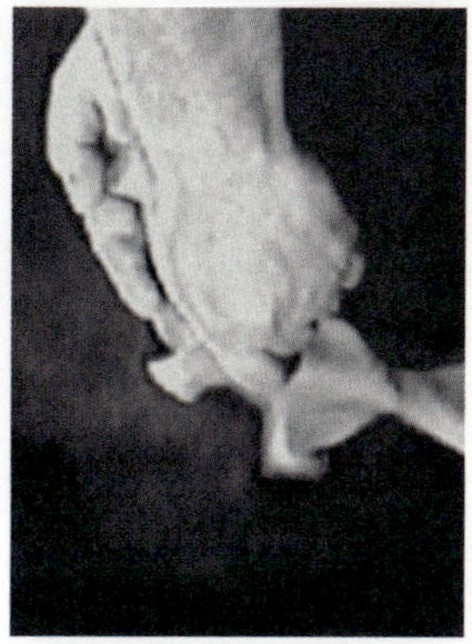
0823G

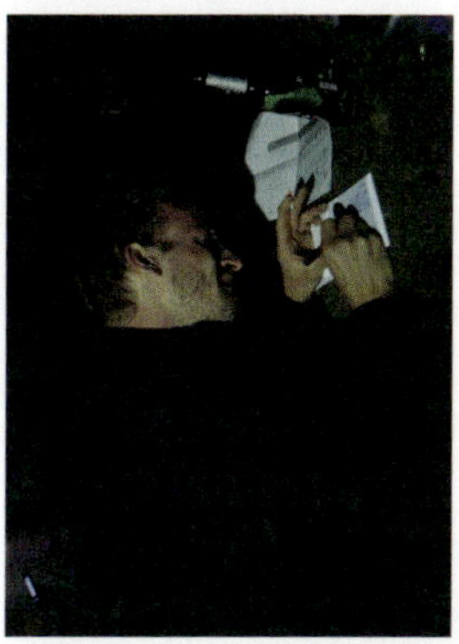
0825P

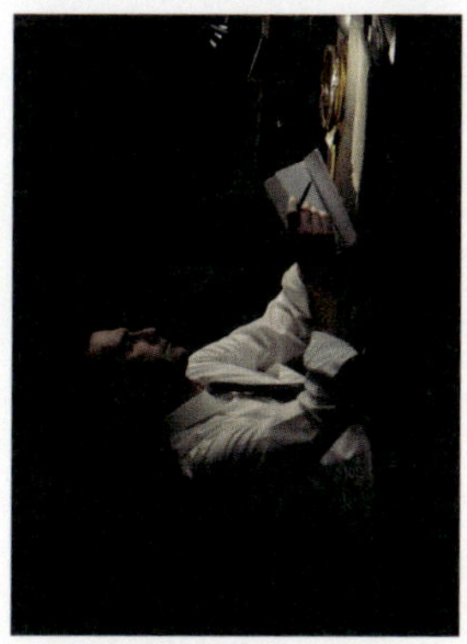
0825G

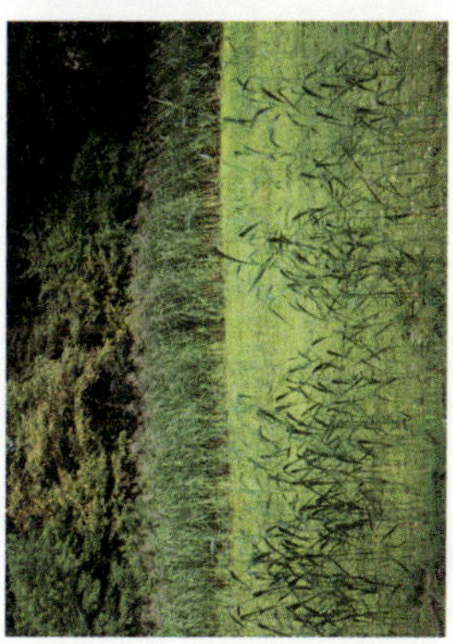
0839P

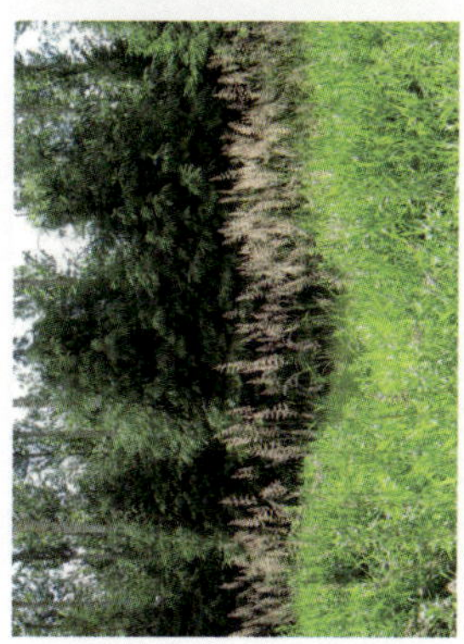
0839G

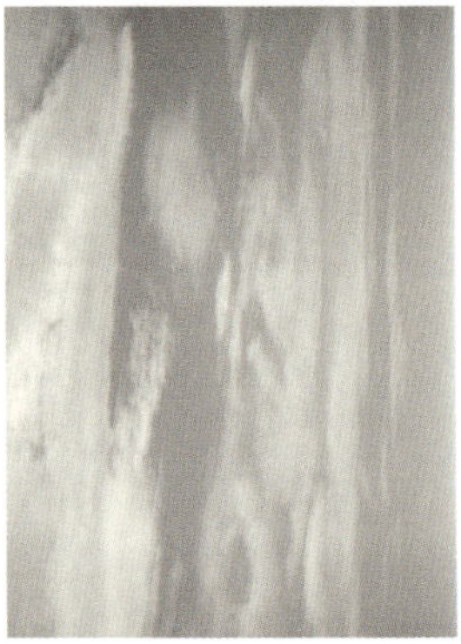
0841P

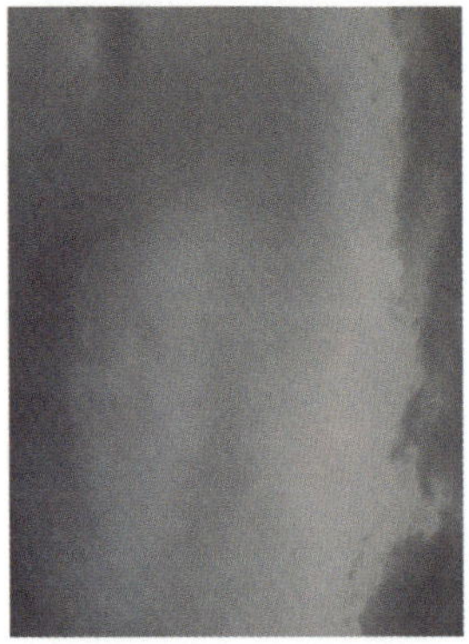
0841G

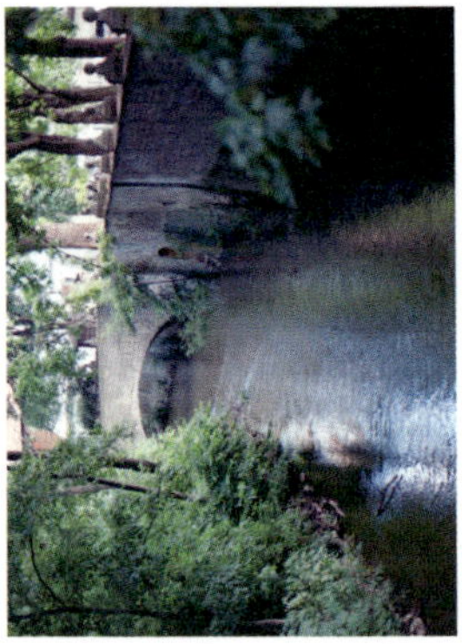
0845P

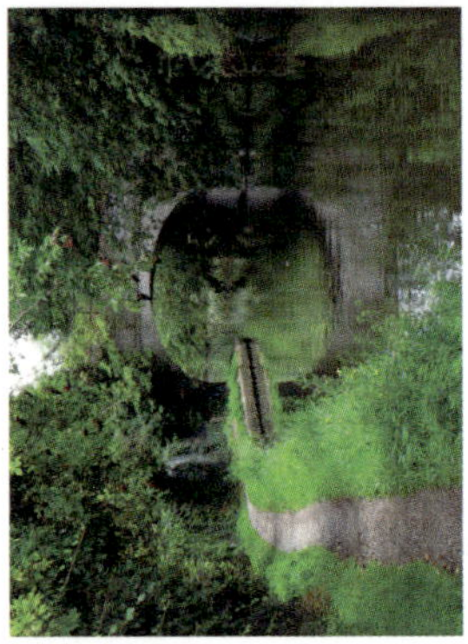
0845G

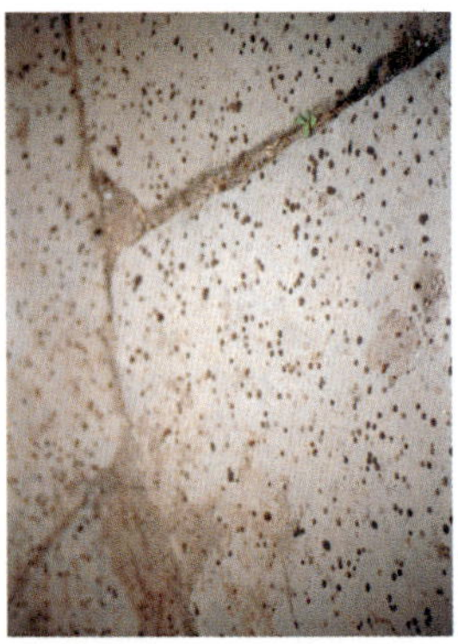
0848P

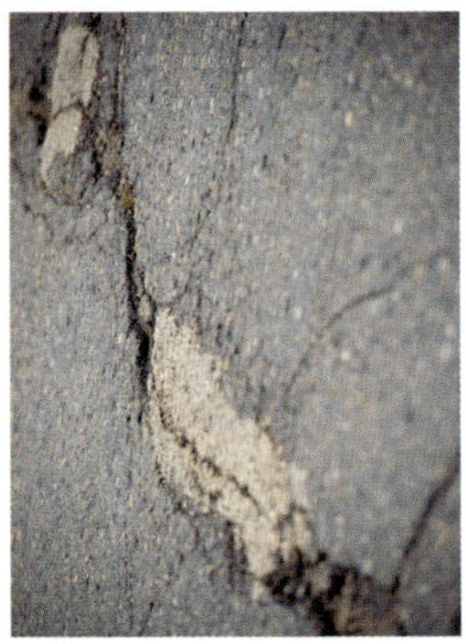
0848G

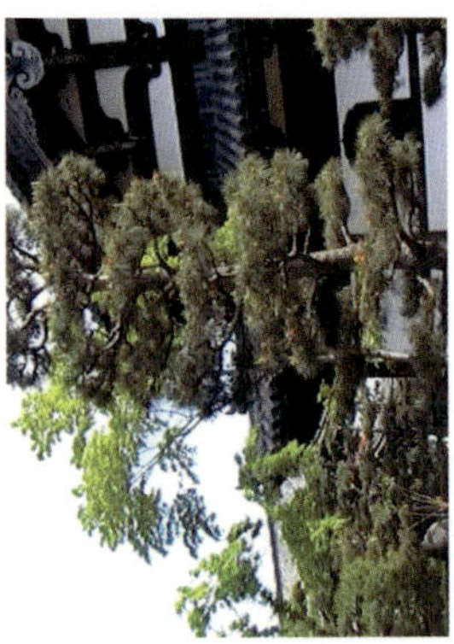
0850P

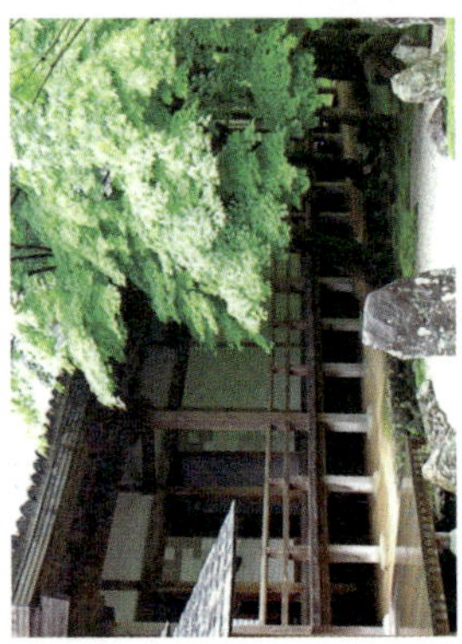
0850G

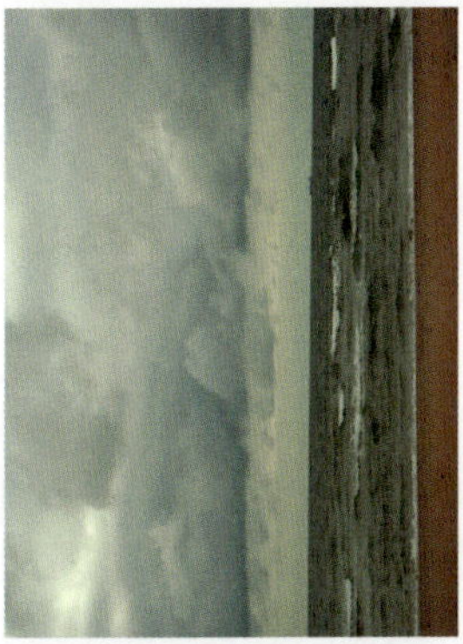

0851P

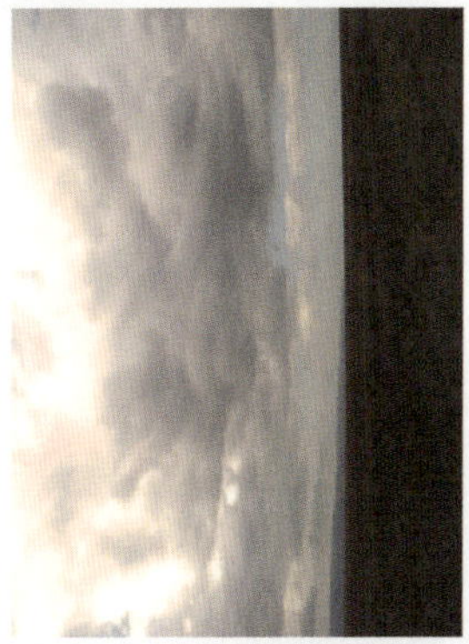

0851G

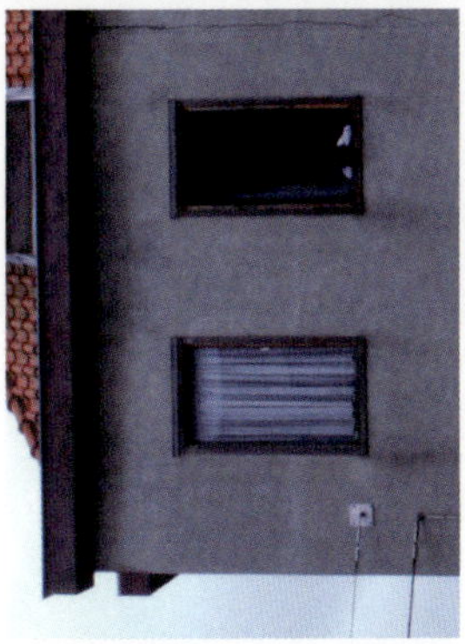

0852P

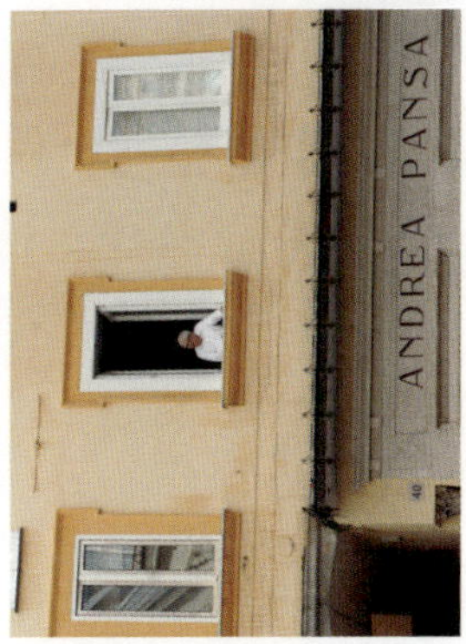

0852G

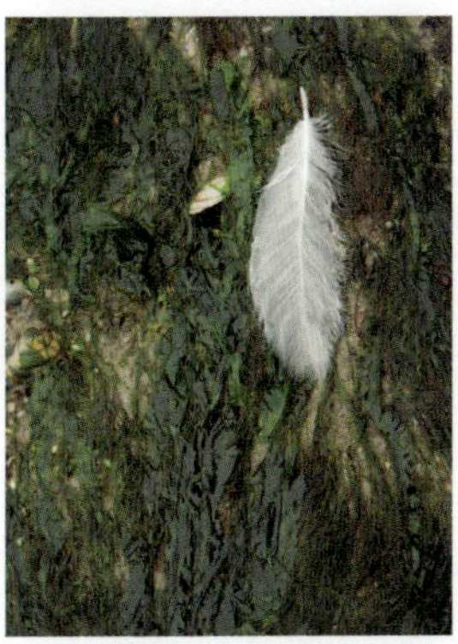

0857P

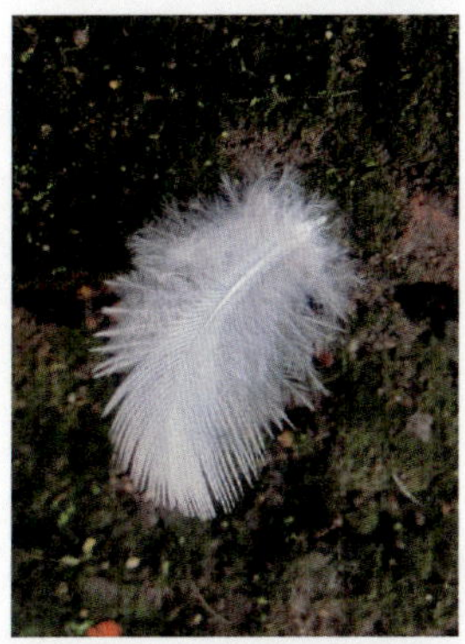

0857G

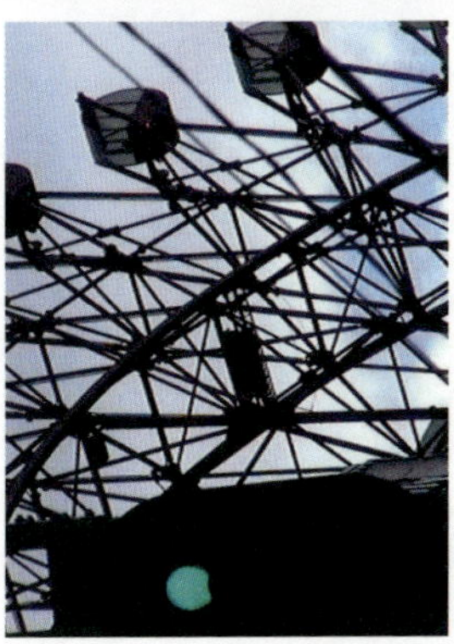

0908P

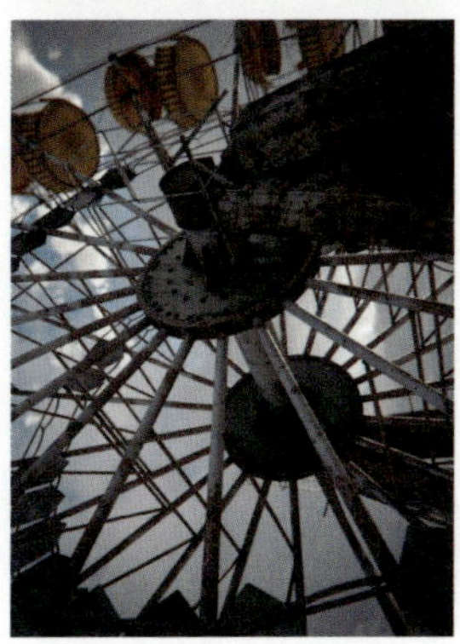

0908G

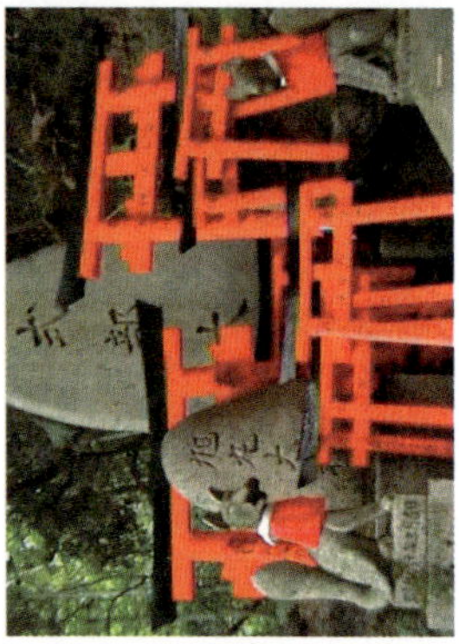

0909P

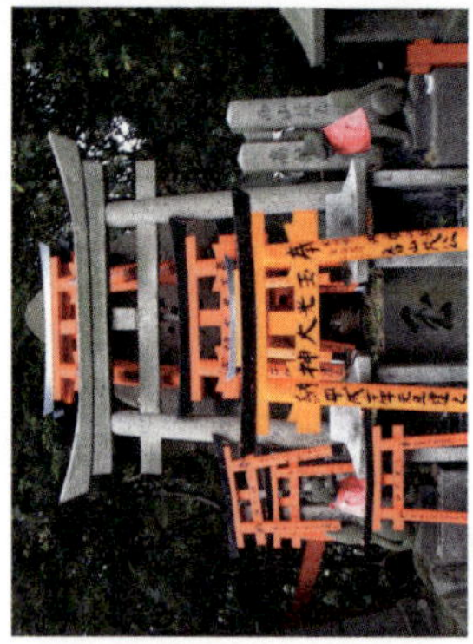

0909G

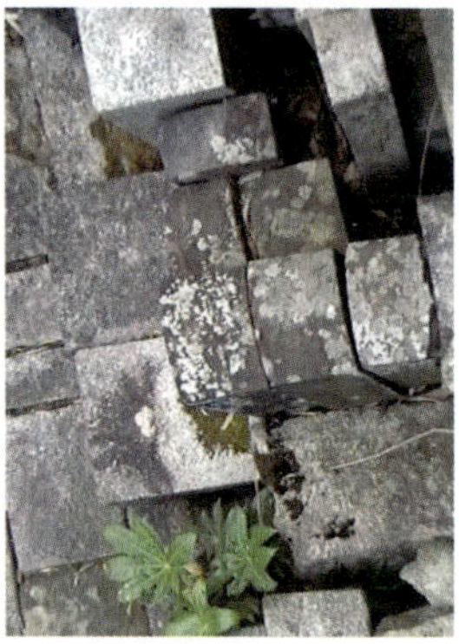

0910P

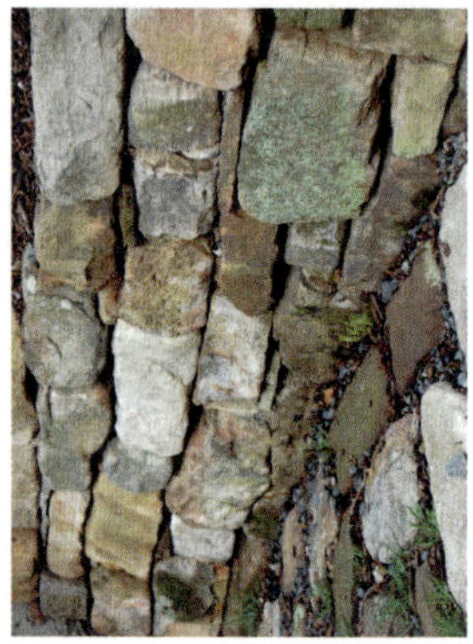

0910G

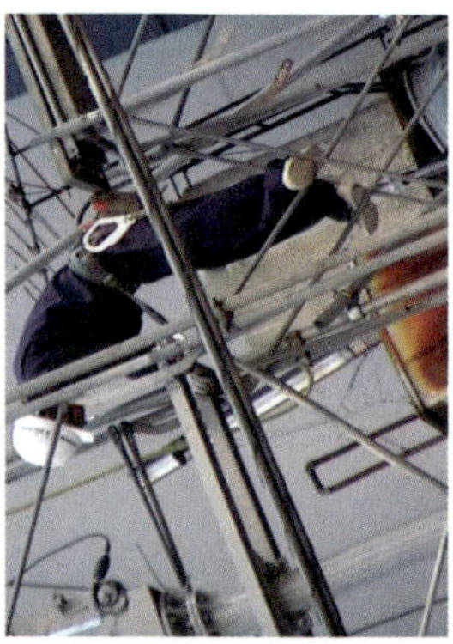

0969P

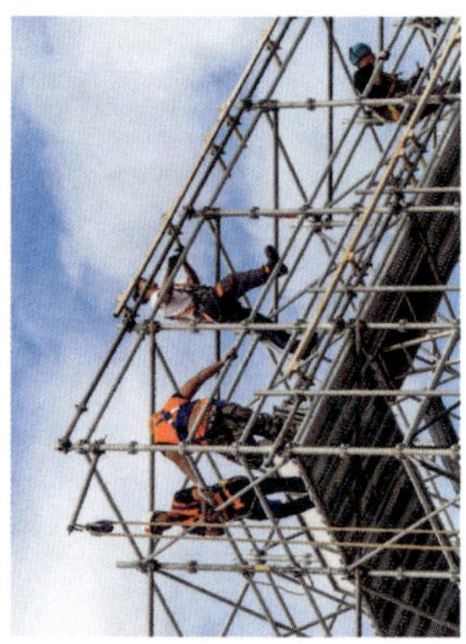

0969G

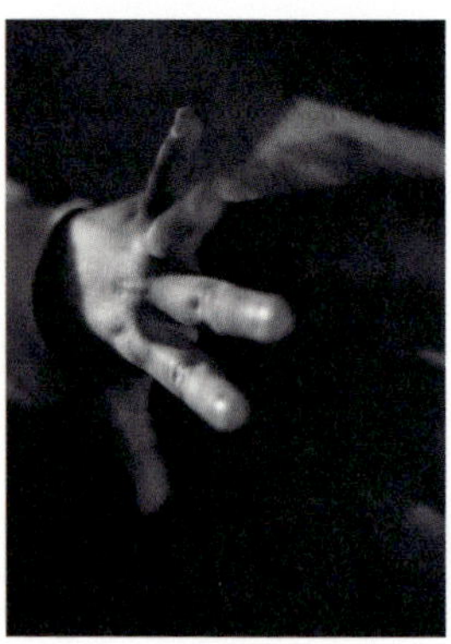

0970P

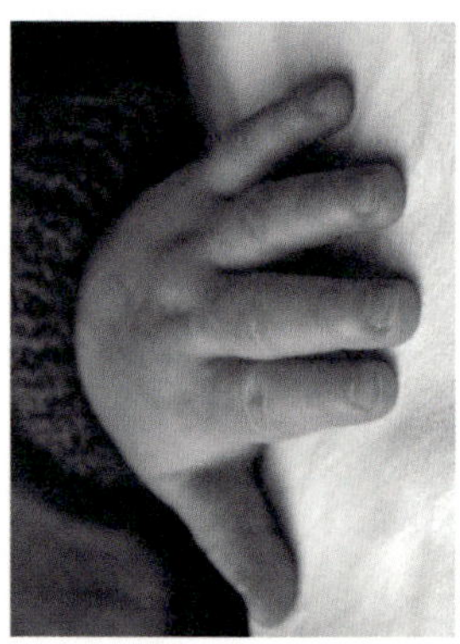

0970G

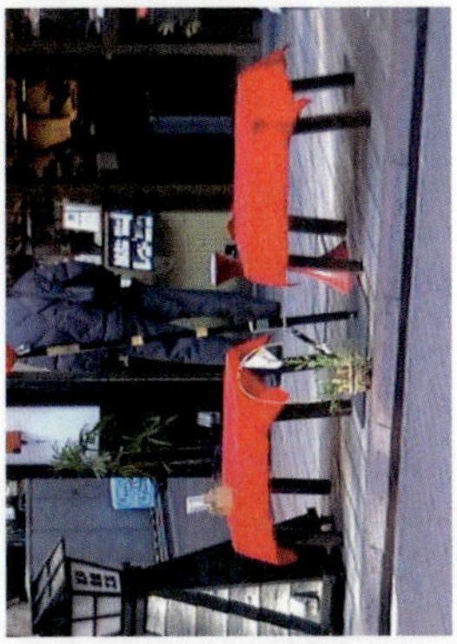
0972P

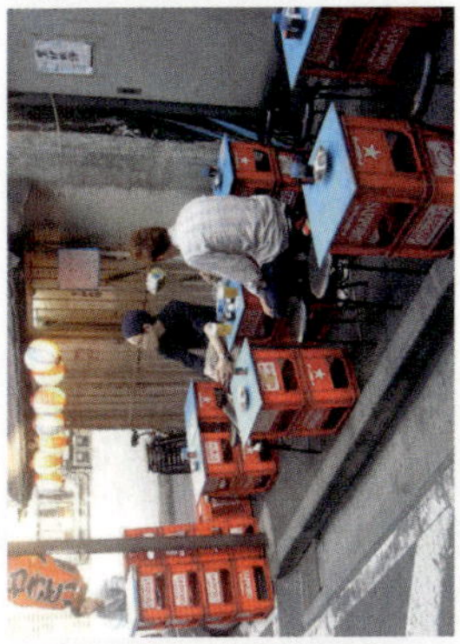
0972G

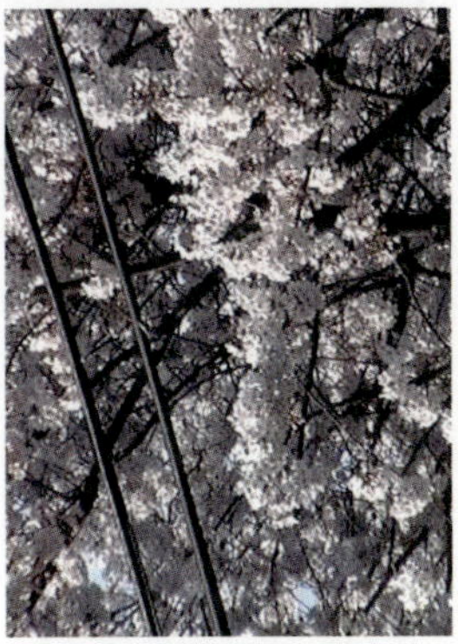
0992P

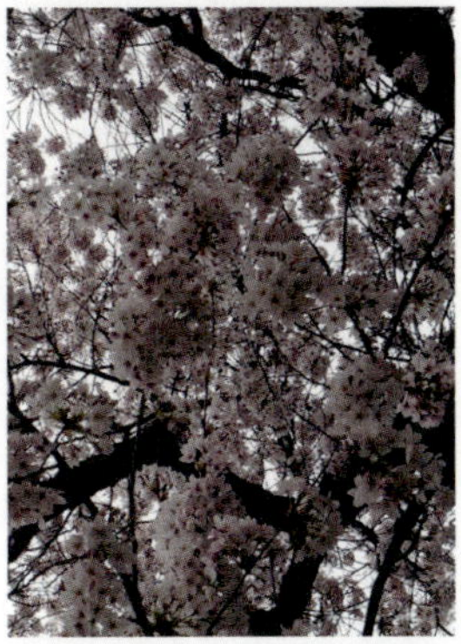
0992G

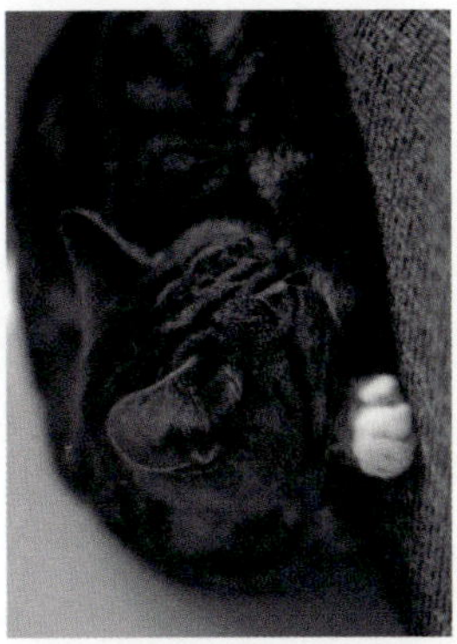
0997P

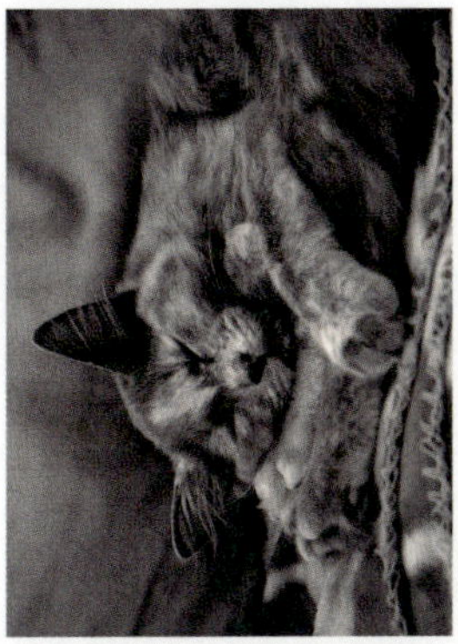
0997G

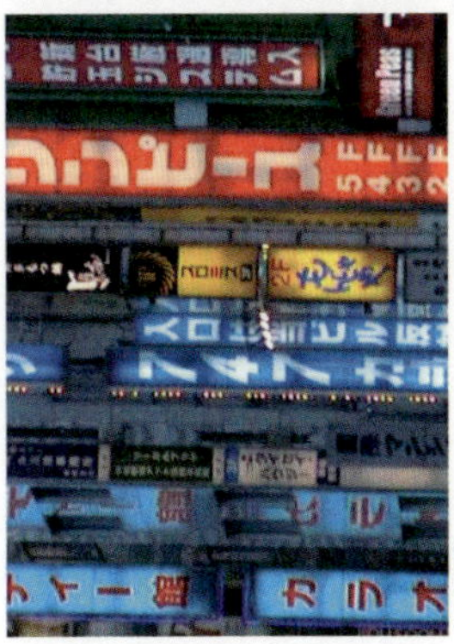
1011P

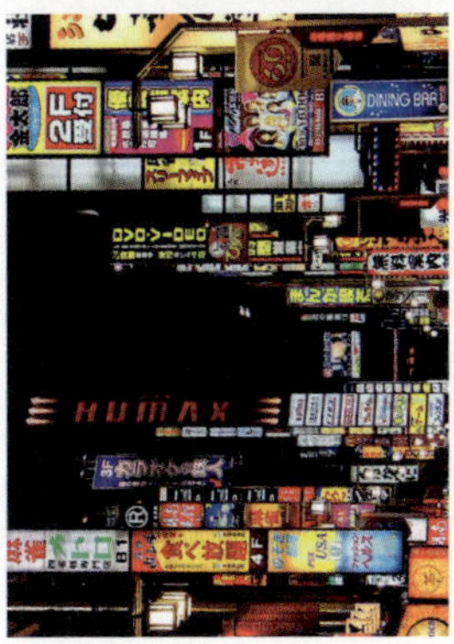

1011G

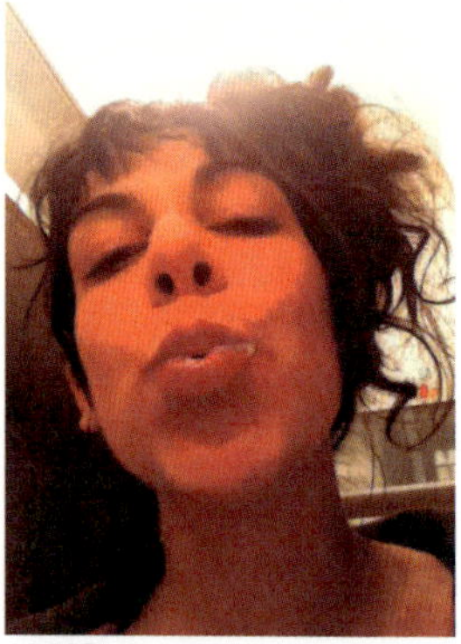
0923P

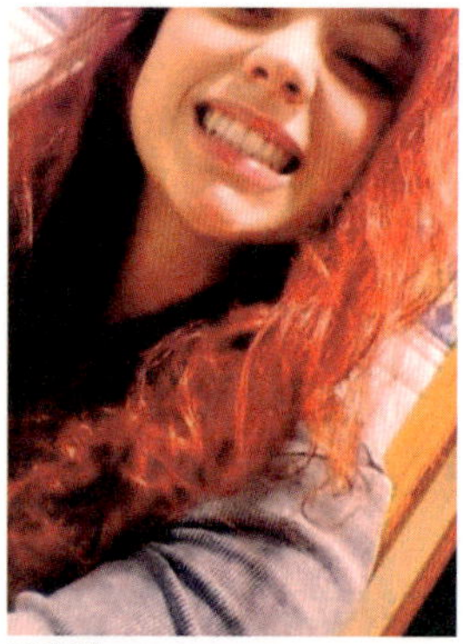
0923G

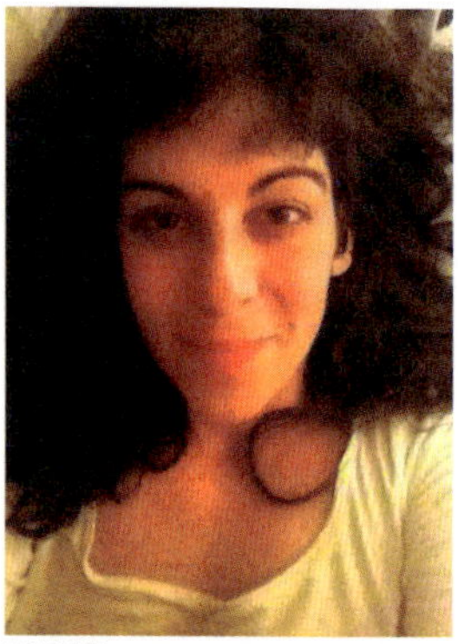
0924P

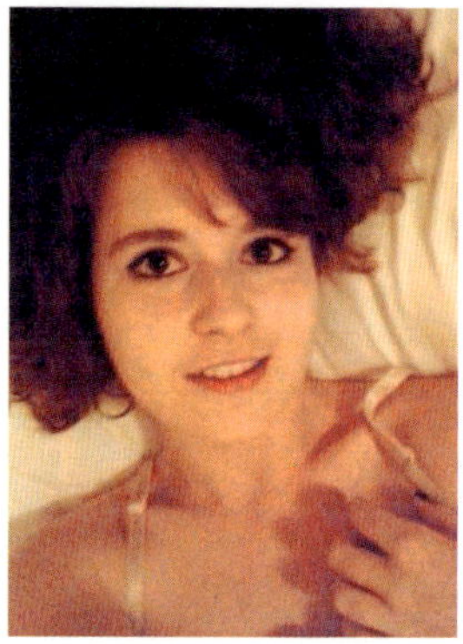
0924G

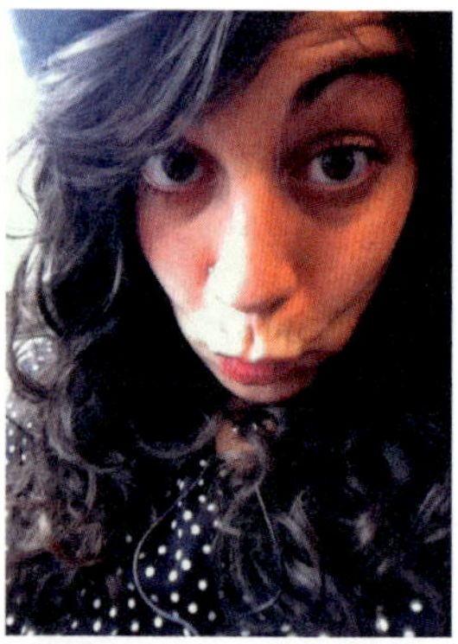
0926P

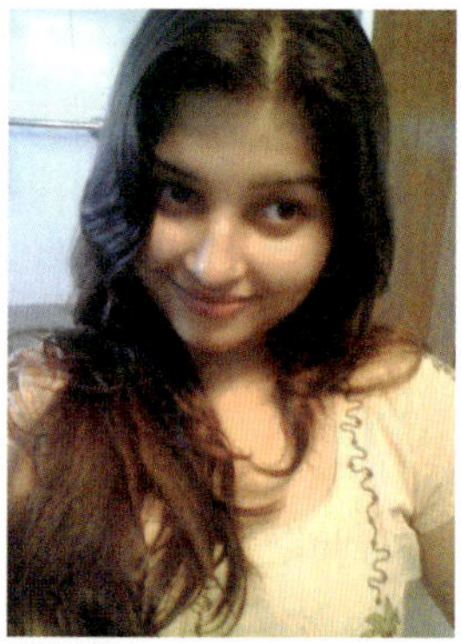
0926G

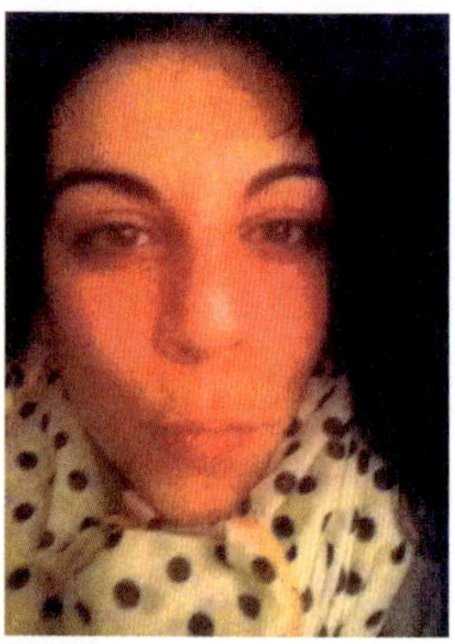
0927P

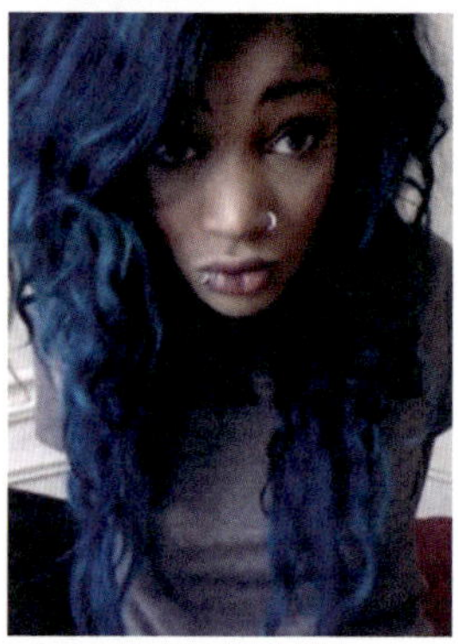
0927G

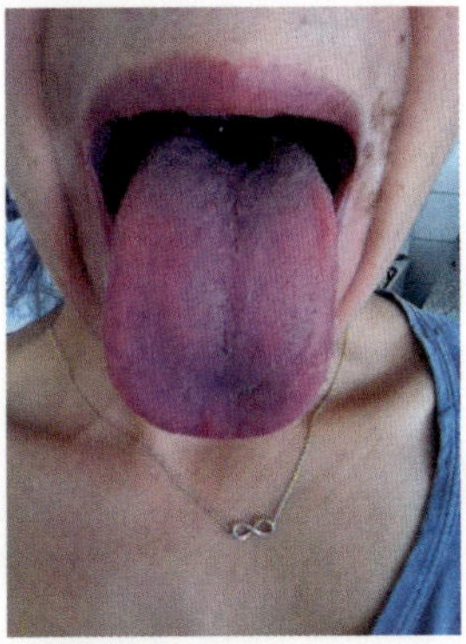

0931P

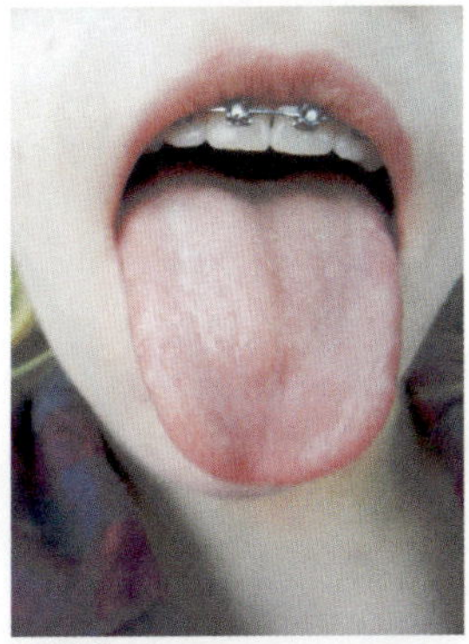

0931G

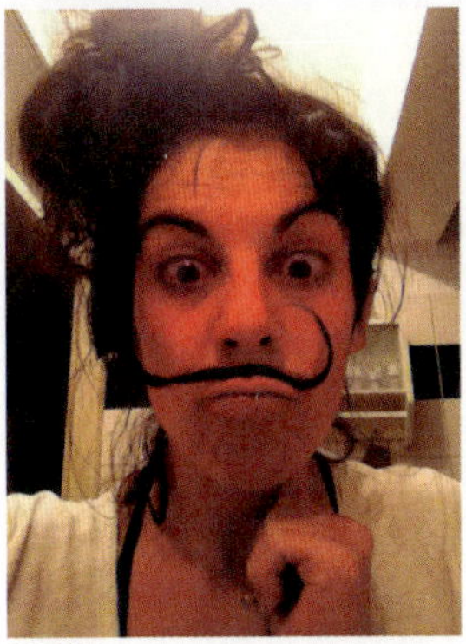

0936P

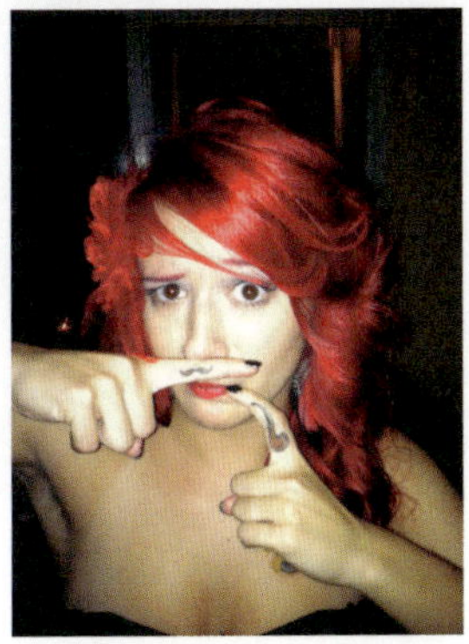

0936G

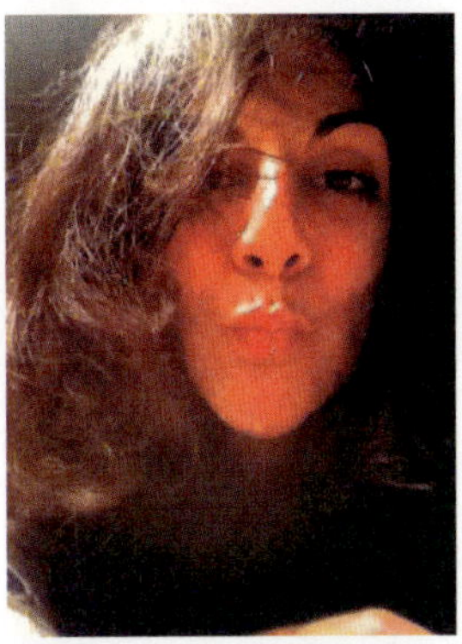

0941P

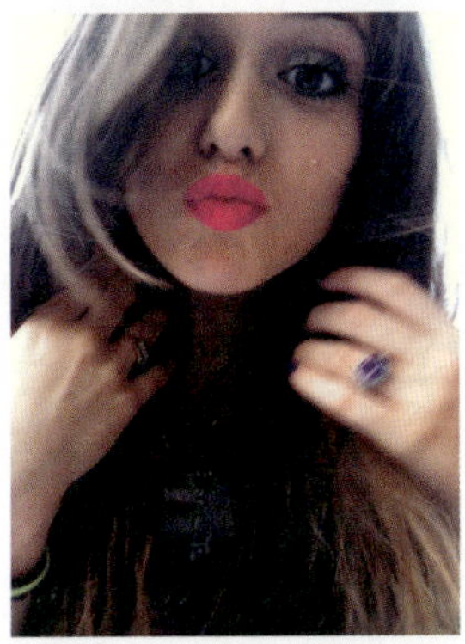

0941G

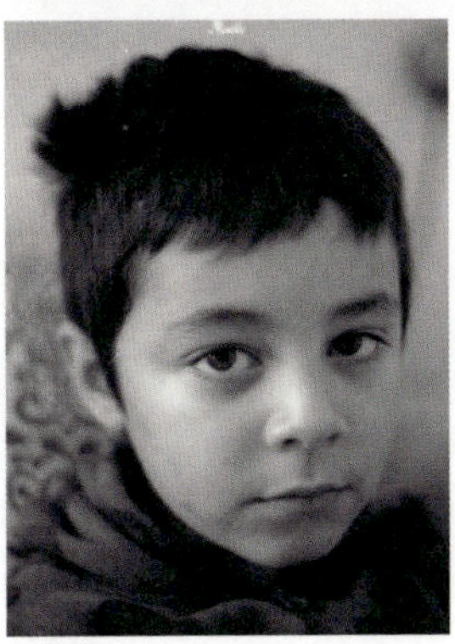

1001P

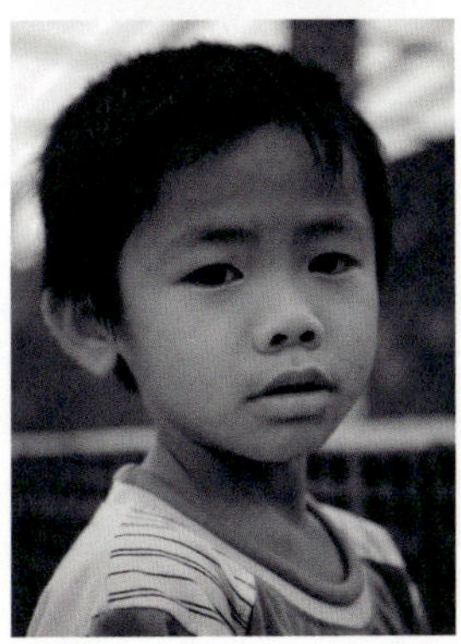

1001G

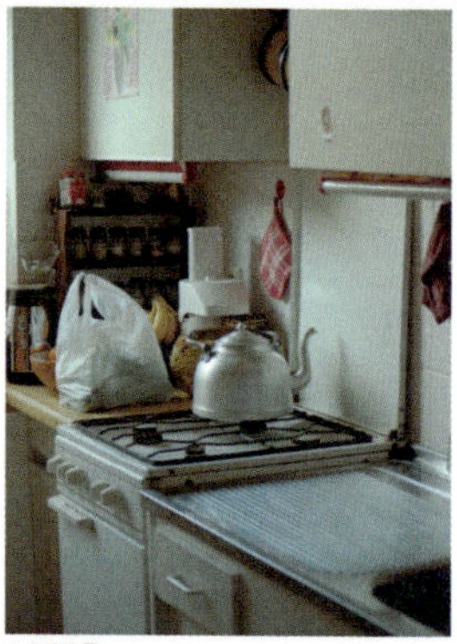

1057P

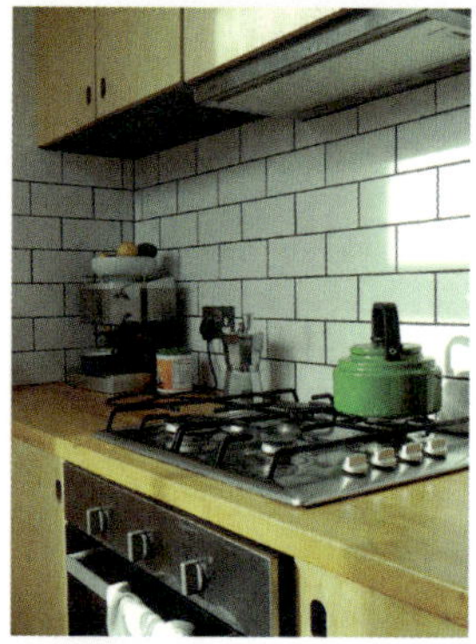

1057G

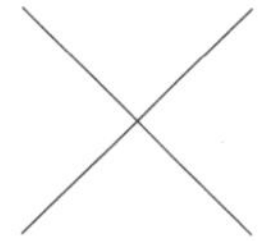

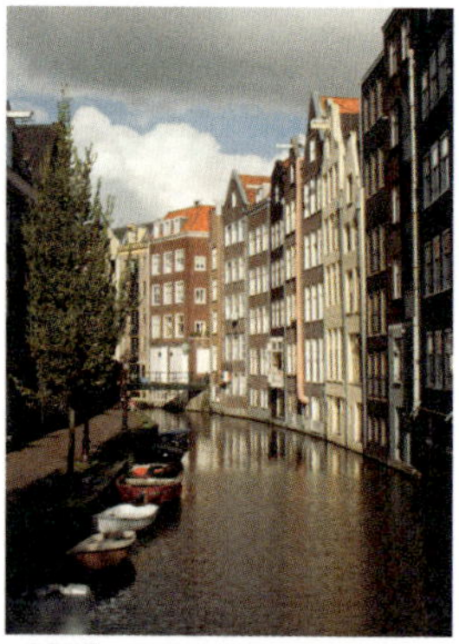

1067P

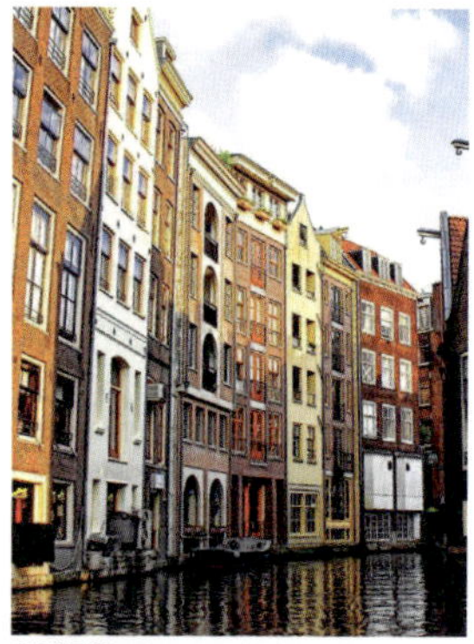

1067G

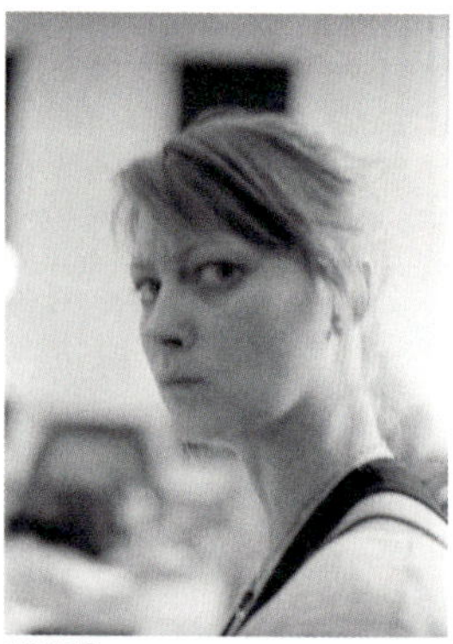

1068P

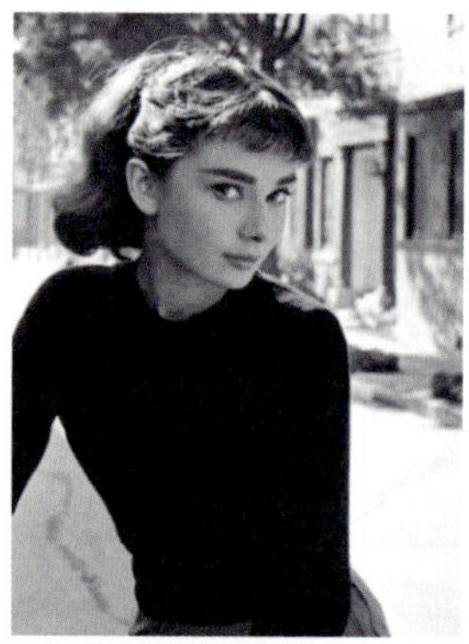

1068G

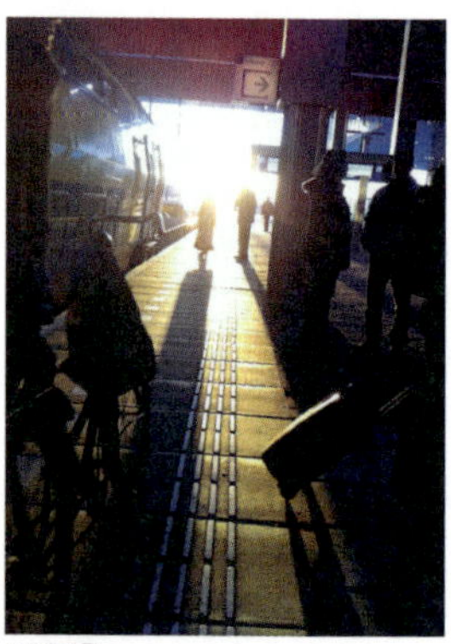

1084P

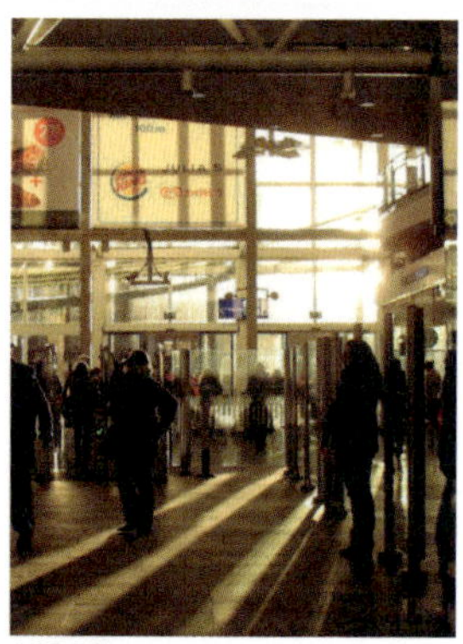

1084G

1085P

1085G

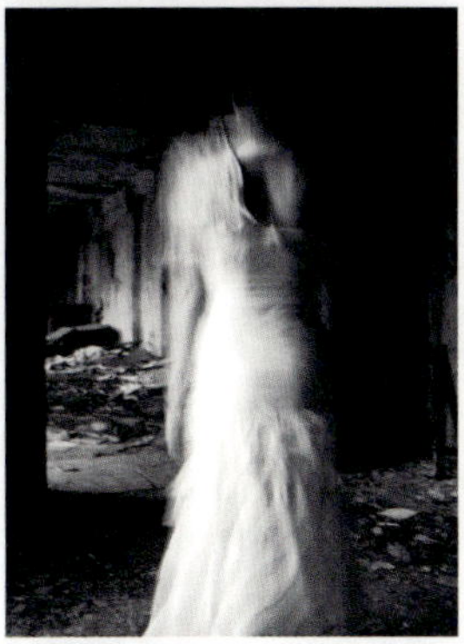
1088P

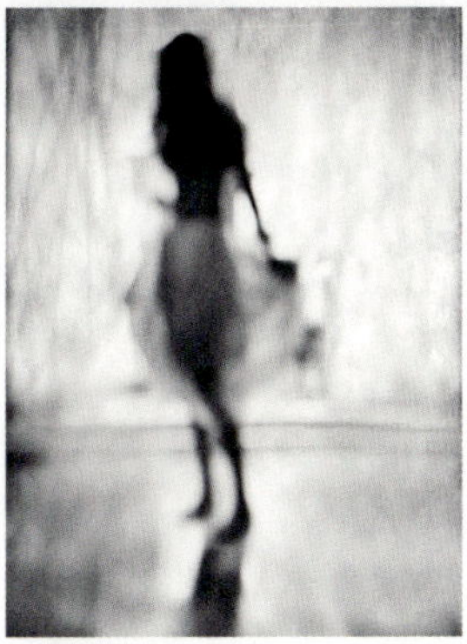
1088G

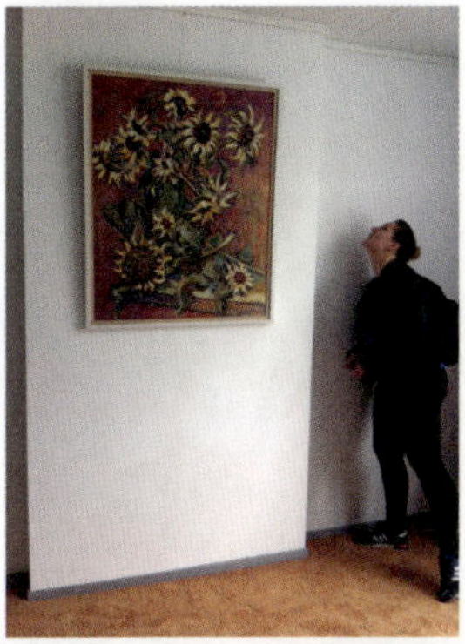
1089P

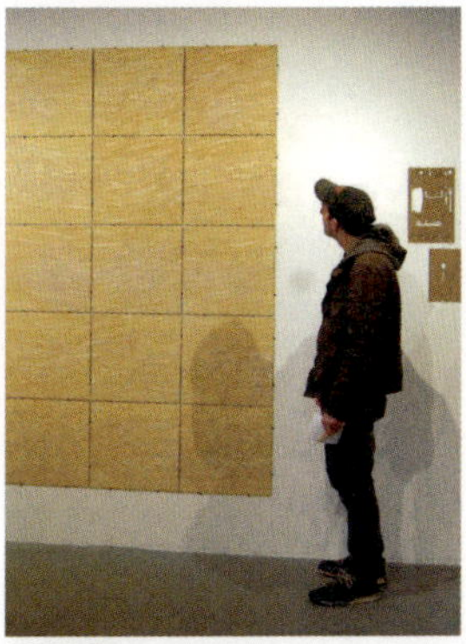
1089G

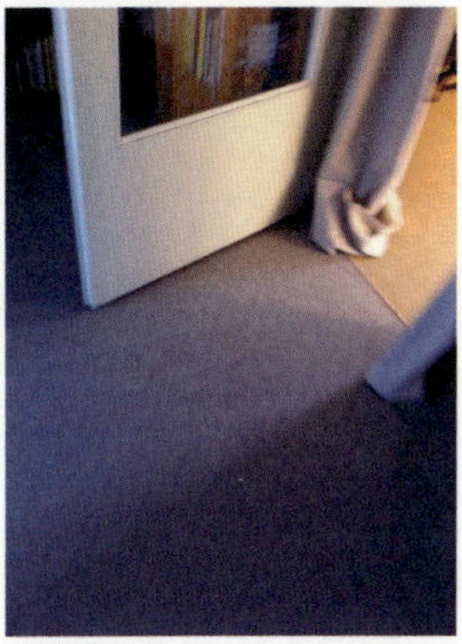
1117P

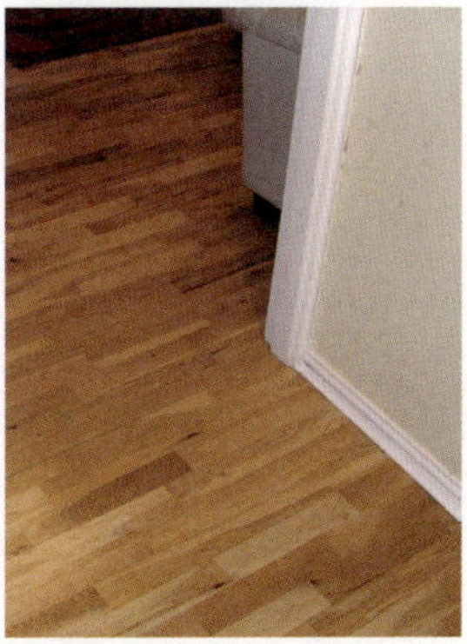
1117G

1123P

1123G

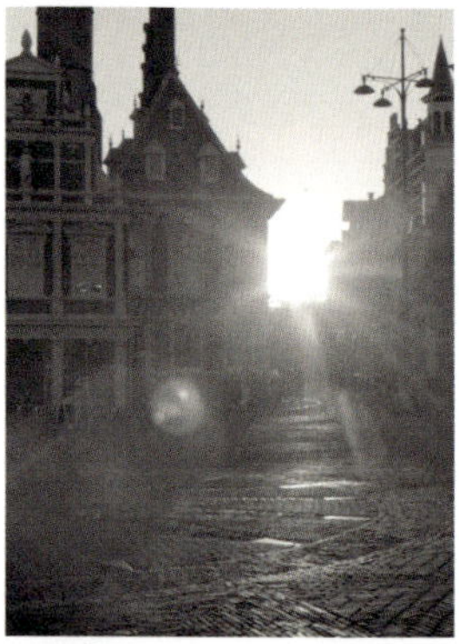
1129P
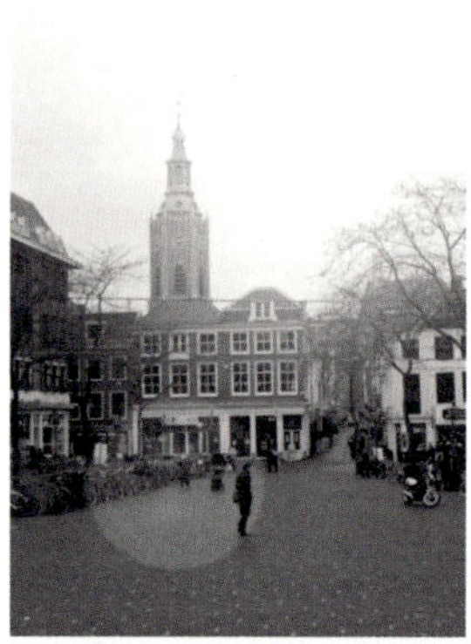
1129G
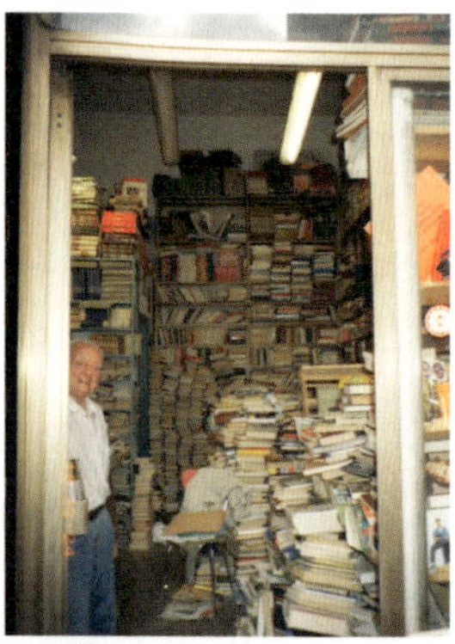
1137P

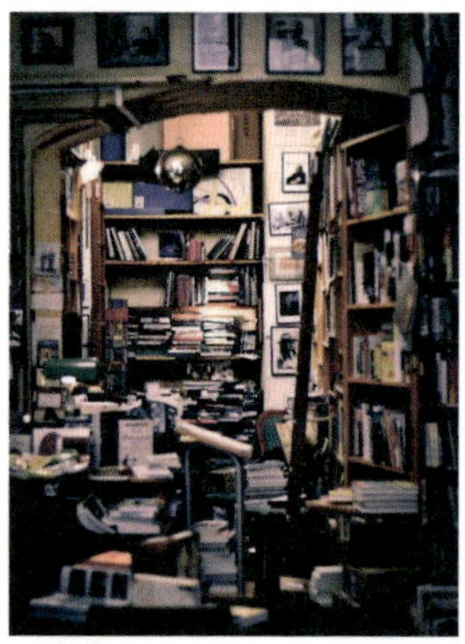
1137G
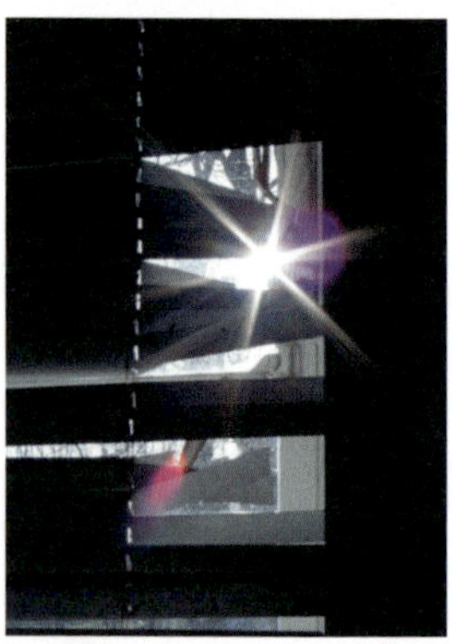
1142P

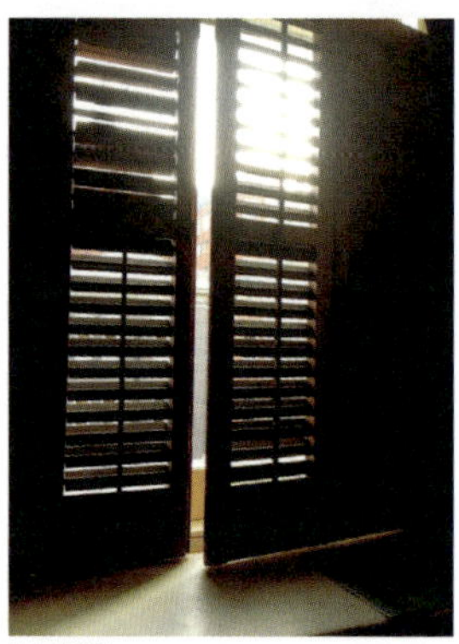
1142G

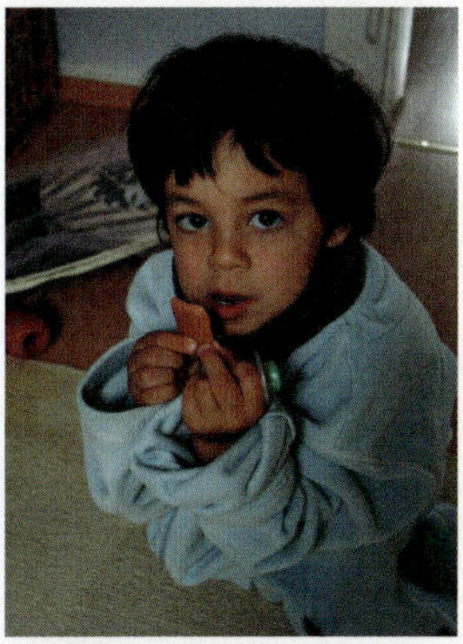

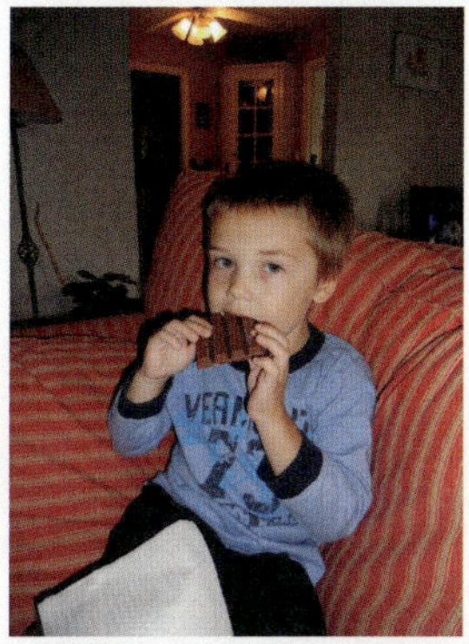

1176P

1176G

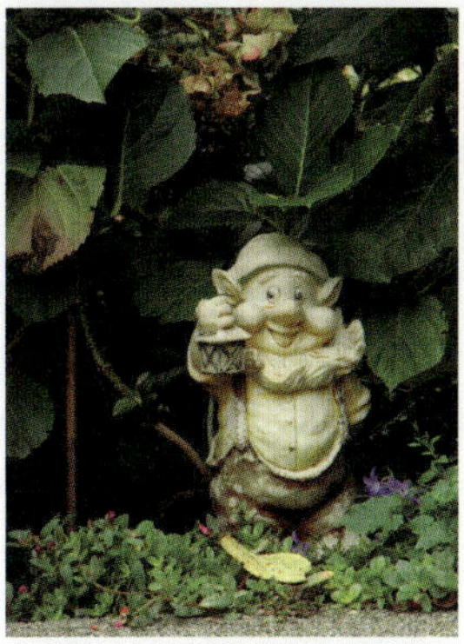

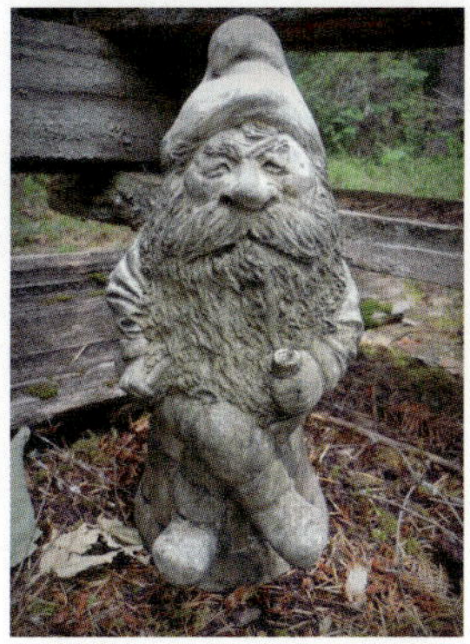

1188P

1188G

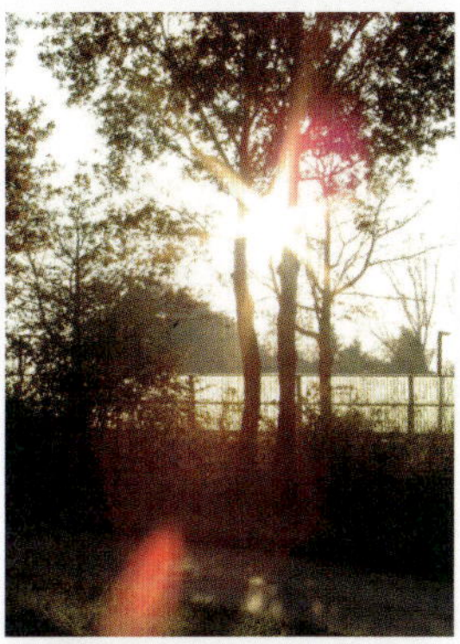

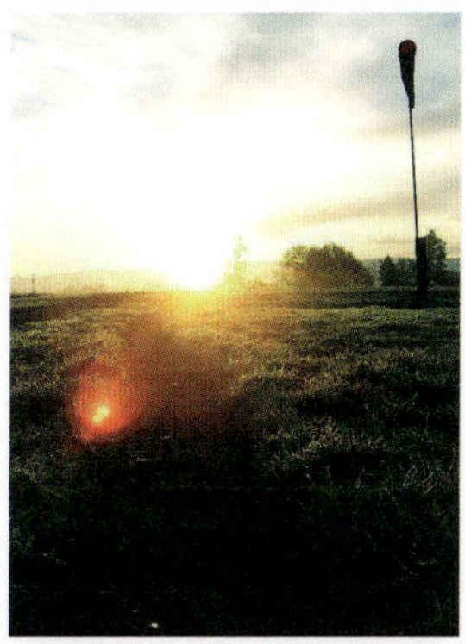

1192P

1192G

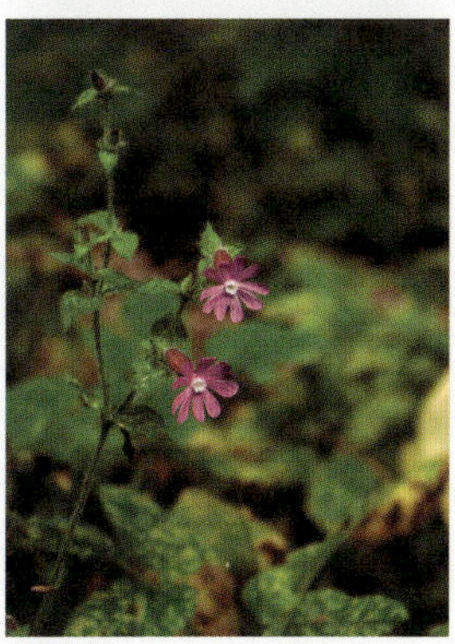
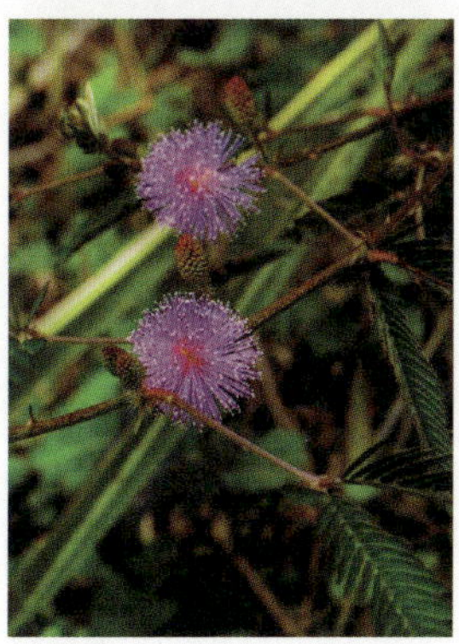

1195P

1195G

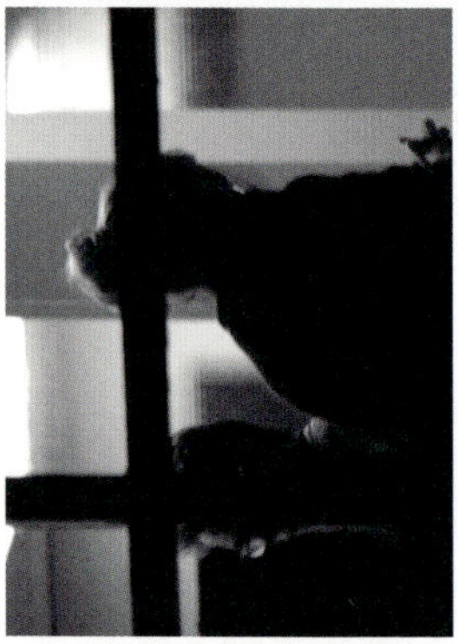

1133P

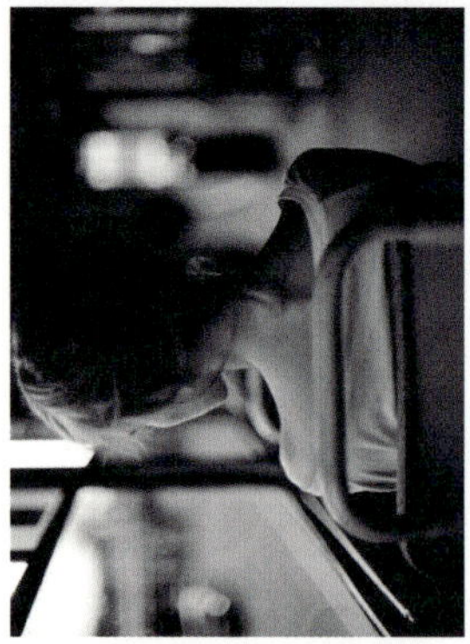

1133G

1323P

1323G

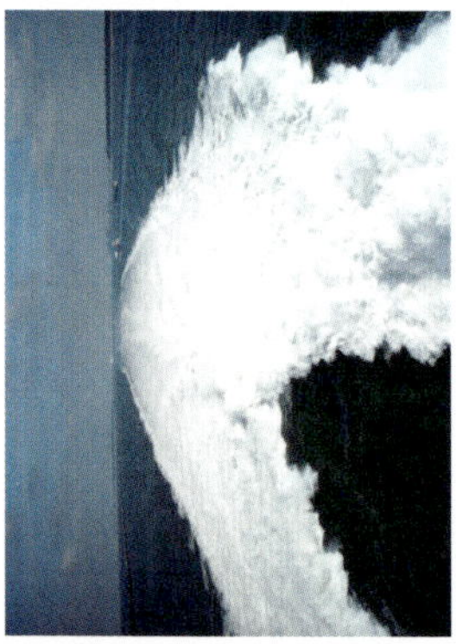

1328P

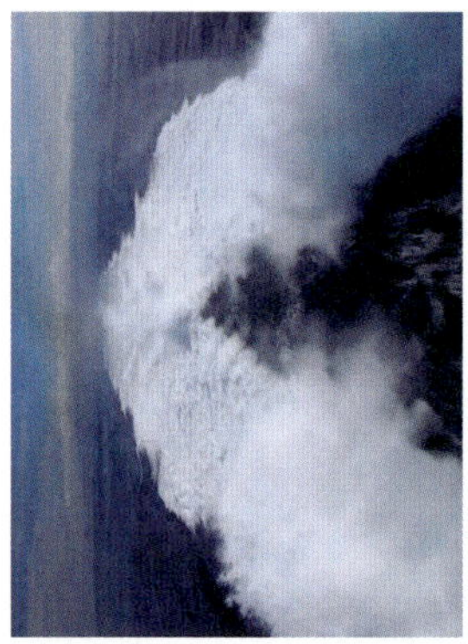

1328G

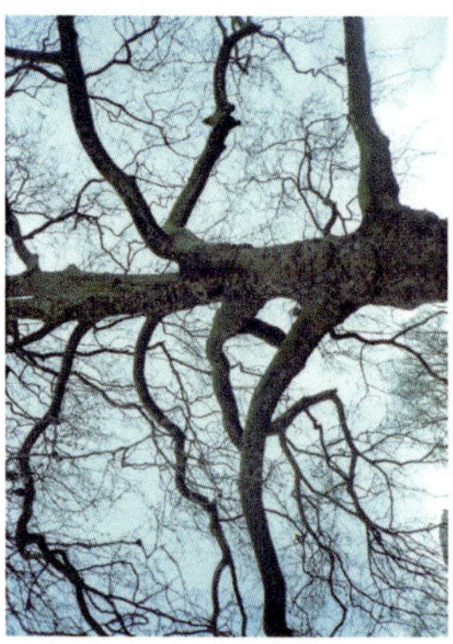

1332P

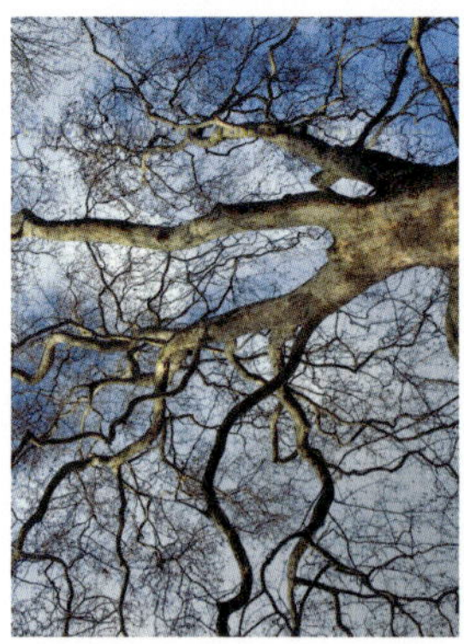

1332G

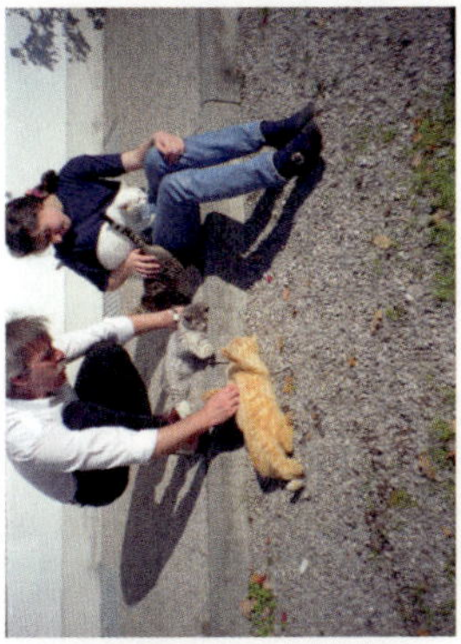

1340P

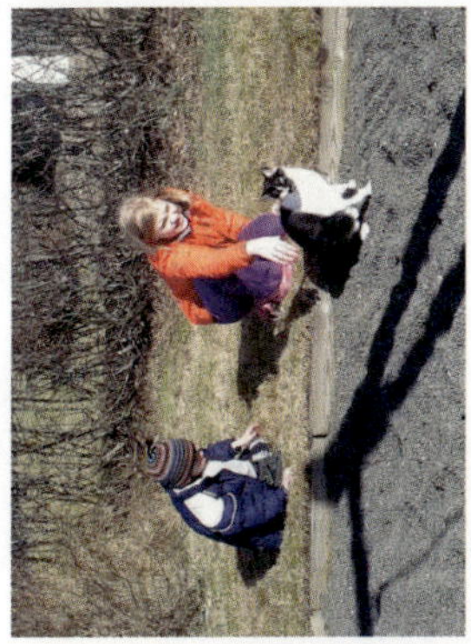

1340G

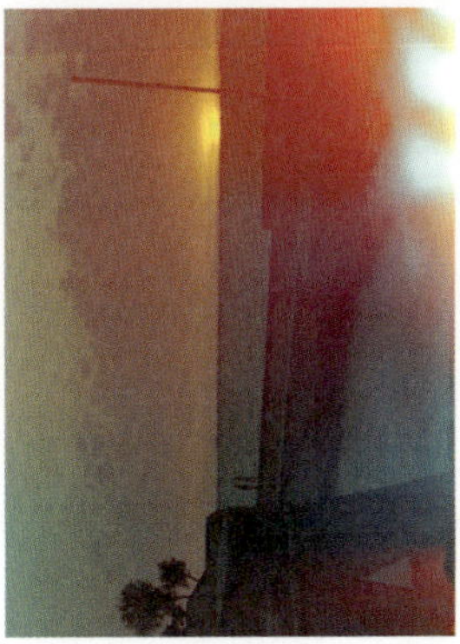

1343P

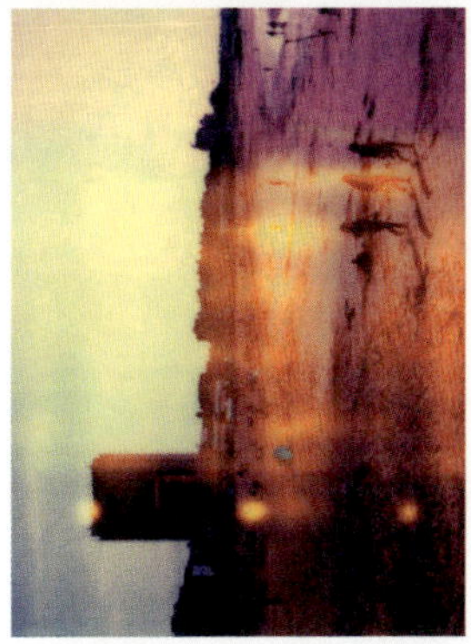

1343G

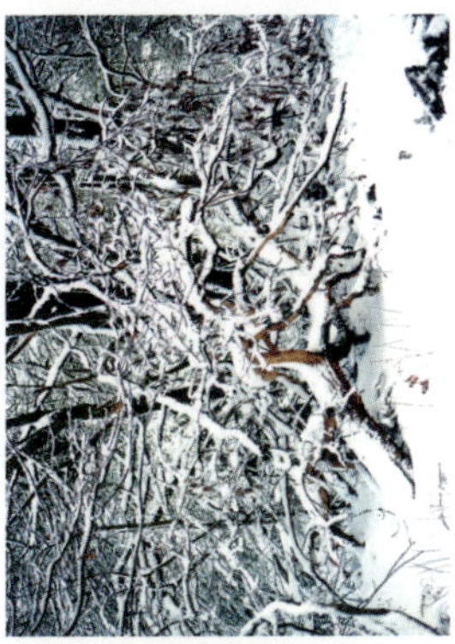

1359P

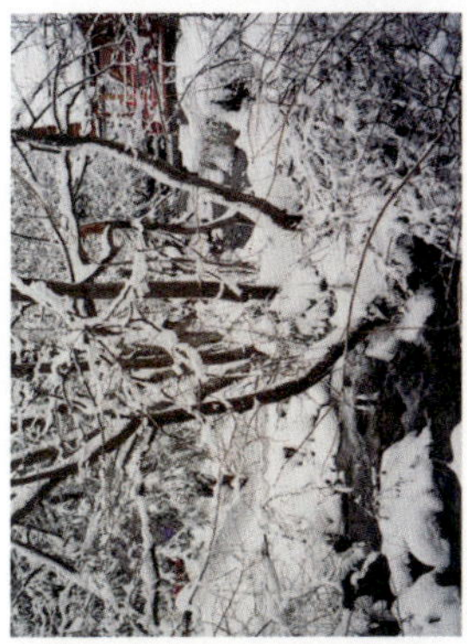

1359G

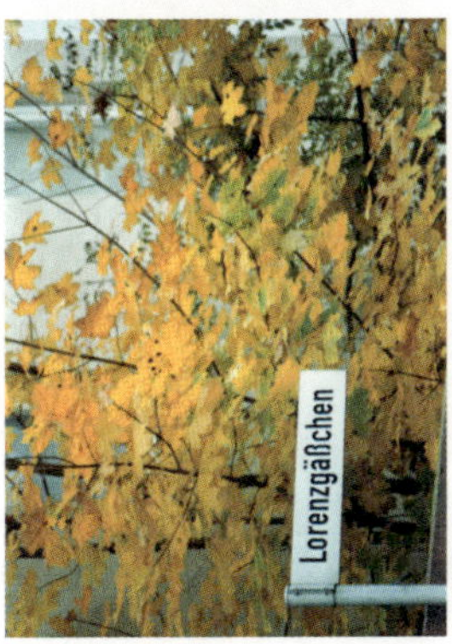

1369P

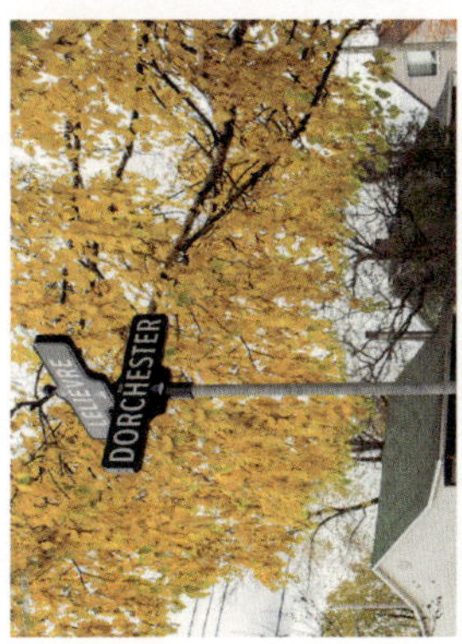

1369G

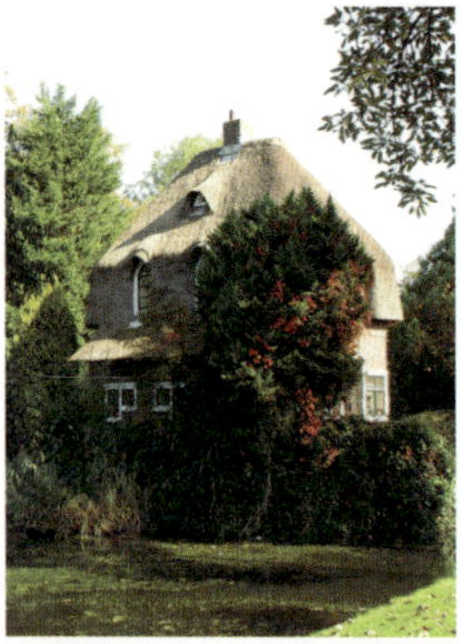

1199P

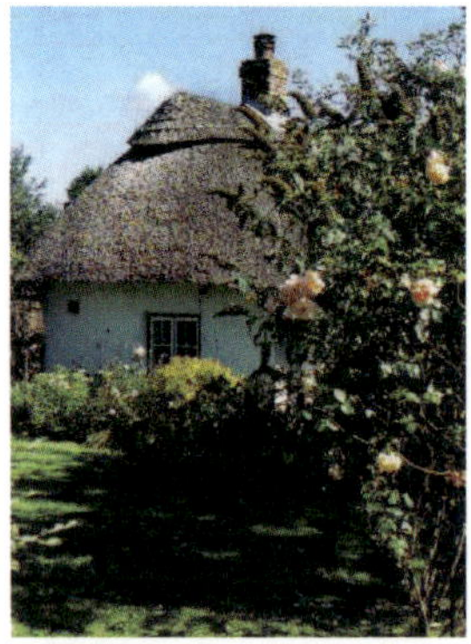

1199G

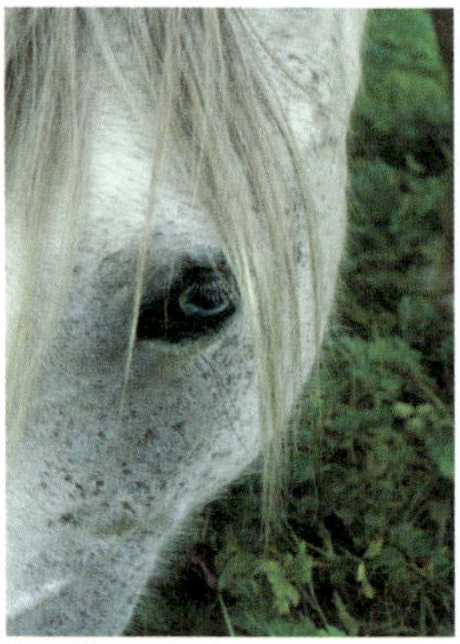

1203P

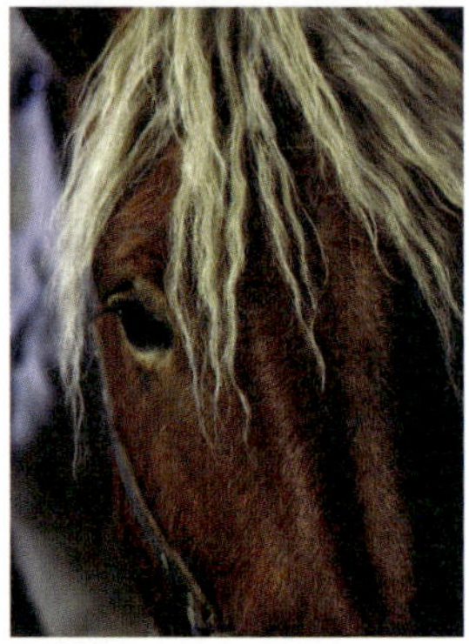

1203G

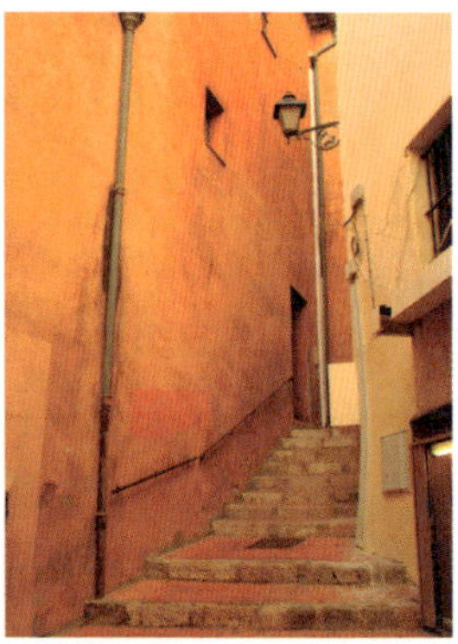

1205P

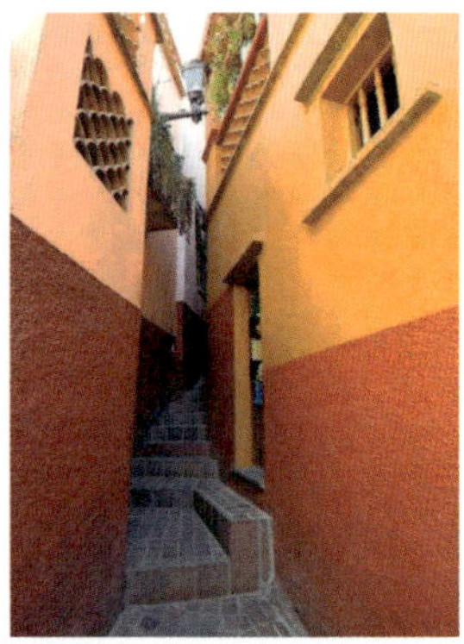

1205G

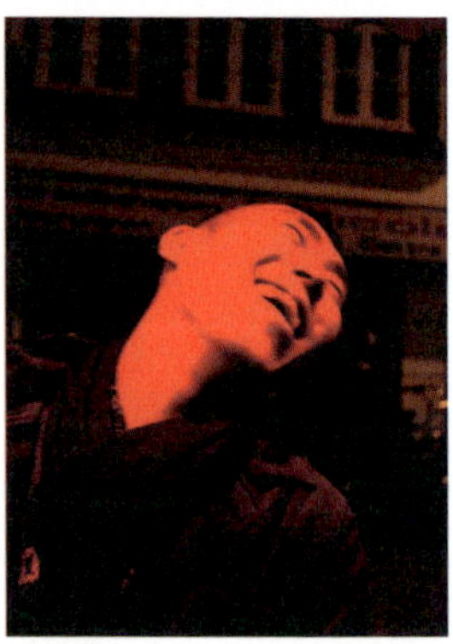

1210P

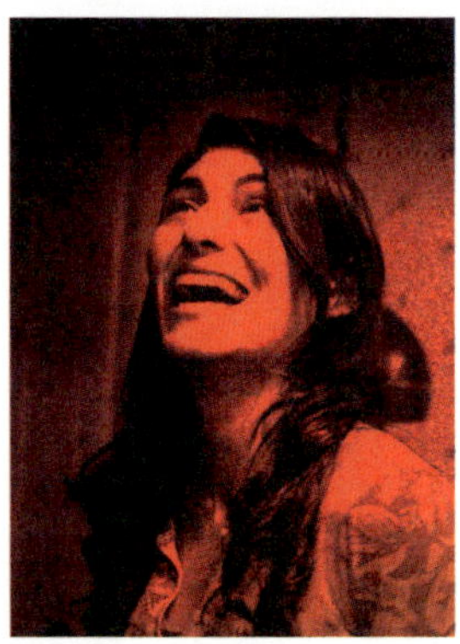

1210G

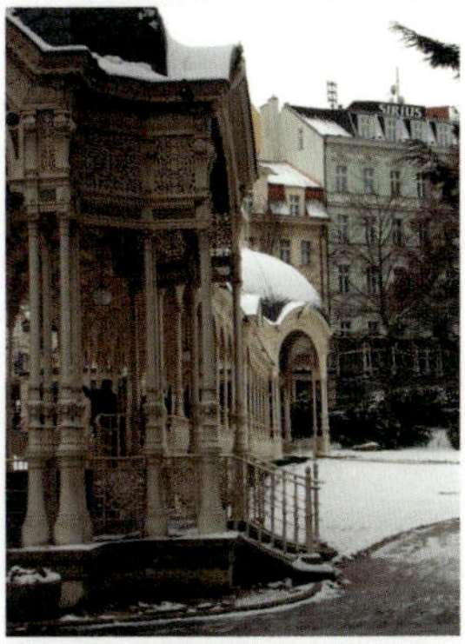
1214P

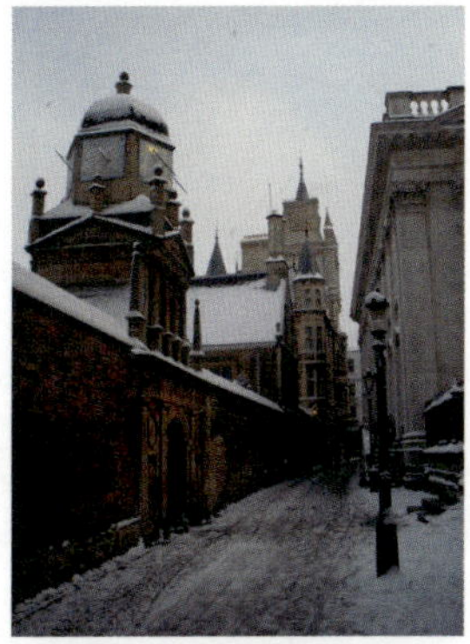
1214G

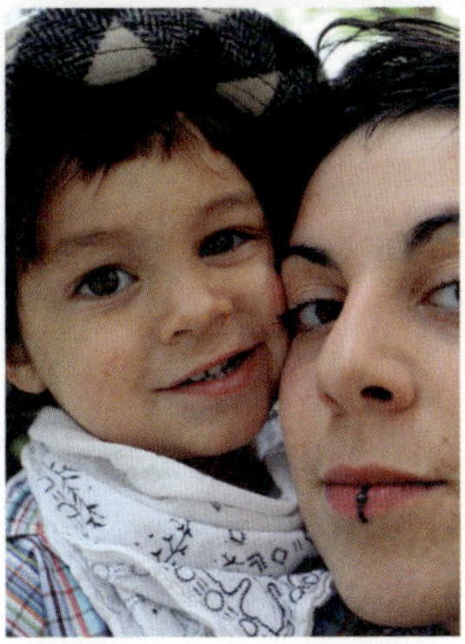
1216P

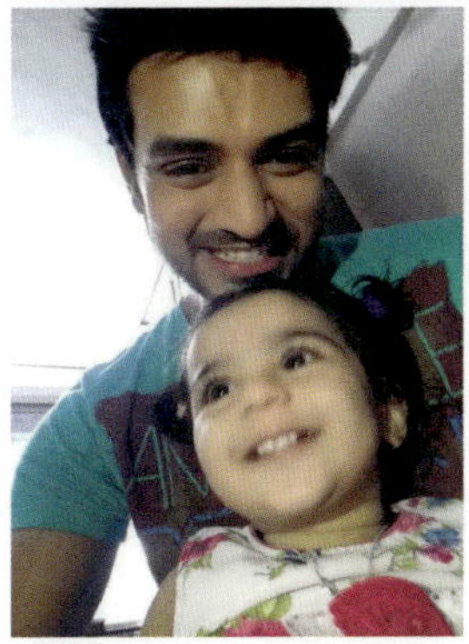
1216G

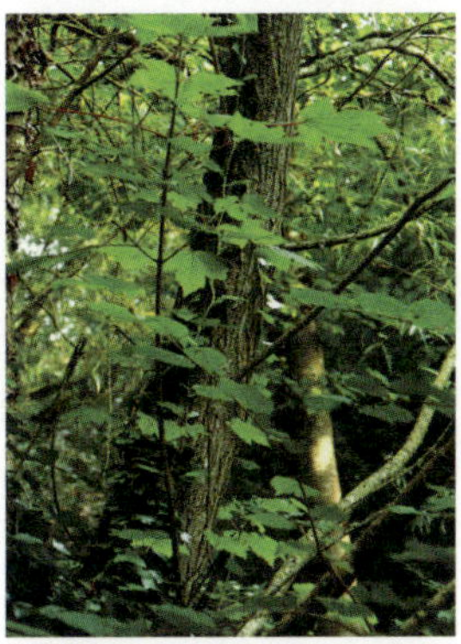
1222P

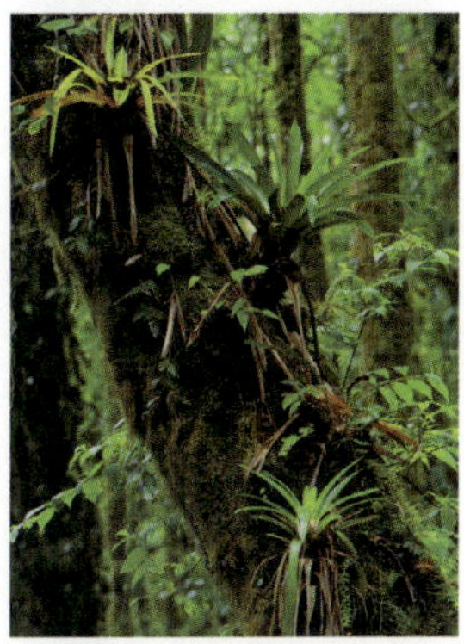
1222G

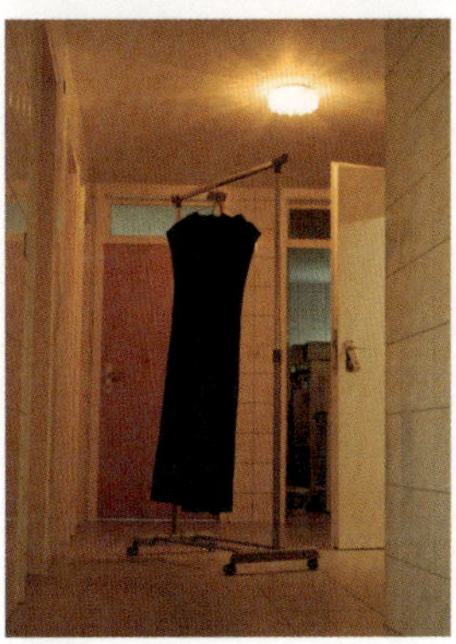
1226P

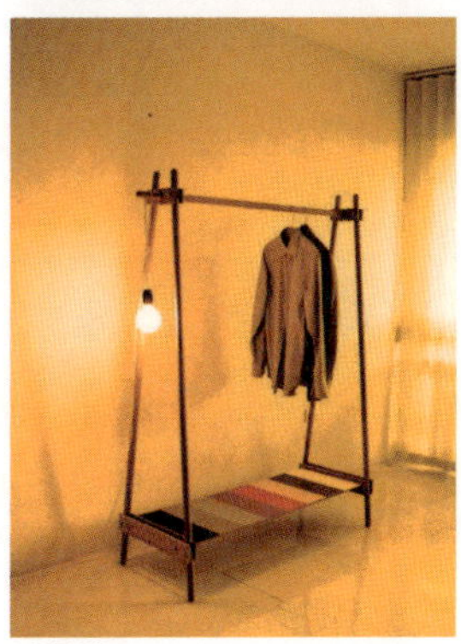
1226G

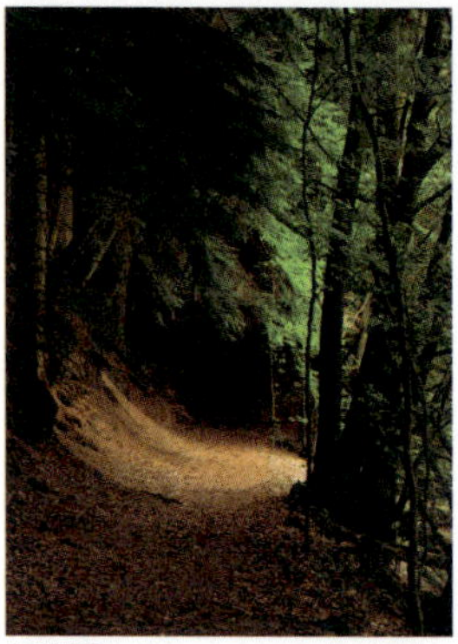
1227P
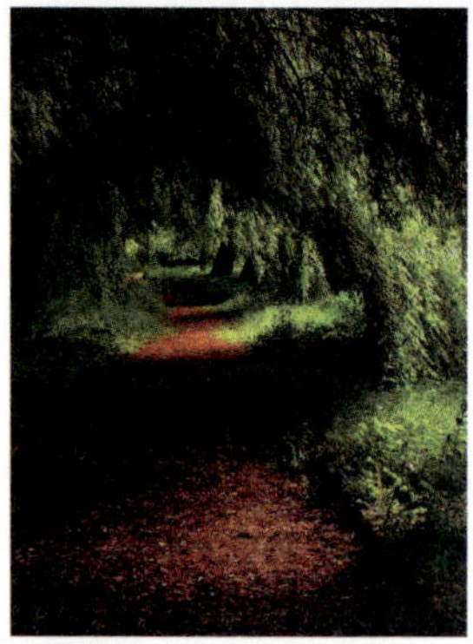
1227G

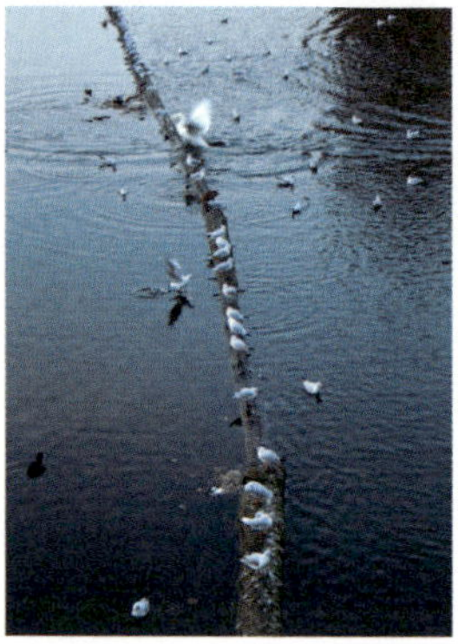
1228P

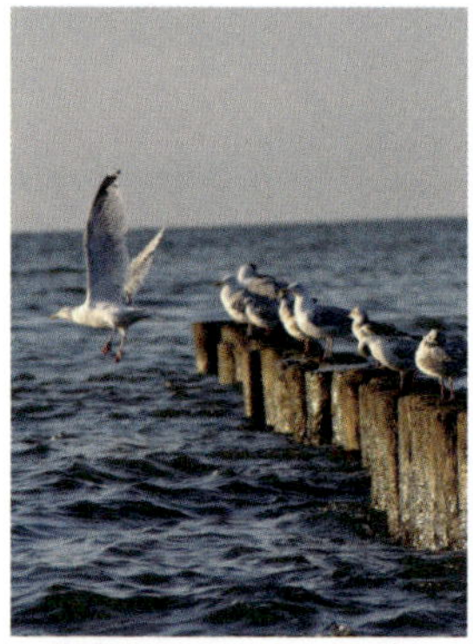
1228G

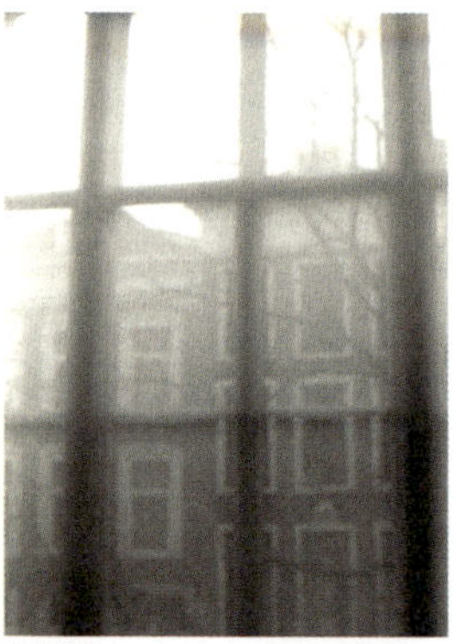
1230P

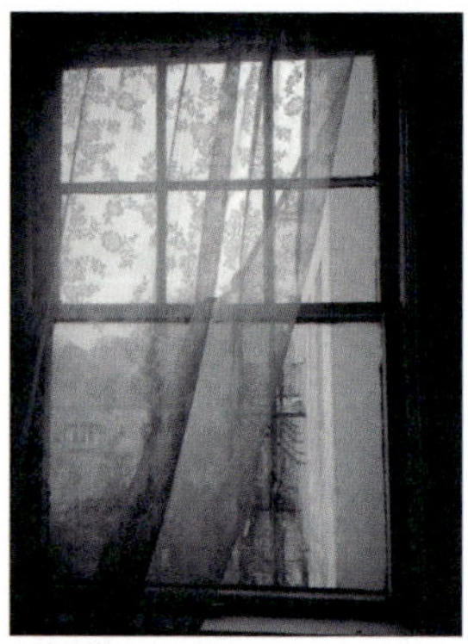
1230G
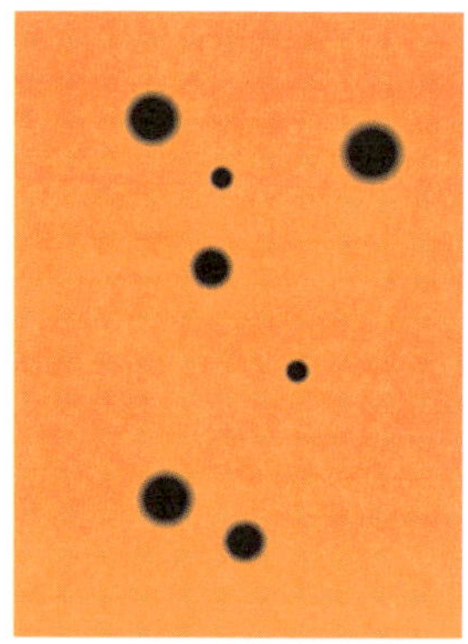

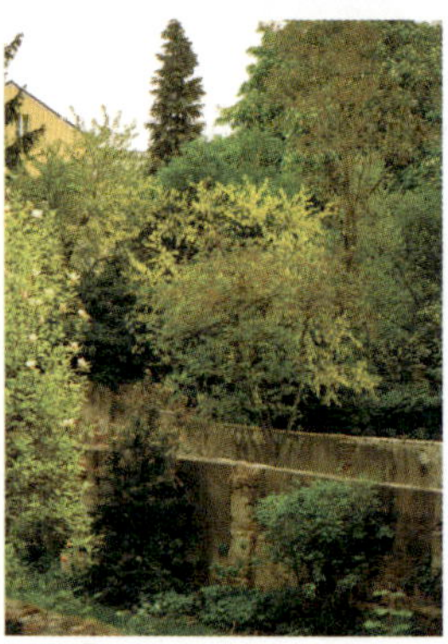
1234P

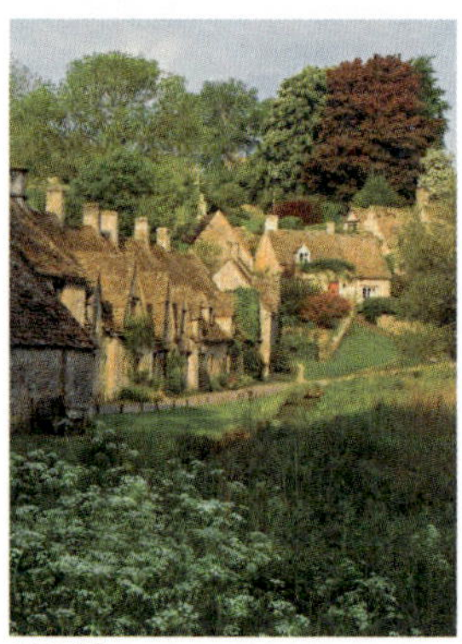
1234G

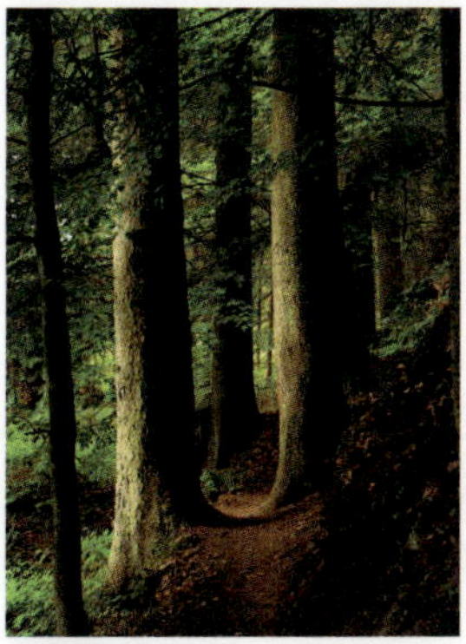

1238P

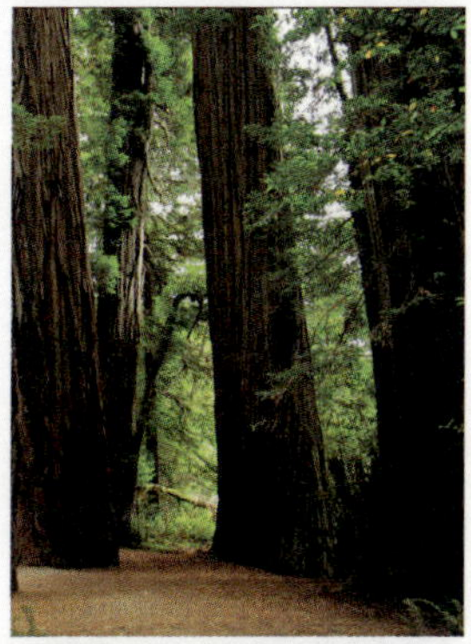

1238G

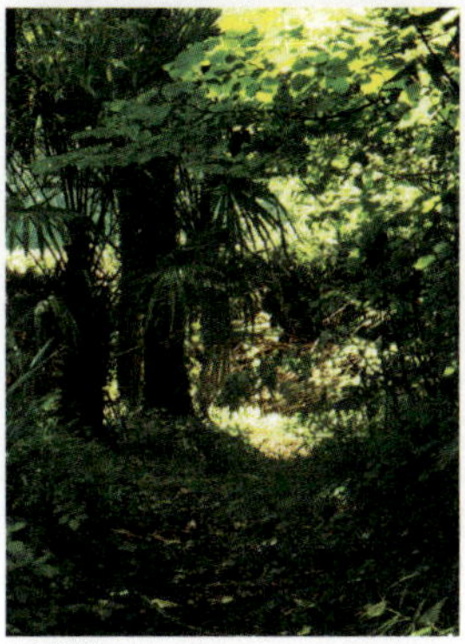

1243P

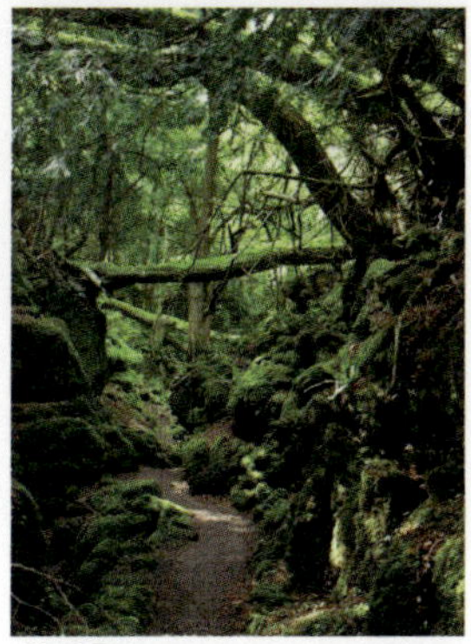

1243G

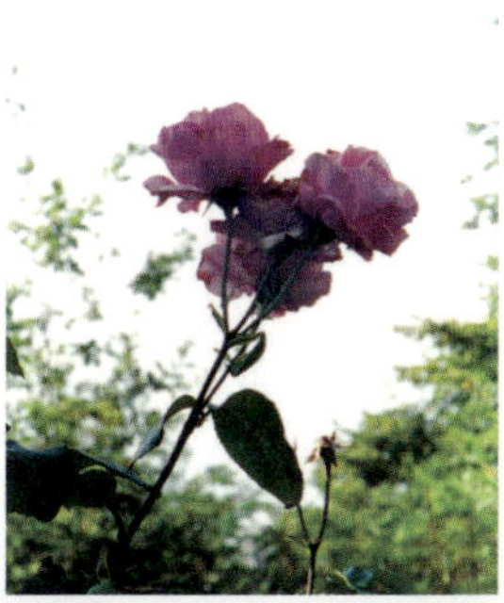

1244P

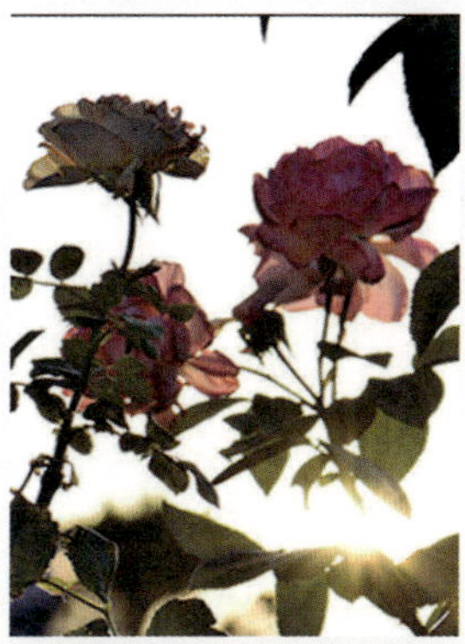
1244G

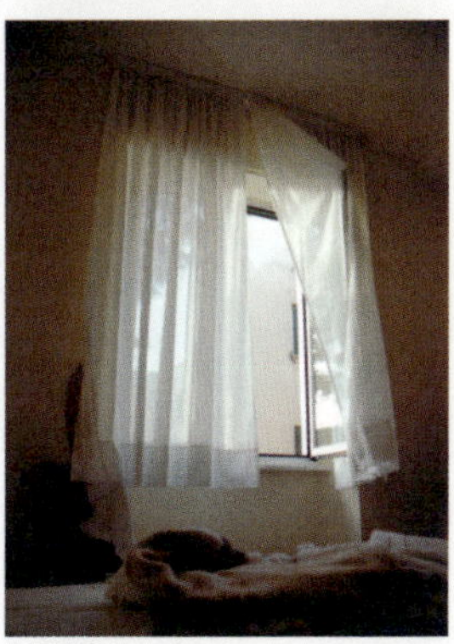

1256P

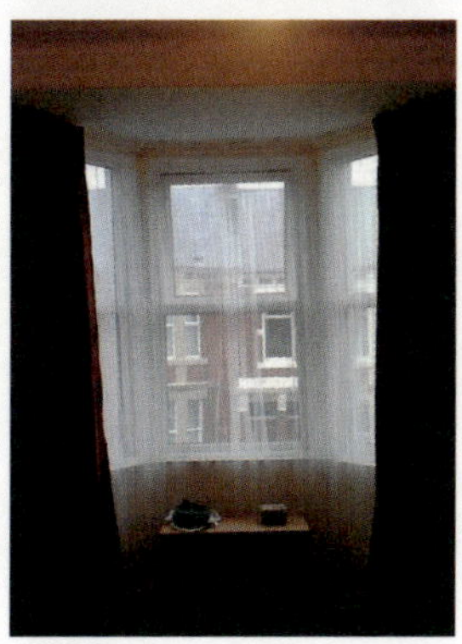
1256G

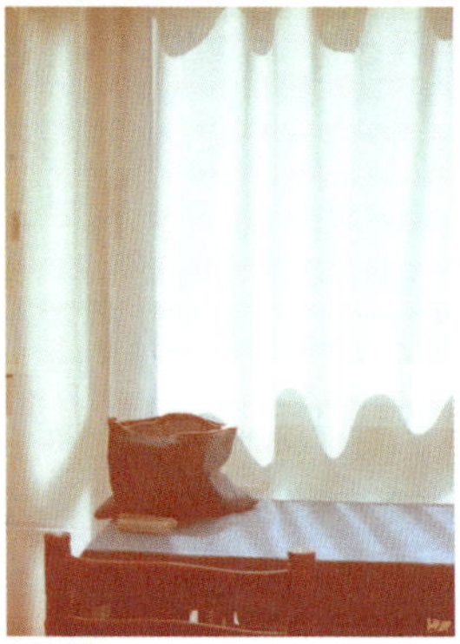

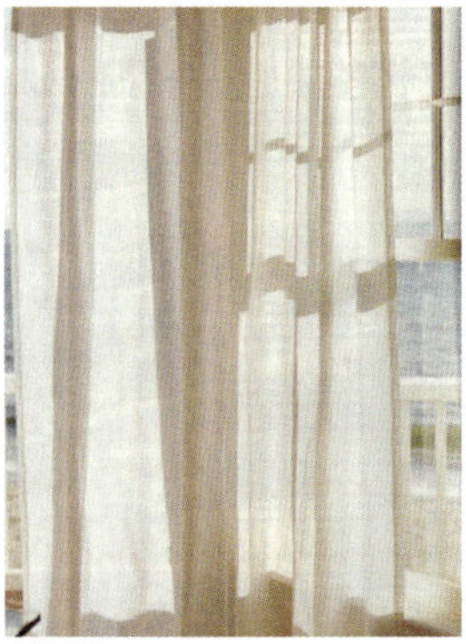

1258P

1258G

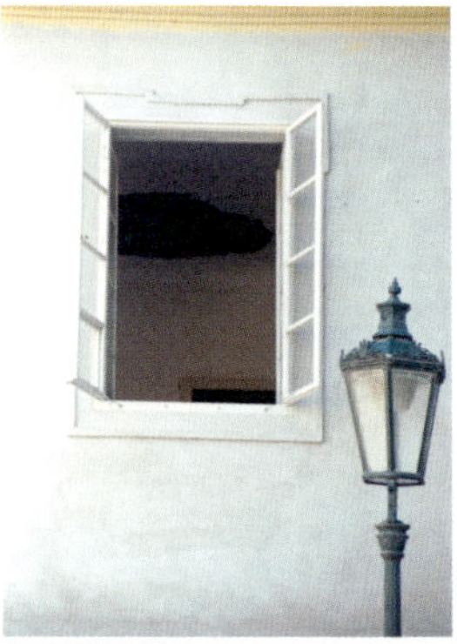

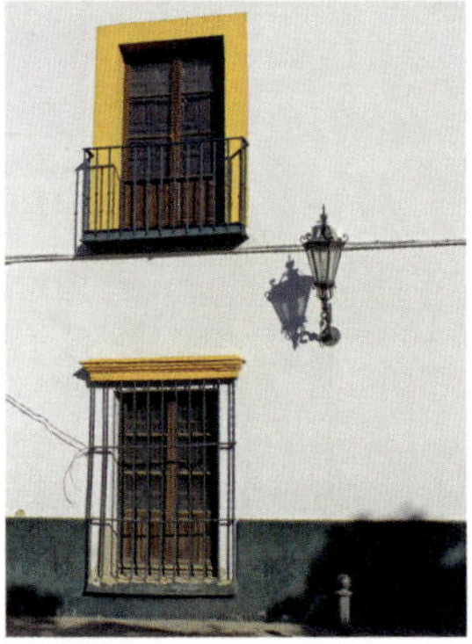

1260P

1260G

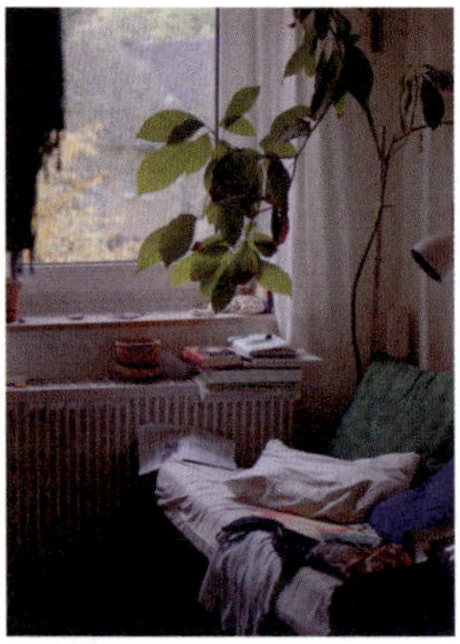
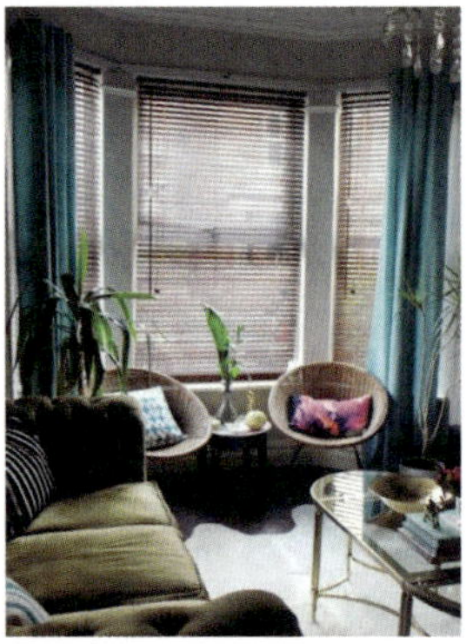

1261P

1261G

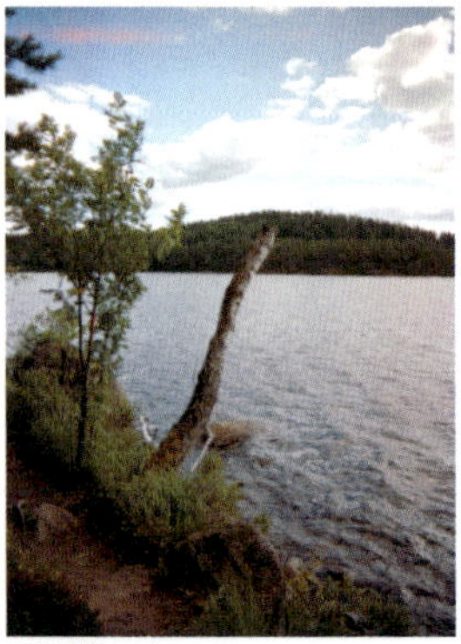

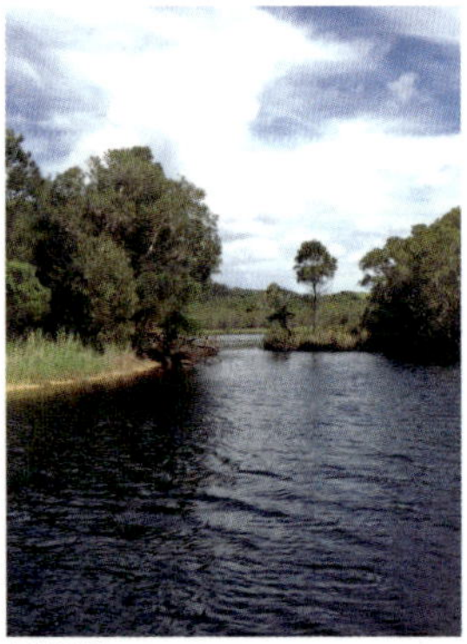

1263P

1263G

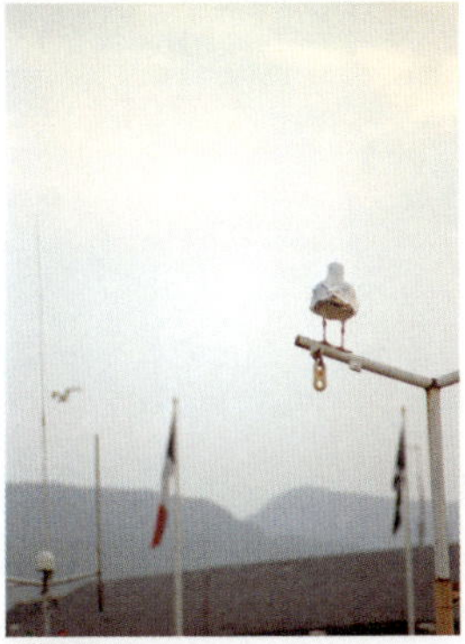
1264P

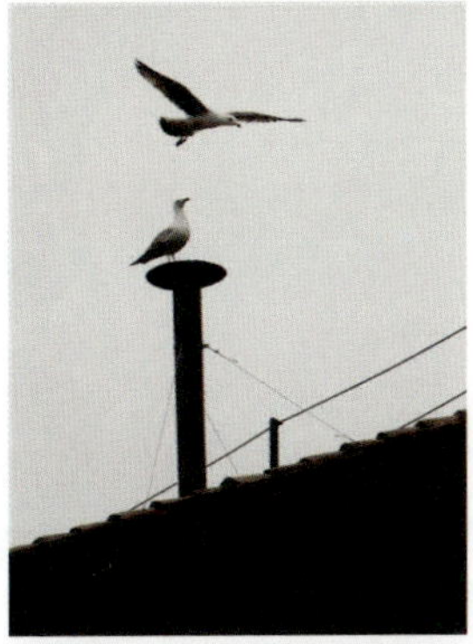
1264G

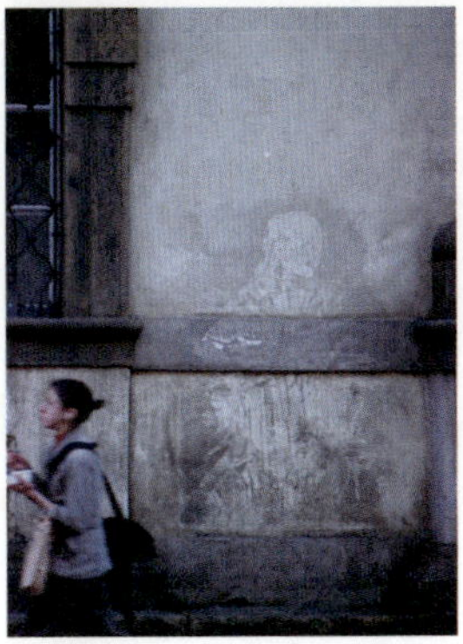
1270P

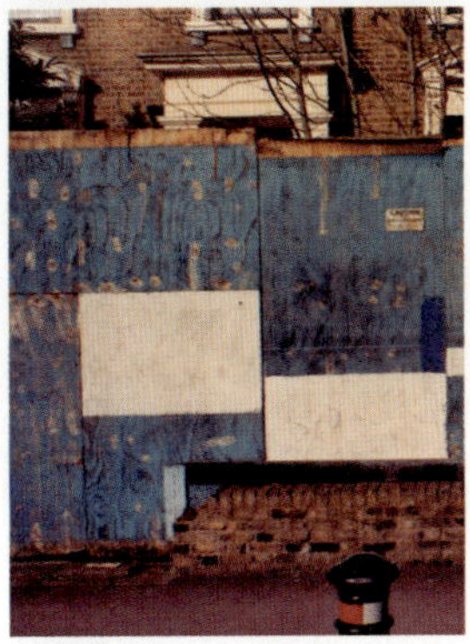
1270G

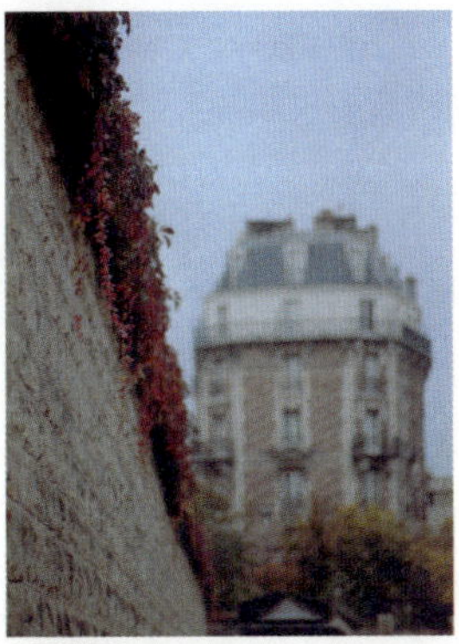
1271P

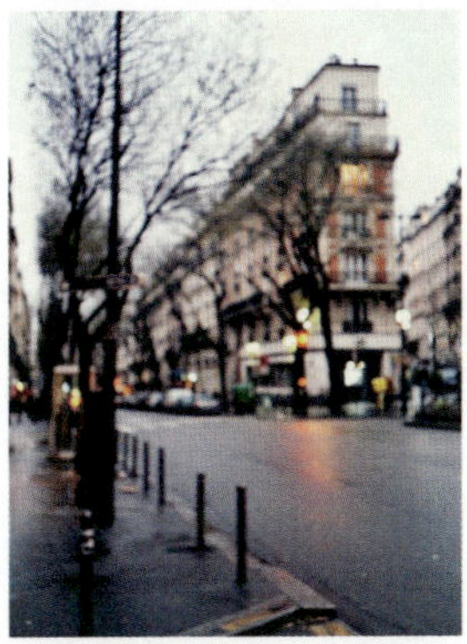
1271G

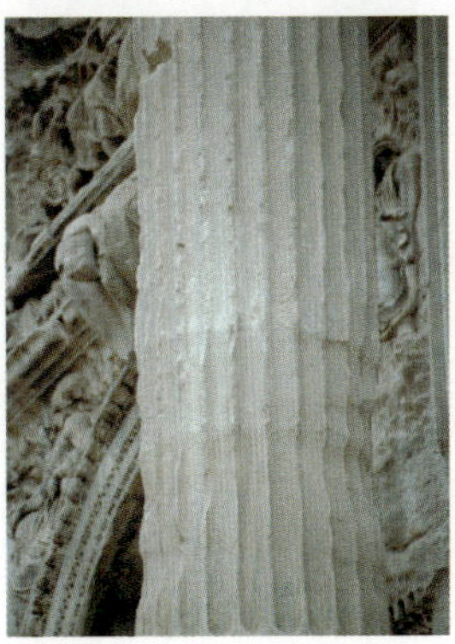
1274P

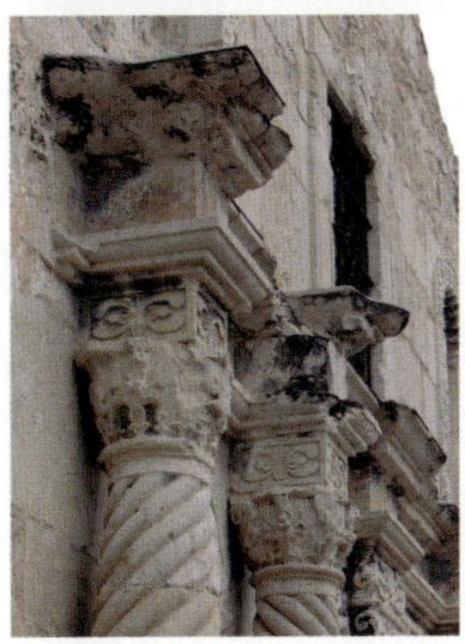
1274G

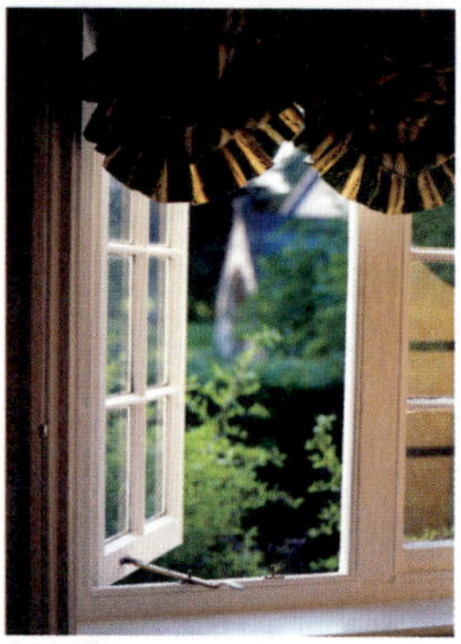

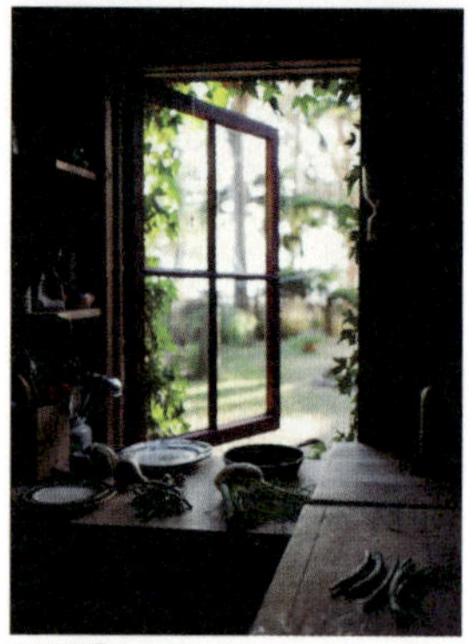

1277P

1277G

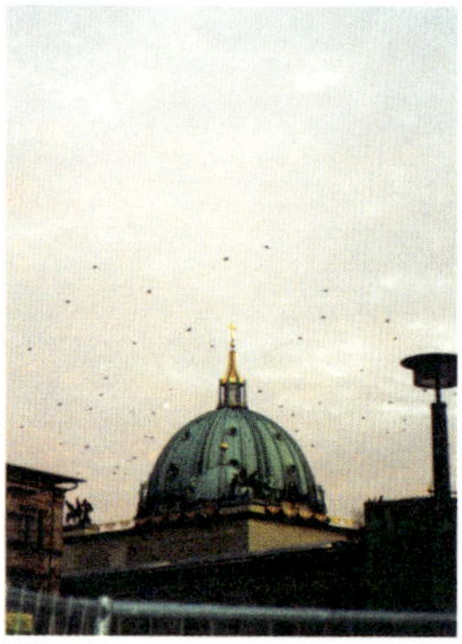

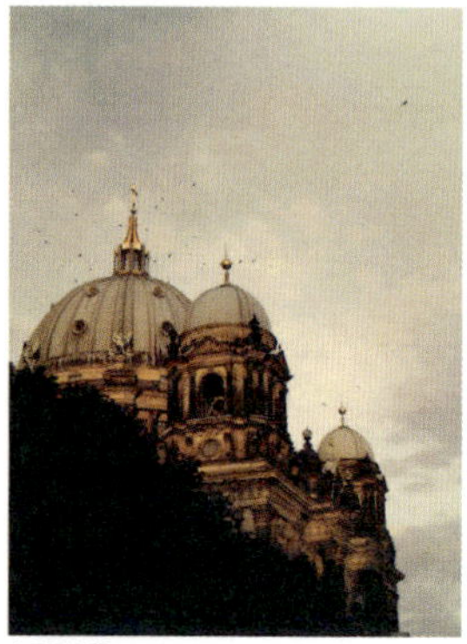

1281P

1281G

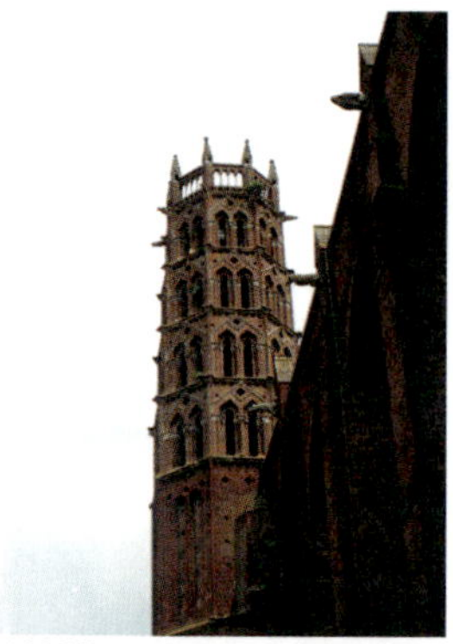

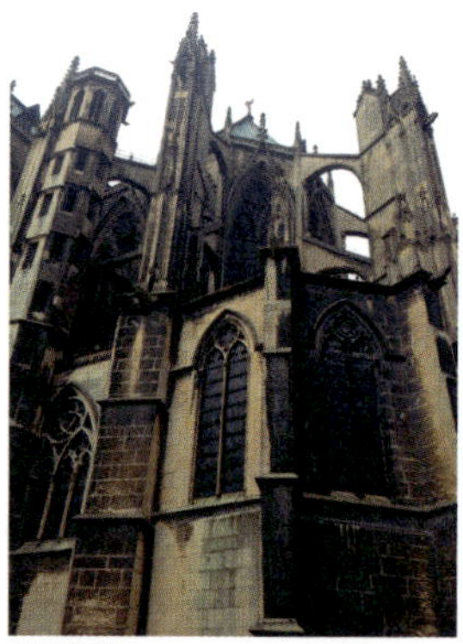

1282P

1282G

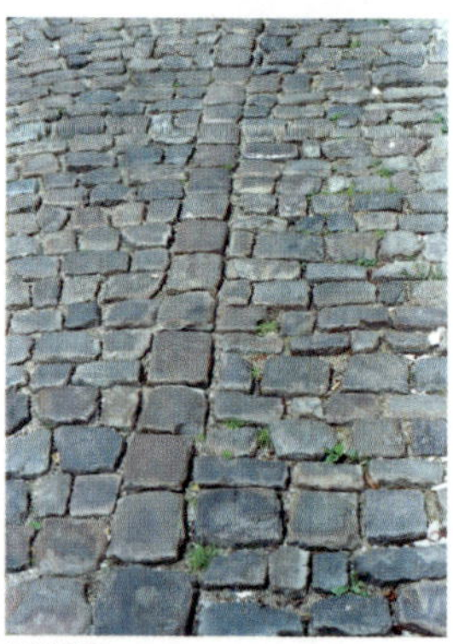

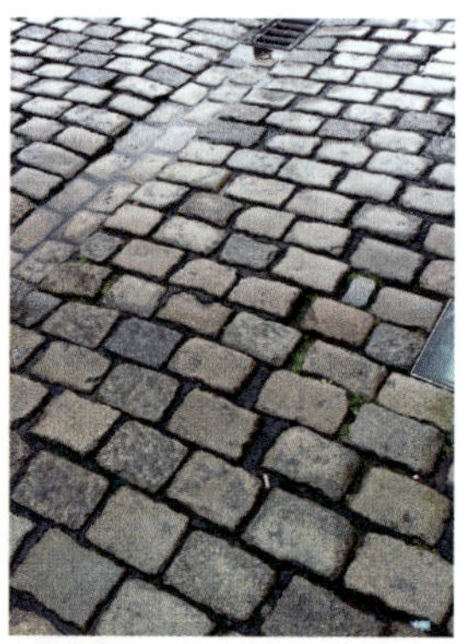

1295P

1295G

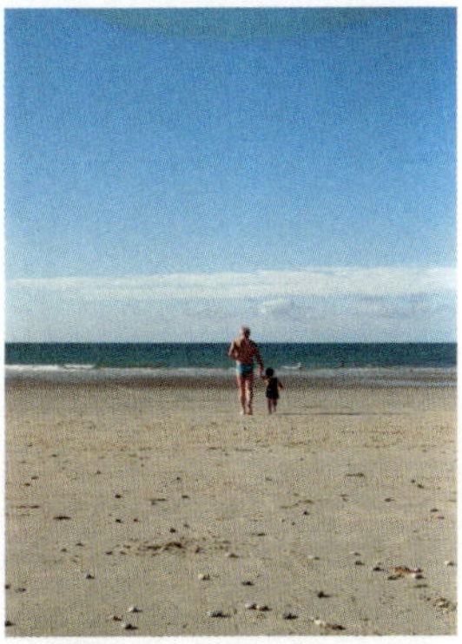
1310P

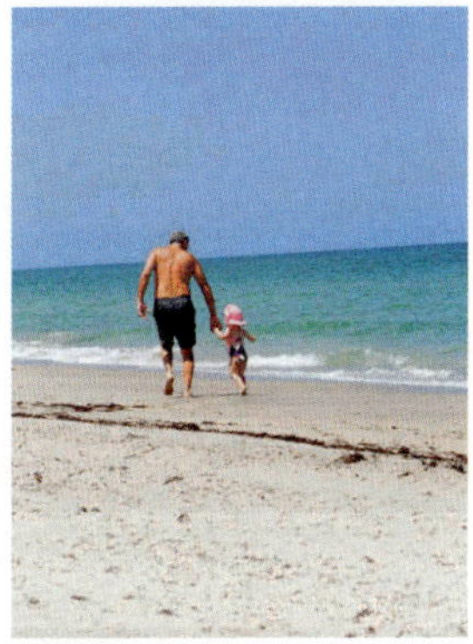
1310G

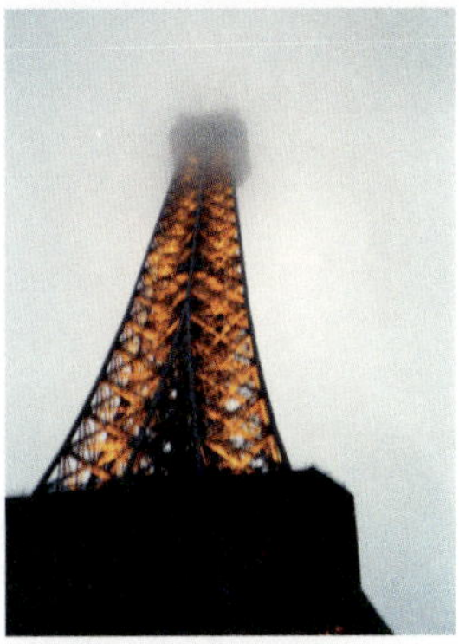
1311P

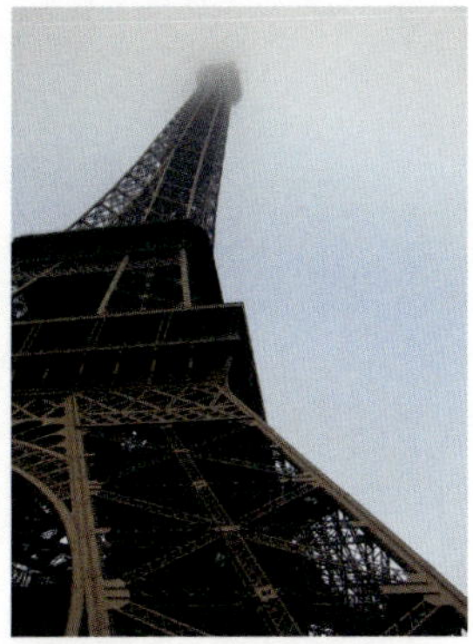
1311G

1463P

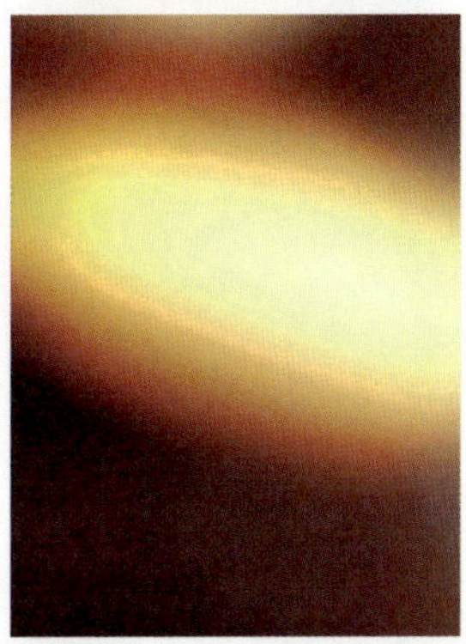
1463G

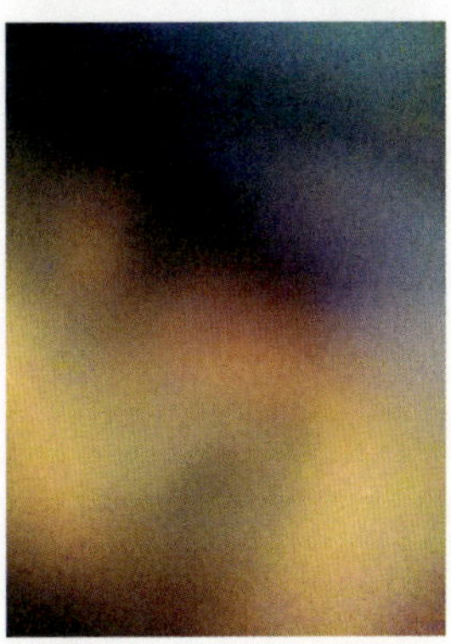
1466P

1466G

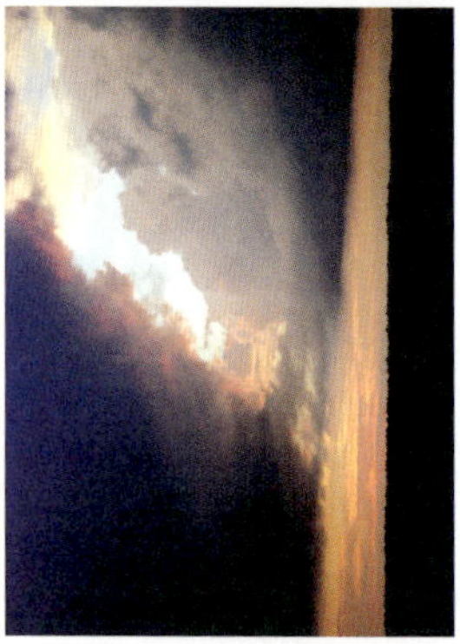
1373P

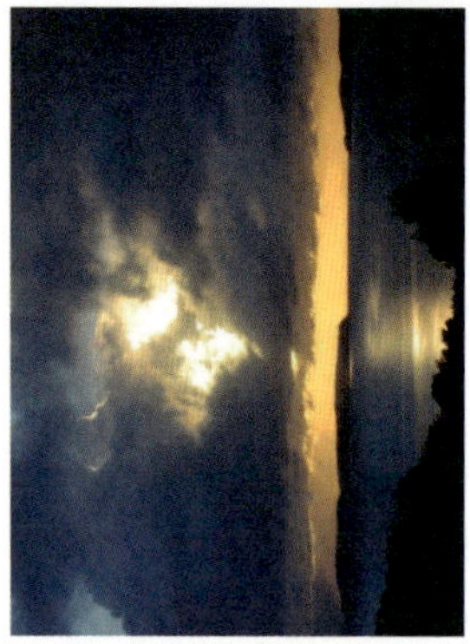
1373G

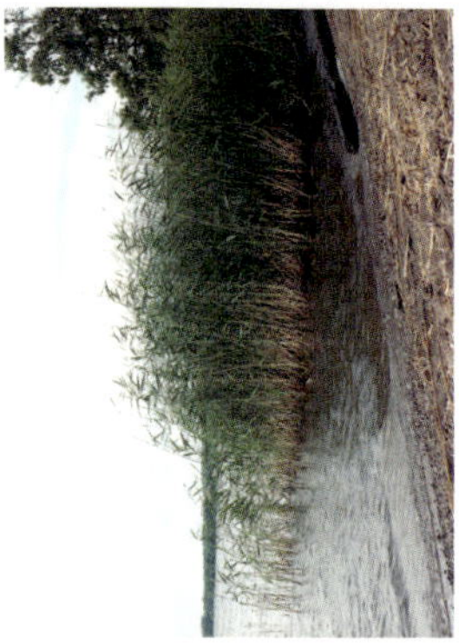
1375P

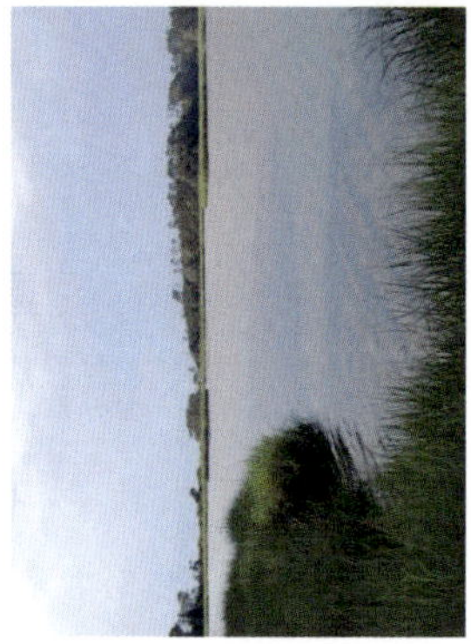
1375G

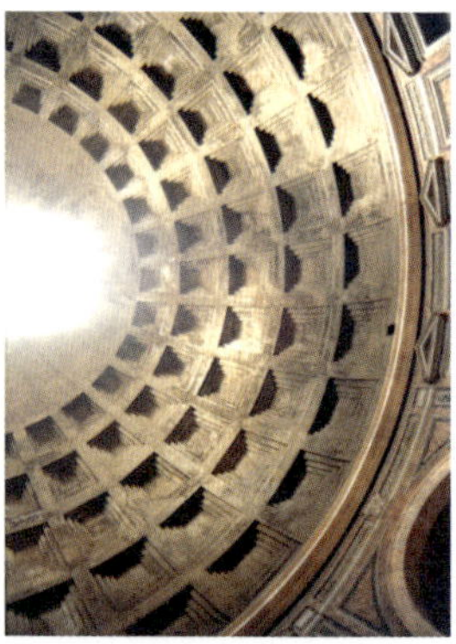
1377P

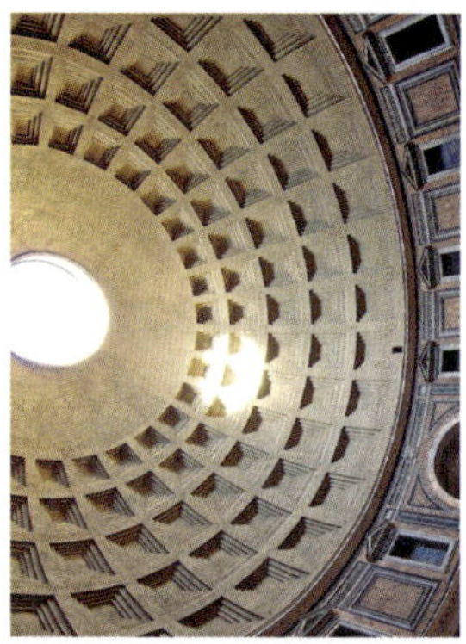
1377G

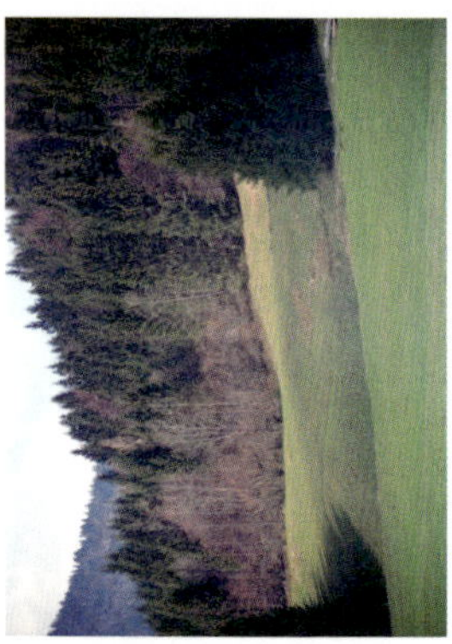
1381P

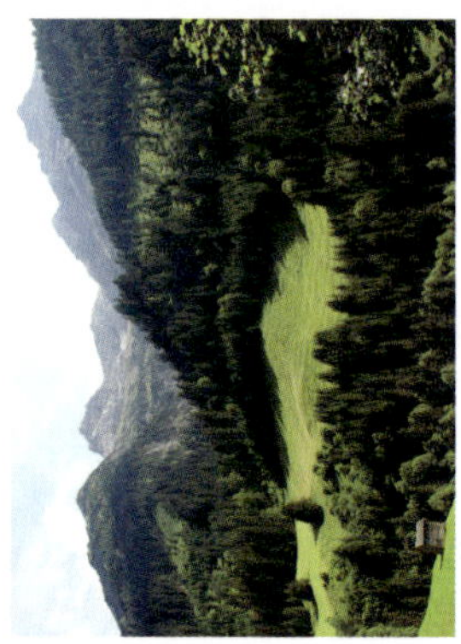
1381G

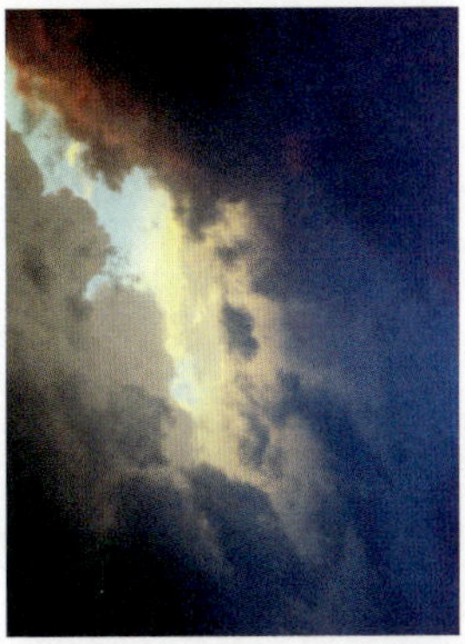
1391P

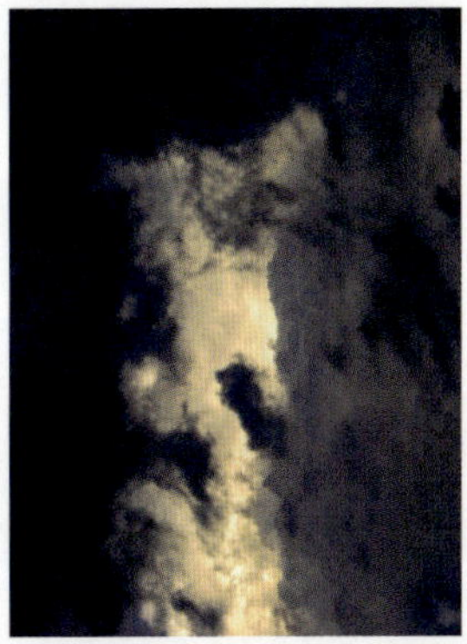
1391G

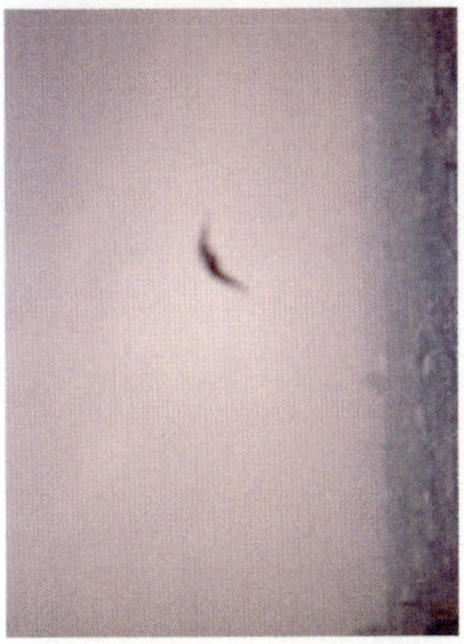
1396P

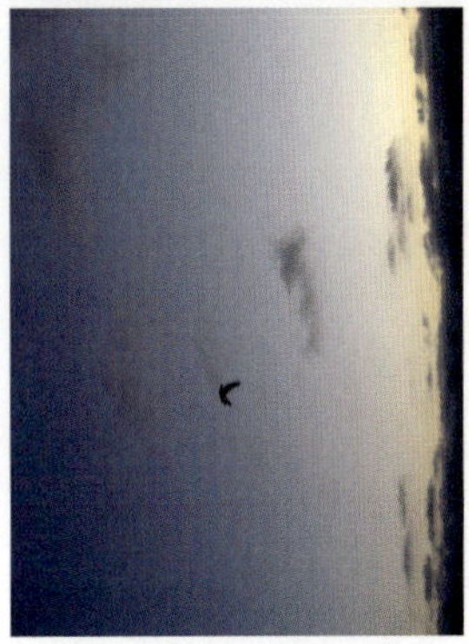
1396G

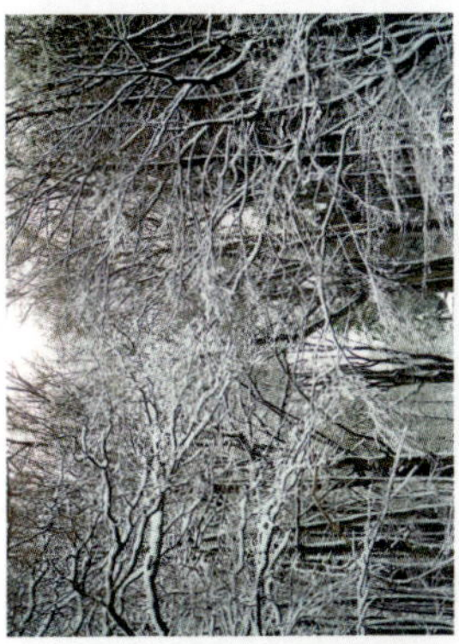
1406P

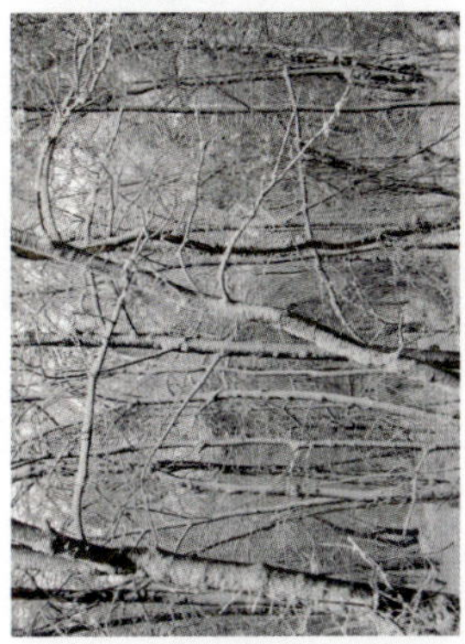
1406G

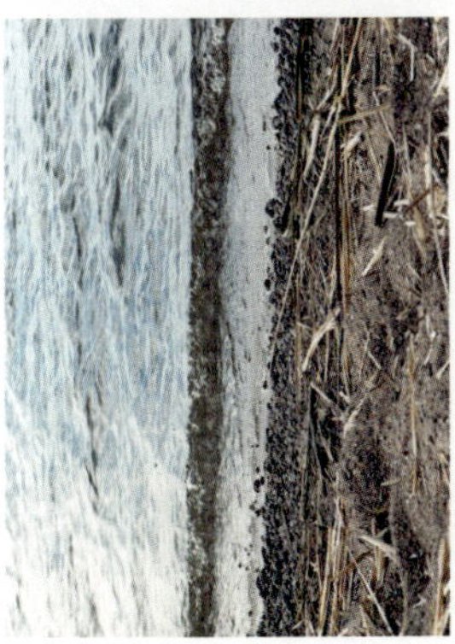
1408P

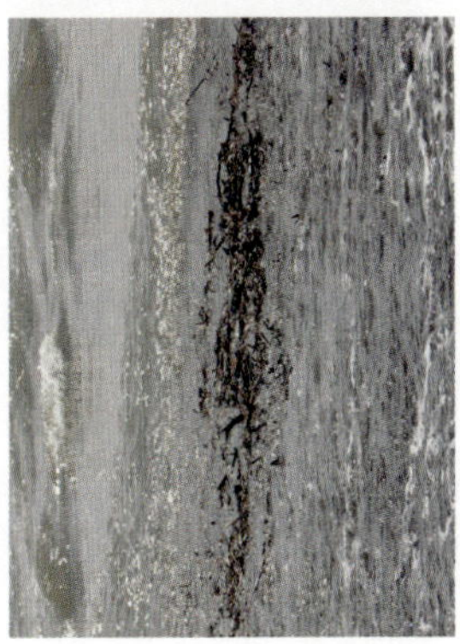
1408G

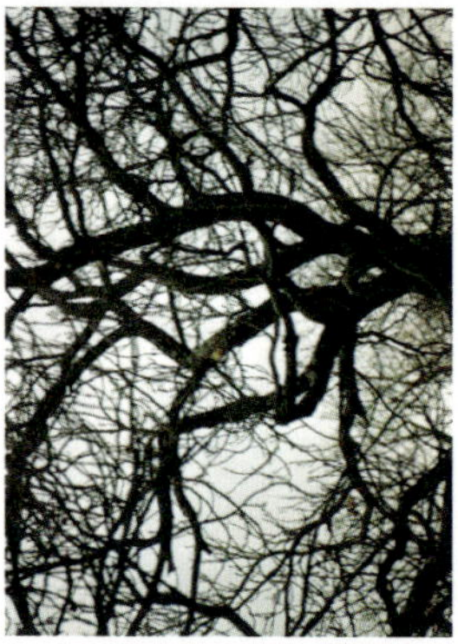

1418P

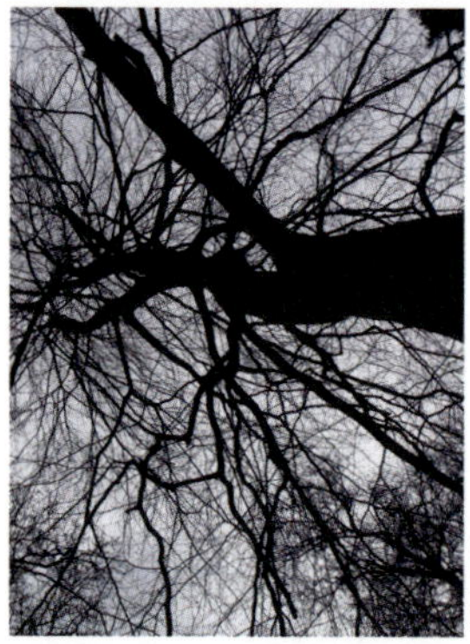

1418G

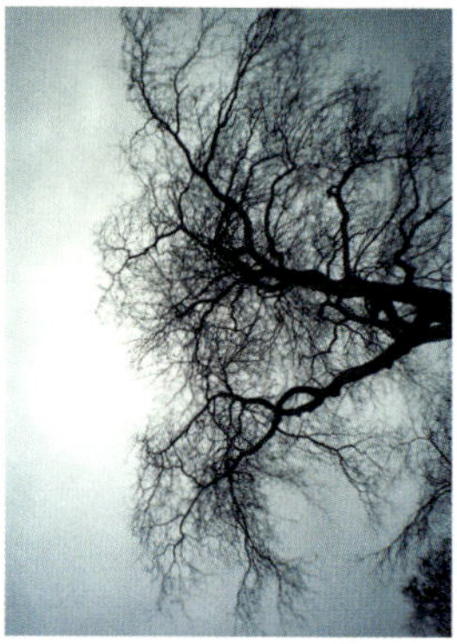

1432P

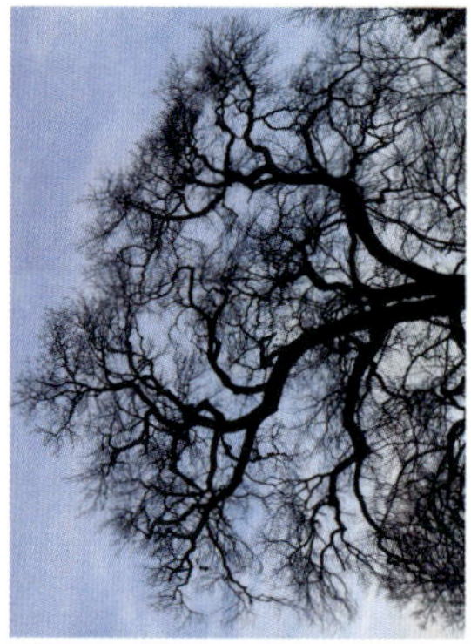

1432G

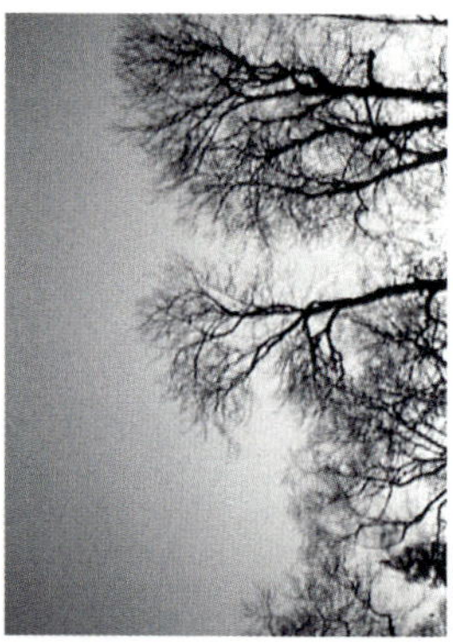

1455P

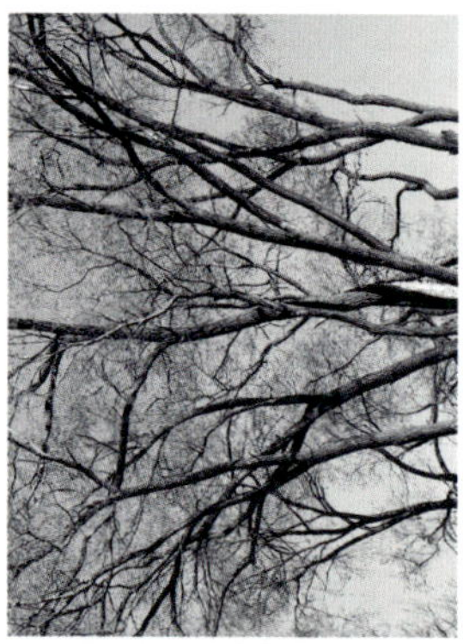

1455G

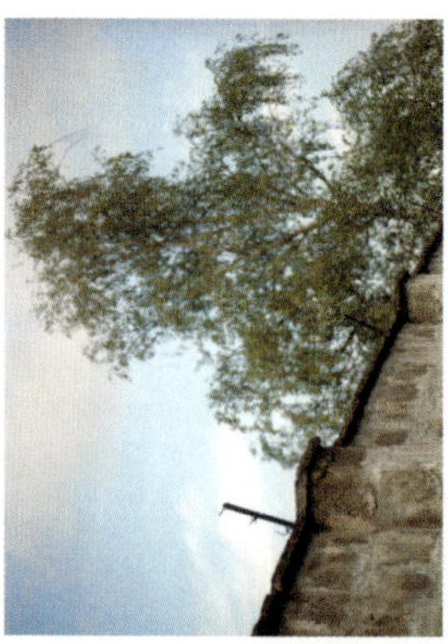

1456P

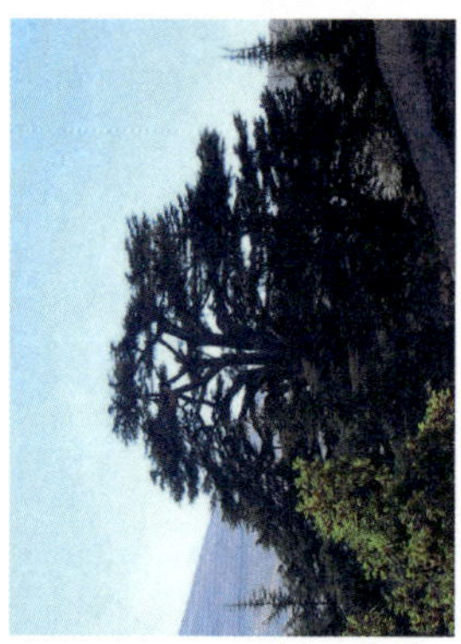

1456G

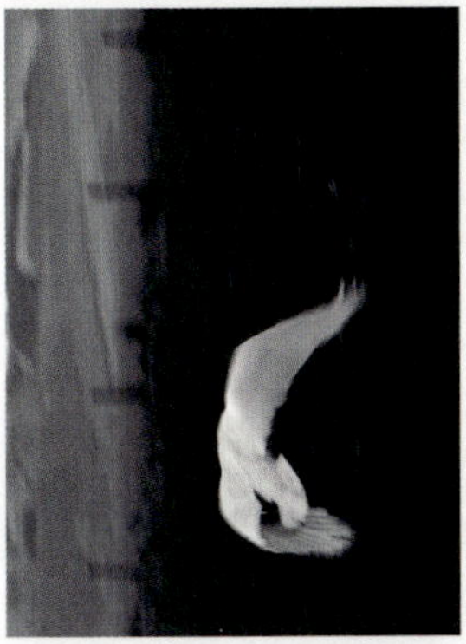
1482P

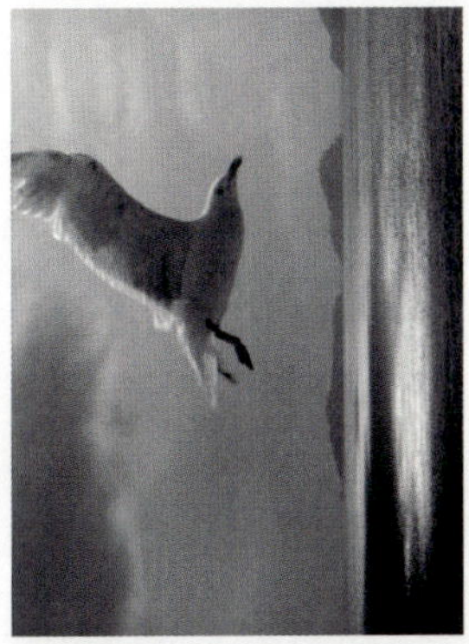
1482G

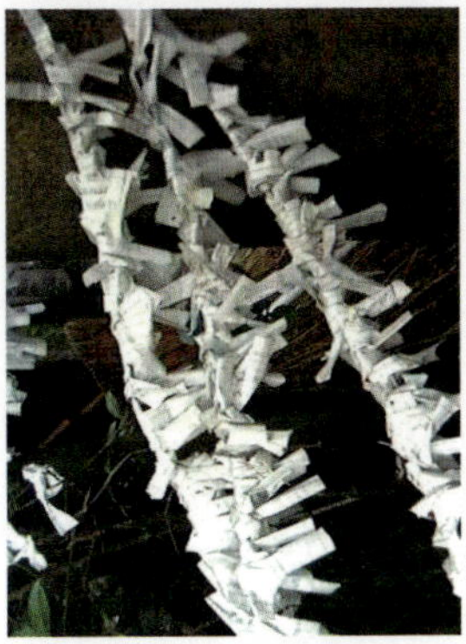
1487P

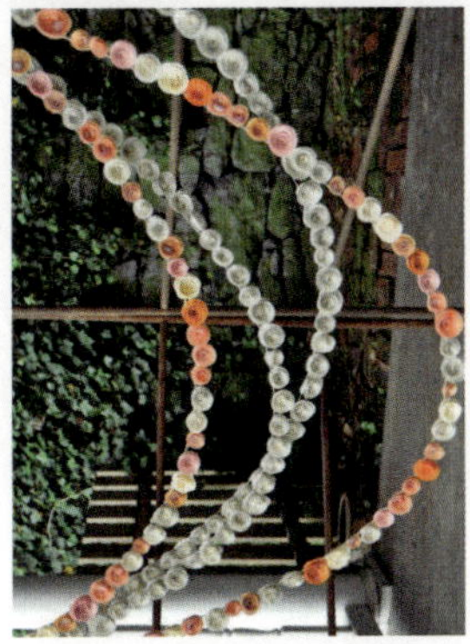
1487G

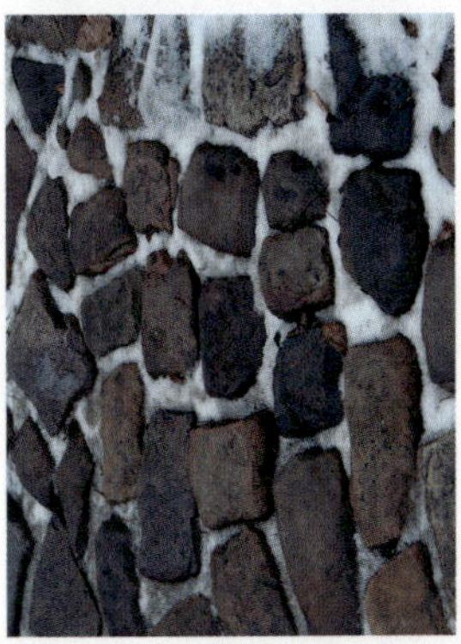
1495P

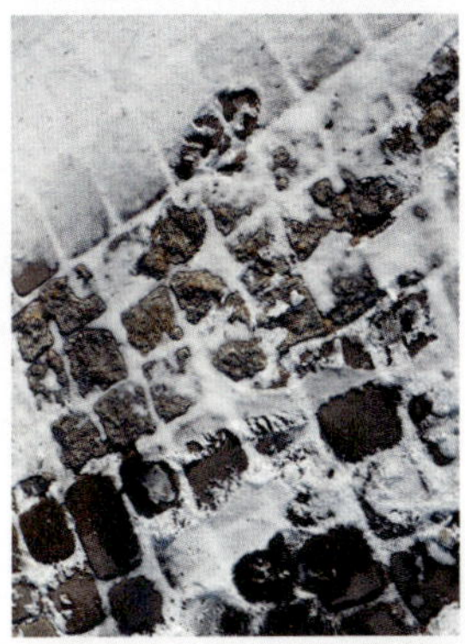
1495G

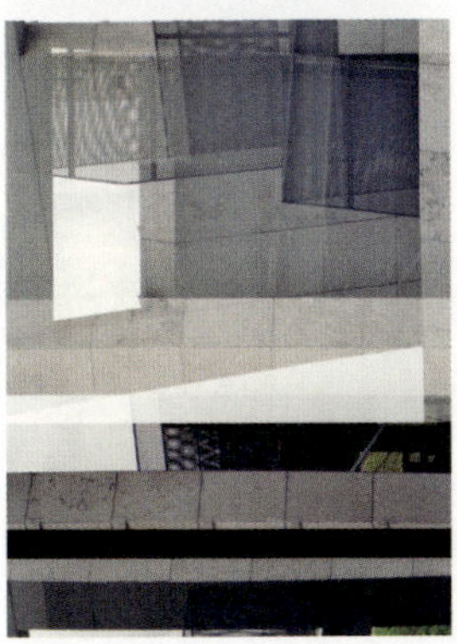
1500P

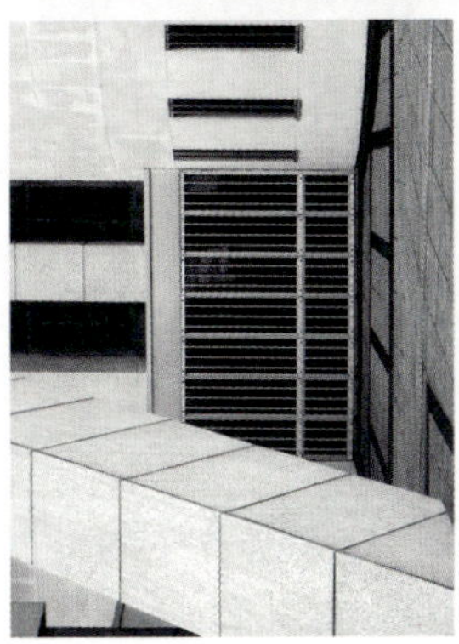
1500G

1

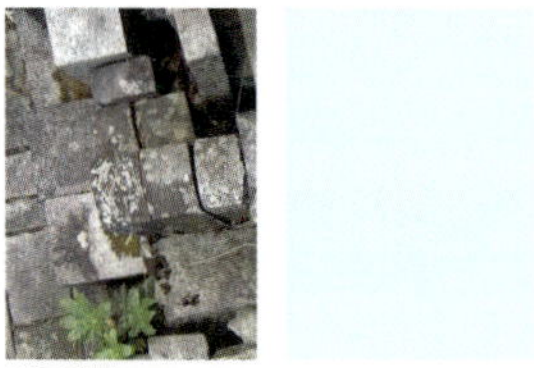
0910P

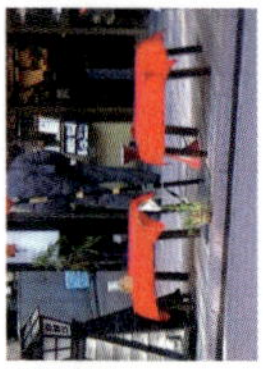
0972P

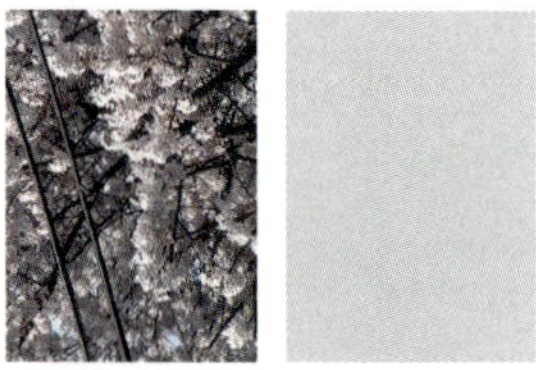
0992P

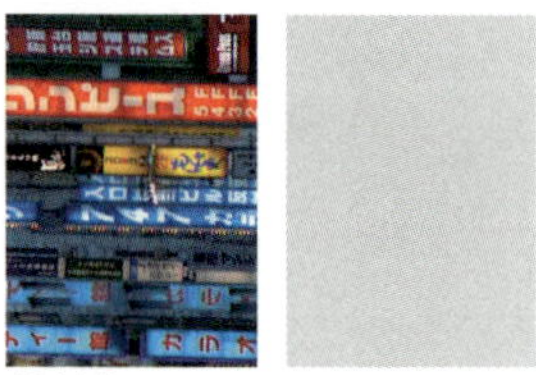
1011P

2

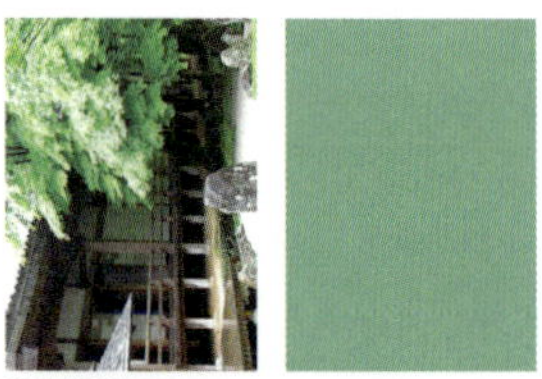
0850G

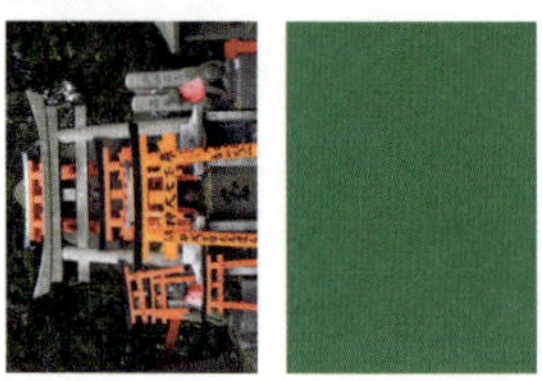
0909G

Most of the photographs I took myself trigger a synaesthetic response, in contrast to their anonymous counterparts. 88% of my own photographs evoke an emotional colour, opposed to a rough 36% of the anonymous ones. I noticed during this research that the strongest colour associations came from the pictures giving me the strongest emotions, from which I deduce that the colour is linked to an emotion and the emotion itself is often perceived as colour.

Some anonymous pictures would give me a colour sensation, but then there would be an element in the picture that resonates with a personal memory or experience. A person who looked to me slightly like Ringo Starr from the Beatles would feel dark blue, like a lot of their songs; a snow-covered landscape would give me a 'wintery' black emotion of withdrawal; or a scene would look like a place I have been to, and then evoke the colour I feel when thinking of that place. An example would be photographs from my journey to Japan.[1] Many of the photographs from Google that show scenes from Japan give me a range of greys and greens similar to the ones that my own Japan photographs evoke.[2]

It seems that photographs that do not 'do' anything to me, that do not trigger an emotional response whatsoever, will not provoke a colour association.

This is most strongly illustrated by the fact that the selfies I took myself to send to my boyfriend while we were apart for months almost all give me a colour; they remind me of him and how I felt back then. Selfies of other people, with one exception[3] (she looks like someone I know), do not seem to do anything to me. Neither do pictures of strangers, not even cute kids — which are supposed to affect me, as a woman in a reproductive age range. The only people I do not know personally but who gave me a colour association are Audrey Hepburn, Jon Hamm — the actor impersonating Don Draper in Mad Men — and that random guy that reminded me of Ringo Starr.[4] It is evident that they can affect me because I have seen them before; I associate films or music pieces with them. Some of my neurons have been preoccupied with them previously, as it were.

Of the 72 pictures from Google that evoked a coloured emotion, only 16 display a colour, a colour range or a single element with a colour that resembles the emotion, roughly 22%.

Of my private pictures, 190 give me a colour; 45 of those pictures (roughly 23%) have colours that correspond with or contain an element linked with the synaesthetic response.

For example, the ladies sitting in a café in Paris might evoke a red colour purely through the coke bottle we see on the table,[5] which is present in all our minds through the red colour of the brand. Concerning the photograph of an old house in Strasbourg,[6] there seems to be a clear link between the window panels displayed and

2 (cont.)

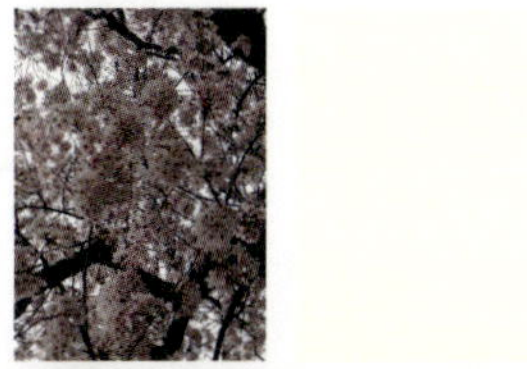

0992G

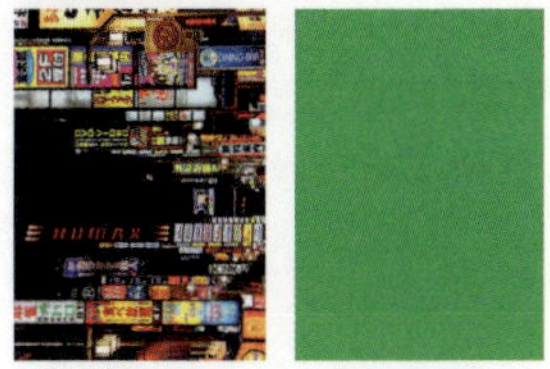

1011G

3

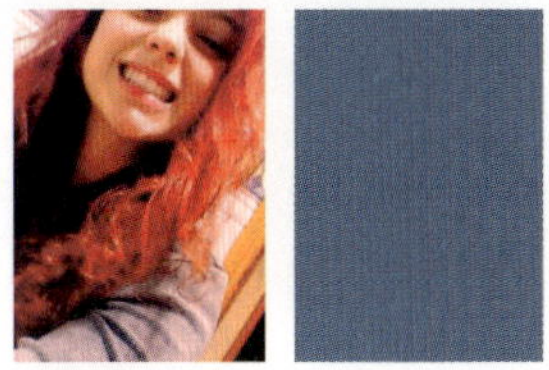

0923G

4

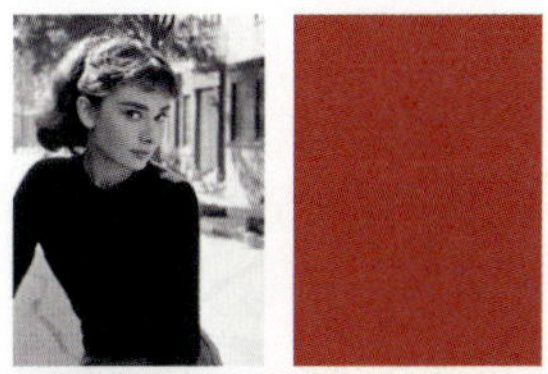

1068G

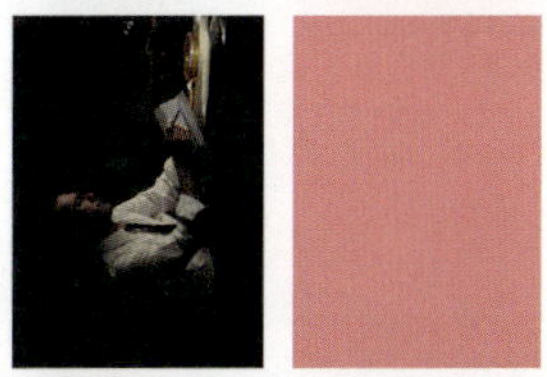

0825G

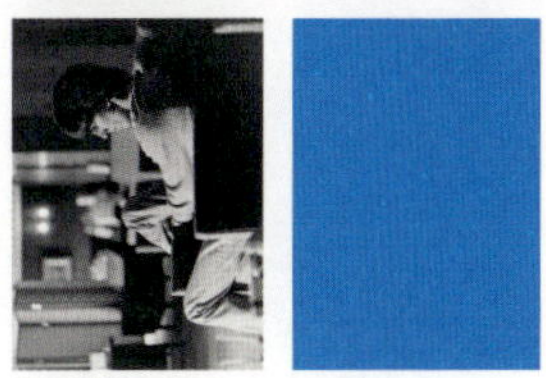

0025G

5

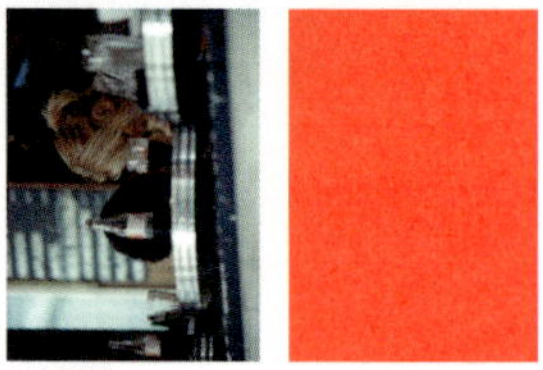

0220P

6

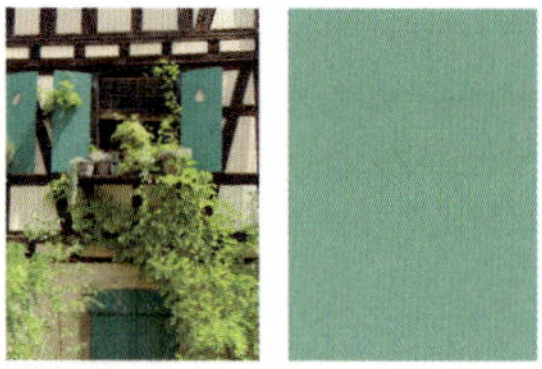

0530P

7

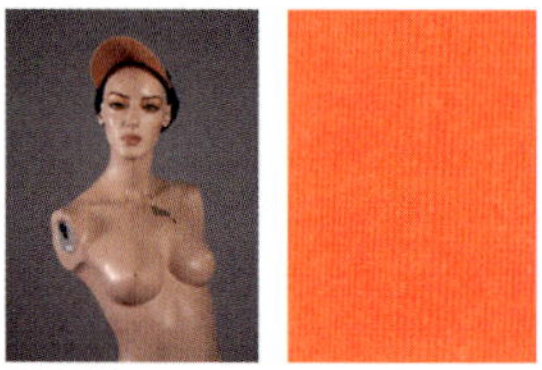

0541P

8

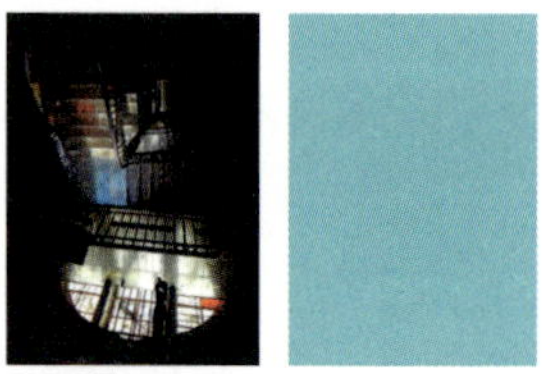

0552P

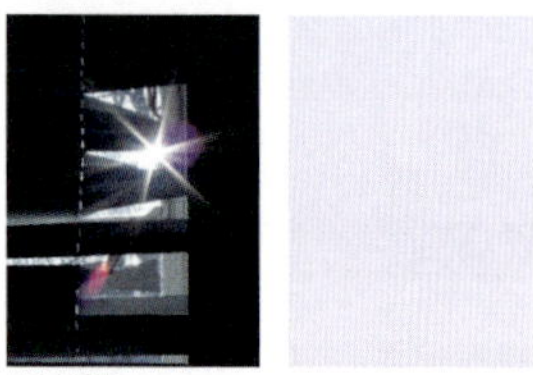

1142P

9

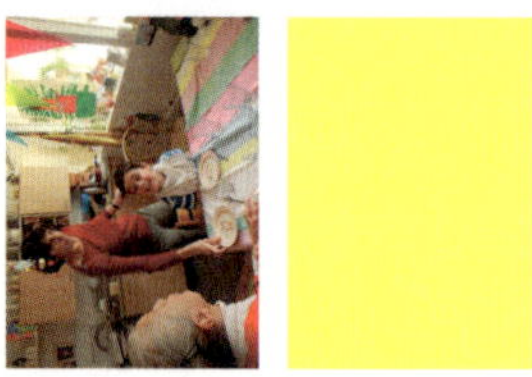

0713P

the colour I feel when seeing that image. The mannequin has an orange element,[7] but I believe that orange shade to be more the colour of that episode of my life, when I was in second grade at art school and things were just starting to sort themselves out a bit. Orange is a mostly positive colour, and often linked to memories of times that were easy. The staircase in the Tacheles building in Berlin has a single blue element, but it is so central that it might evoke the blue feeling; same for the capture of a sunbeam creeping through some slats.[8] As well as my son's birthday party table[9] — is it the decoration that creates the colour I feel? We will see more about that in chapter three.

Another question investigated was: were there parallels between the colour associations? When both pictures, the one from my archive and the corresponding one from found footage, evoked a colour, were they the same?

In most cases, they were not. In 62 cases, my photograph and the Google equivalent both had an emotional response, but the colours were matching or at least corresponded vaguely in only 12 cases (approximately 19%).

An interesting detail was that some pictures from Google felt almost familiar; sometimes I was not sure whether it was not one that I had shot and might have forgotten. Since the images were randomised and renamed for the research, I would not know. I can see that some pictures resemble mine so much, in motif and style, that they trigger strong emotions as well. Examples for this phenomenon would be two black and white photographs of hands[10] — I photograph hands a lot, and often in this particular way. Or a couple of rail tracks that look very similar[11] — even to me, no distinction was possible between my own photograph and the one from Google. I considered both of them to be mine.

So apparently, sometimes I thought that photographs I did not take myself were mine, quite certainly because they evoked a familiar colour, but it also happened that I did not identify my own photographs and missed a synaesthetic reaction because no memory occurred. For example, I did not remember shooting a picture of blurred saris,[12] or a photo of a random cat.[13] It was only later — when I saw the graphic design finished for this chapter — that I realised that they were part of my personal archive. I ended up remembering where and when they were made, which instantly gave them a shade of orange and beige.

It appears thus that only the photographs that trigger an emotional response would evoke a colour. After doing a little more research on that observation, I found that there is a form of synaesthesia called emotion–colour synaesthesia and that, amongst people experiencing synaesthesia, it is rare but not undocumented. A related form of synaesthesia is personality–colour synaesthesia, in which people are perceived either directly as having coloured auras around them or with an indirect colour

10

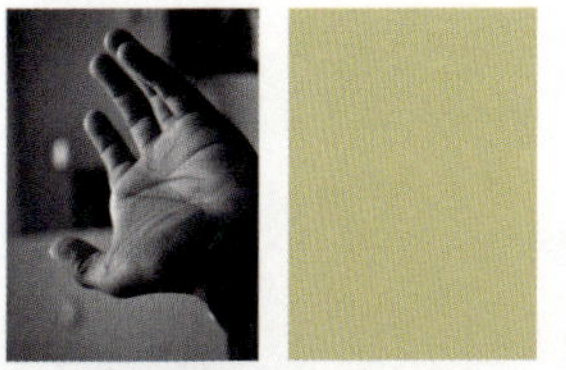

0116G

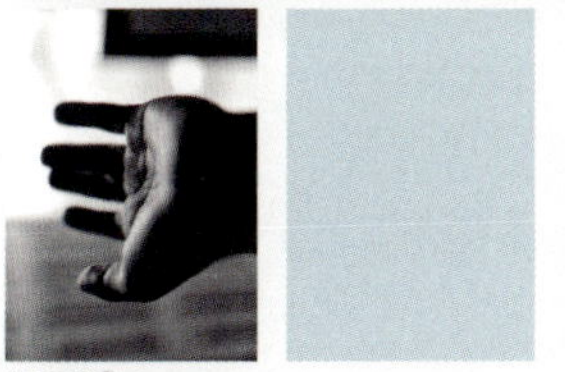

0179G

11

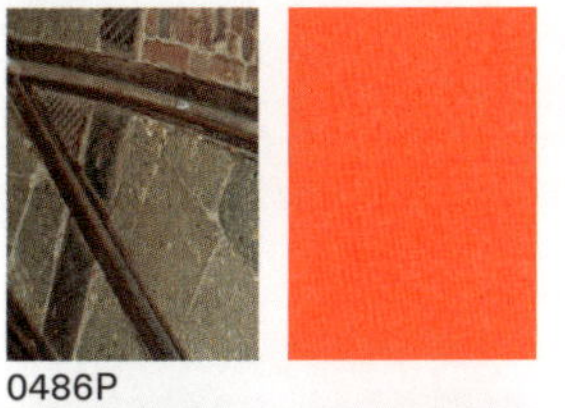

0486P

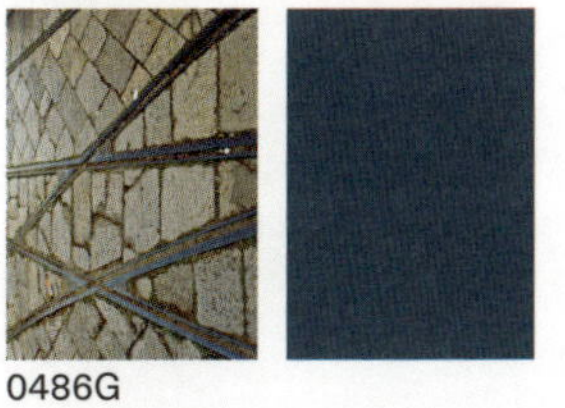

0486G

12

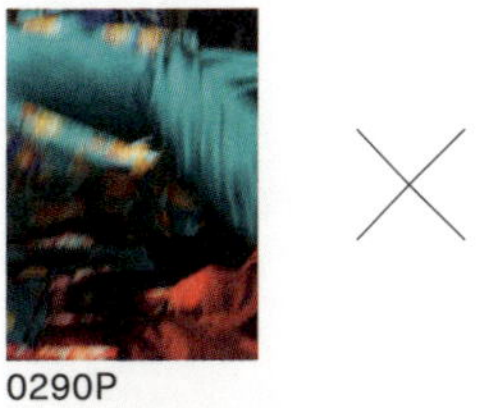

0290P

13

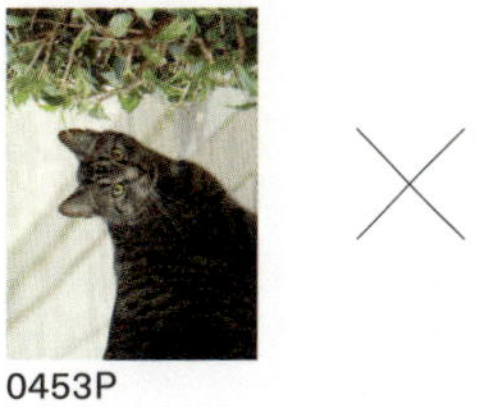

0453P

association when the synaesthete thinks of them. As stated above, I experience quite often that people have a colour which can change over the years, and especially when my emotional relationship to that person evolves.

I conclude that the colours I feel can be categorised as emotion–colour synaesthesia and that they are most often triggered by scenes of familiarity or motifs evoking an emotional response. The next step is to find out whether certain subjects — whether I am emotionally related to them or not — trigger a consistent colour range. For example, does a subject such as hands or a certain place consistently evoke similar emotional colours?

Chapter 2
Motif

To figure out whether the subject of the image could determine the colour, I isolated several topics, like clouds, or trees in winter, and compared the associated colours to see whether there would be similarities. I was curious to see whether one person or a certain subject would always feel the same and thus consistently evoke the same colour.

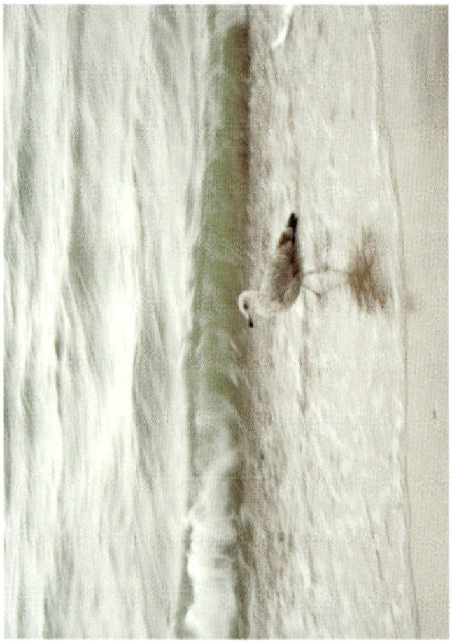
0014P

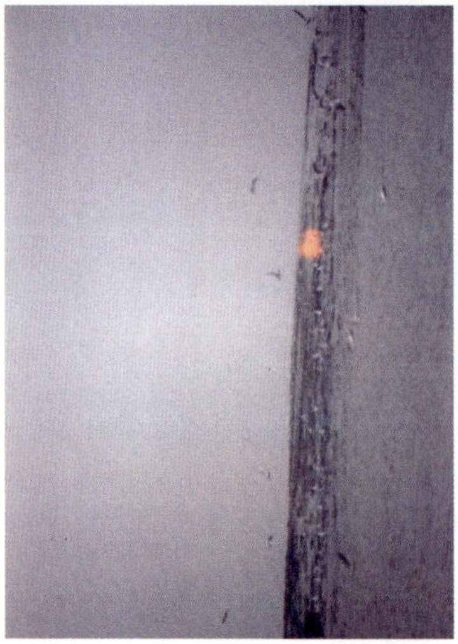
0015P

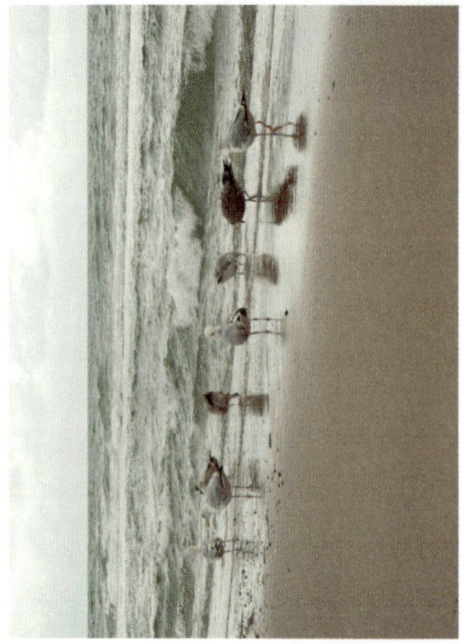
0048P

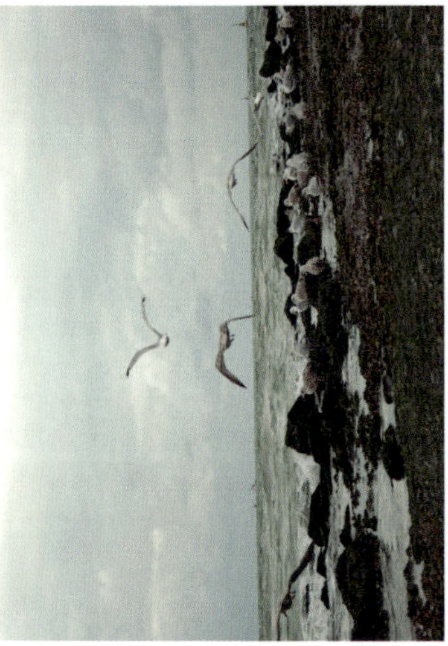
0068P

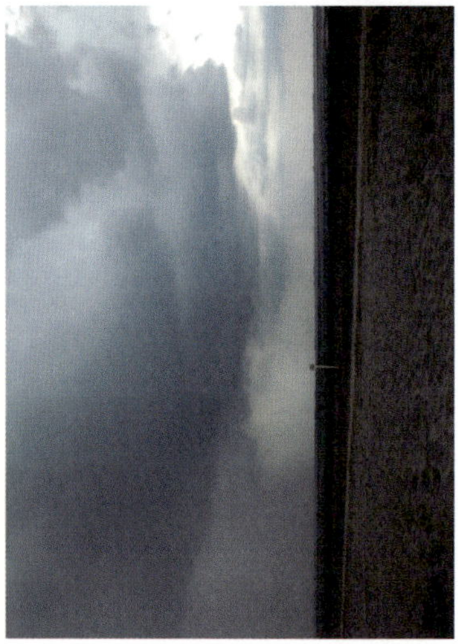
0764P

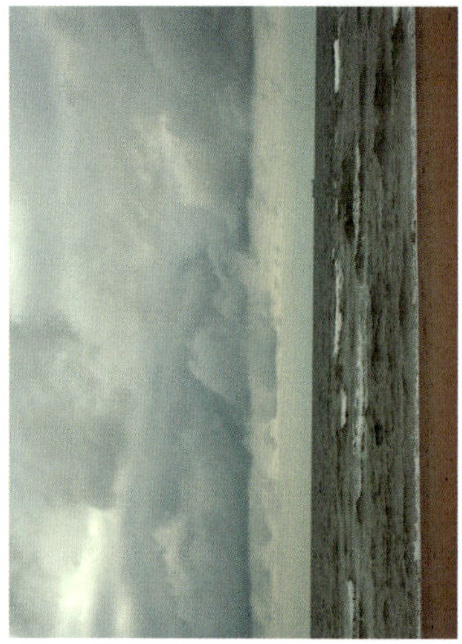
0851P

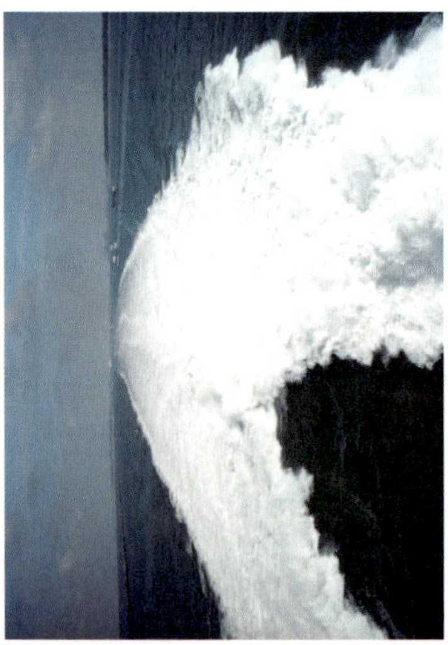
1328P

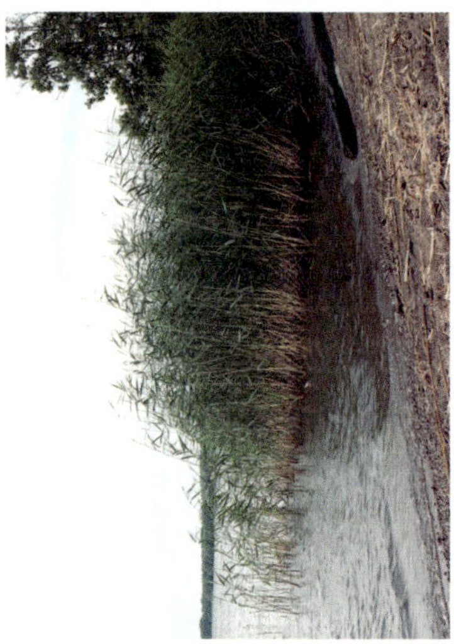
1375P

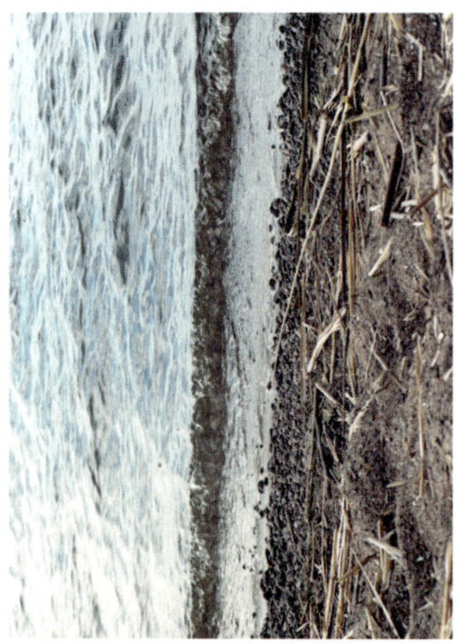
1408P

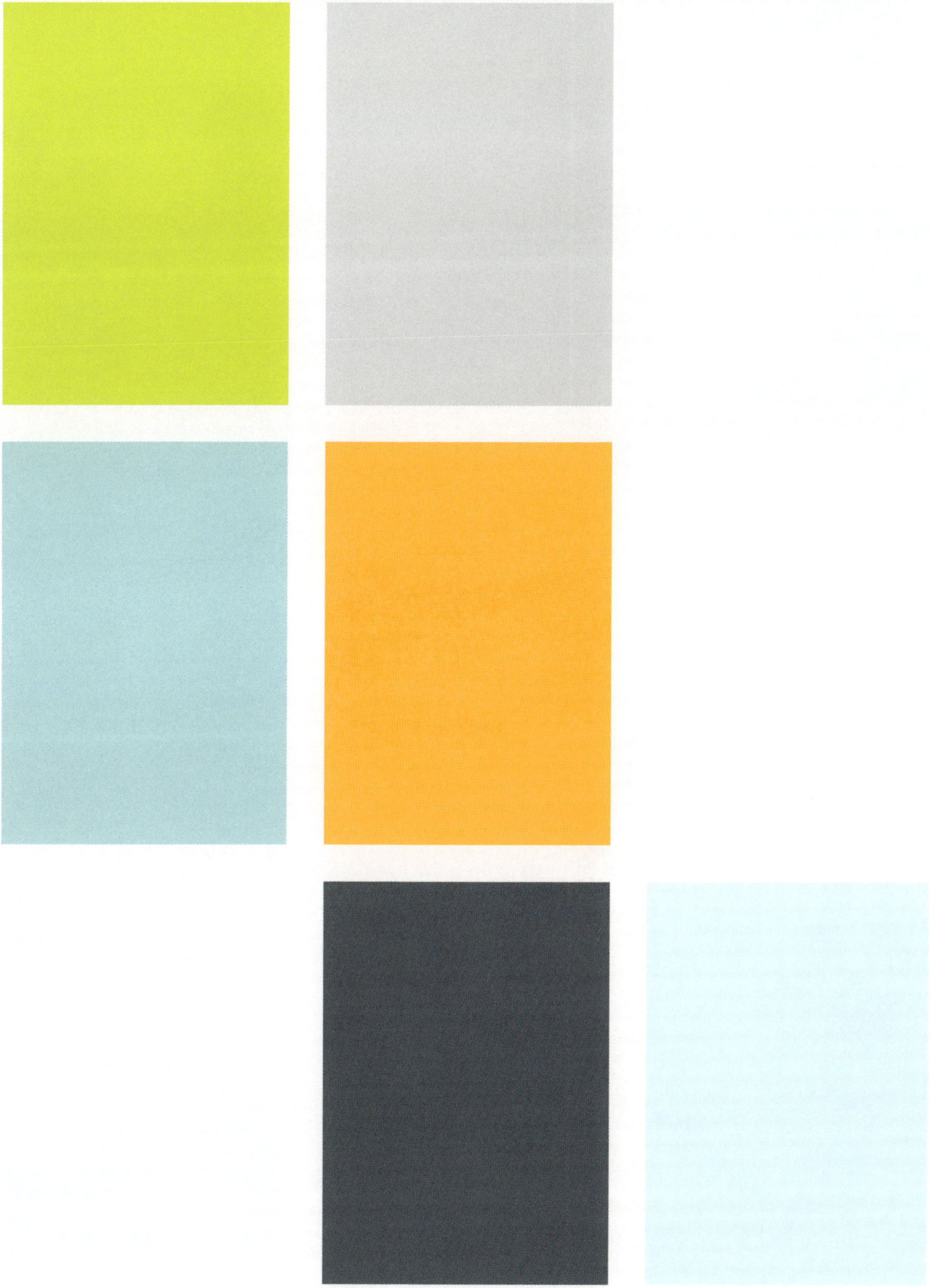

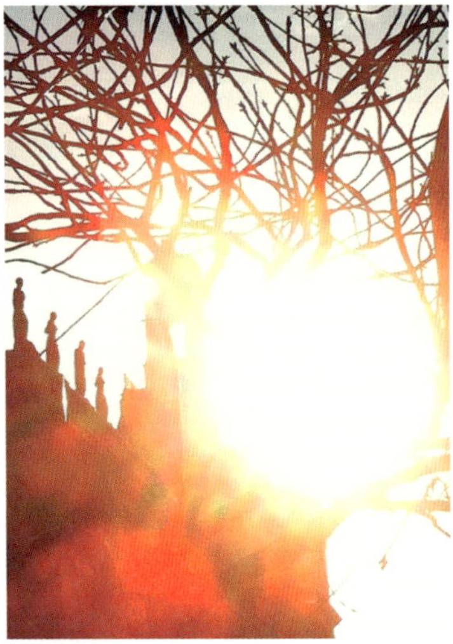
0026P

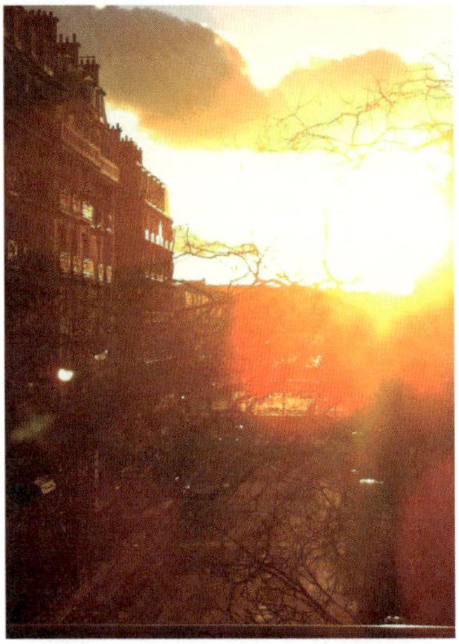
0028P

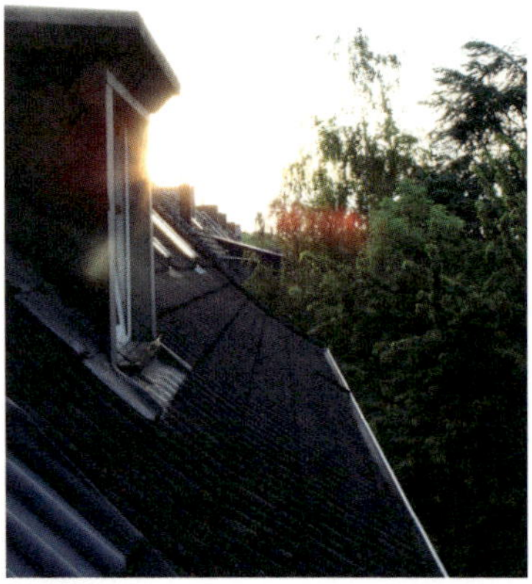
0240P

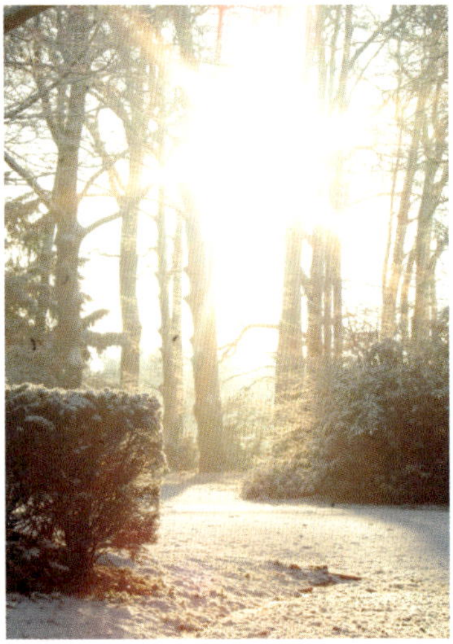
0472P

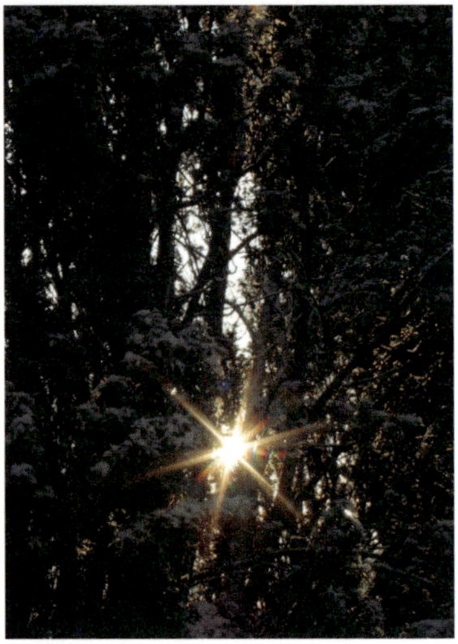
0482P

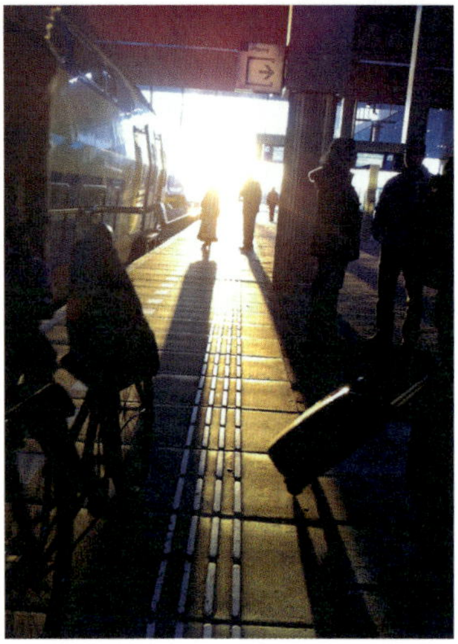
1084P

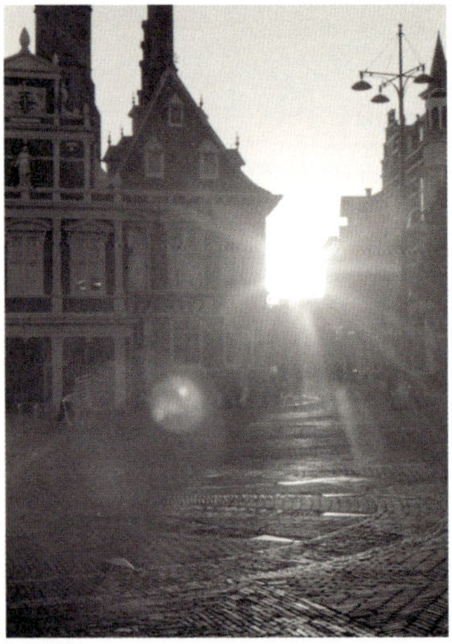
1129P

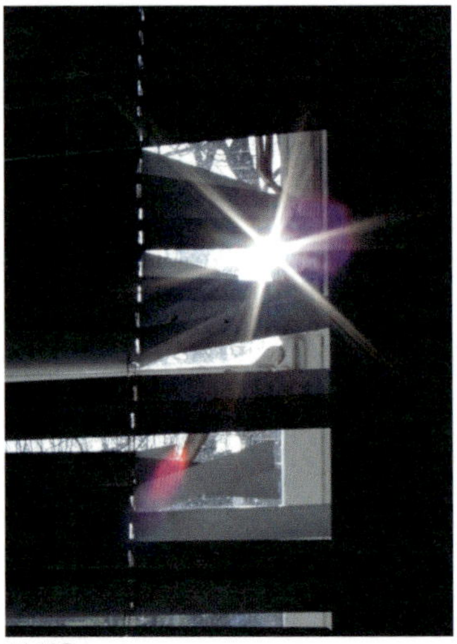
1142P

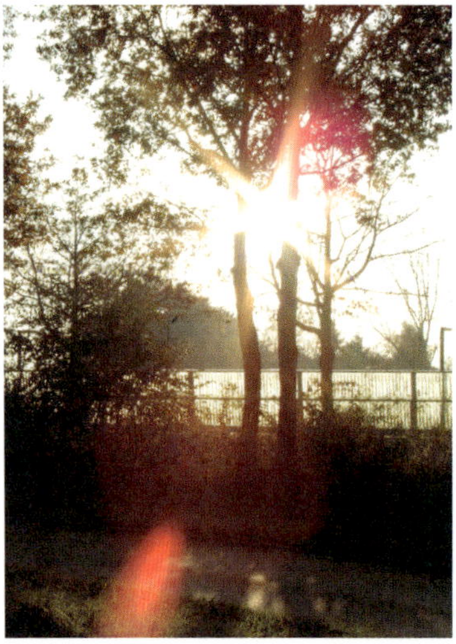
1192P

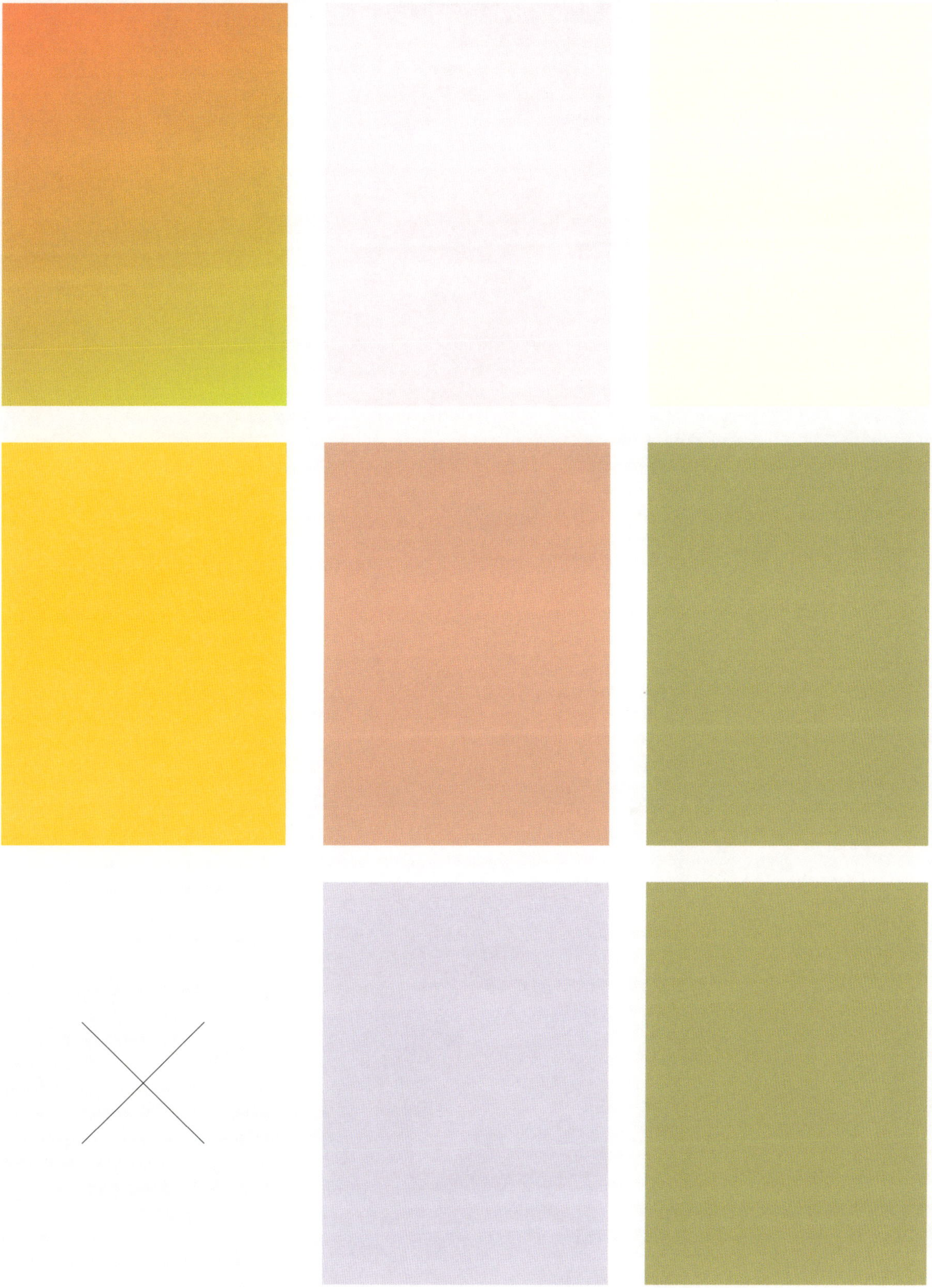

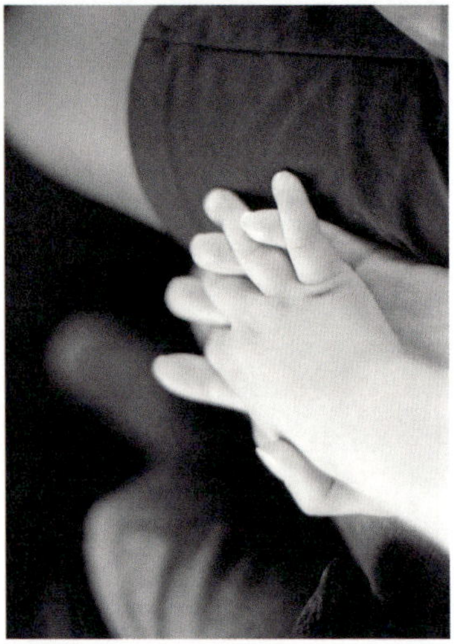
0058P

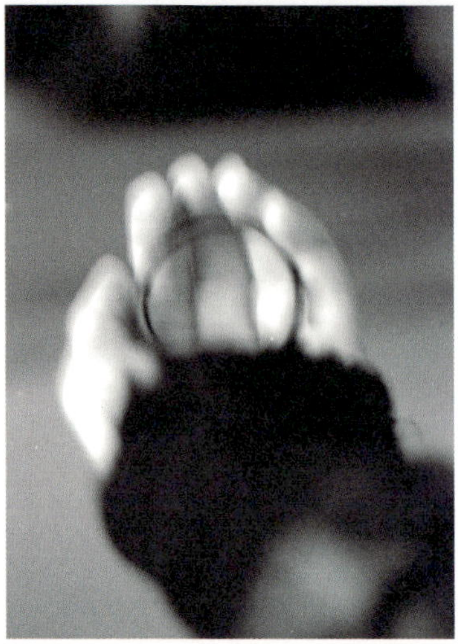
0116P

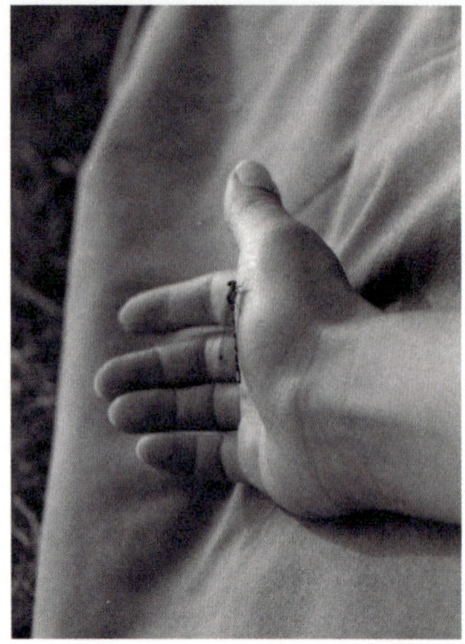
0179P

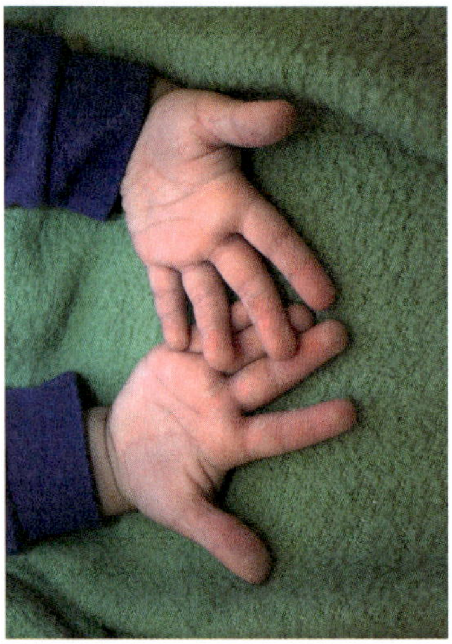
0380P

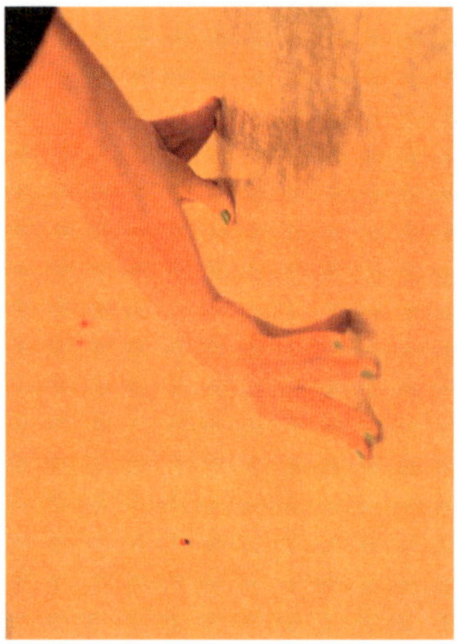
0399P

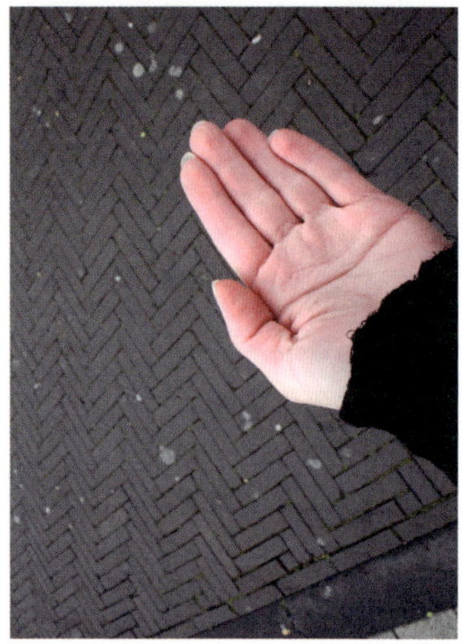
0671P

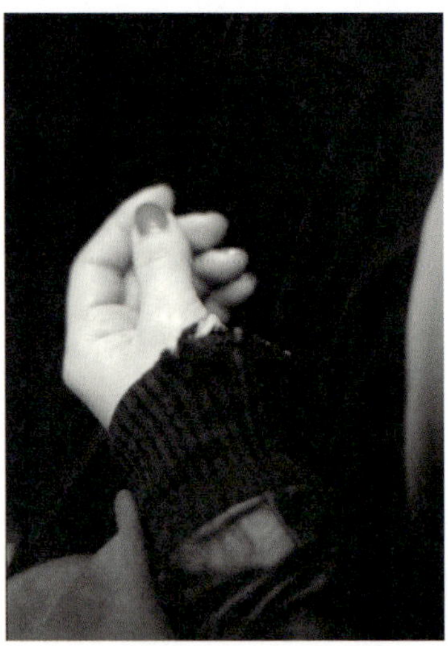
0767P

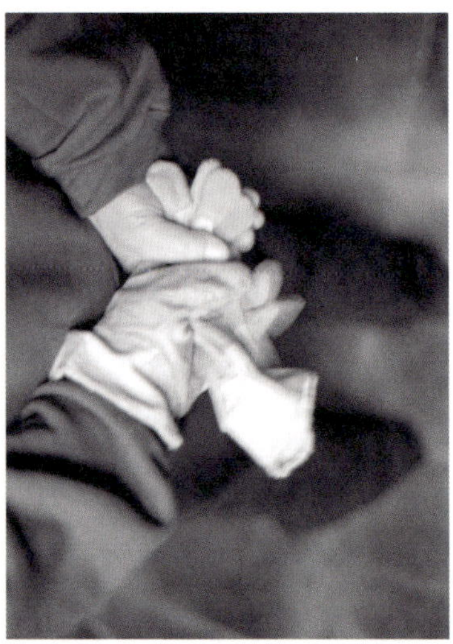
0823P

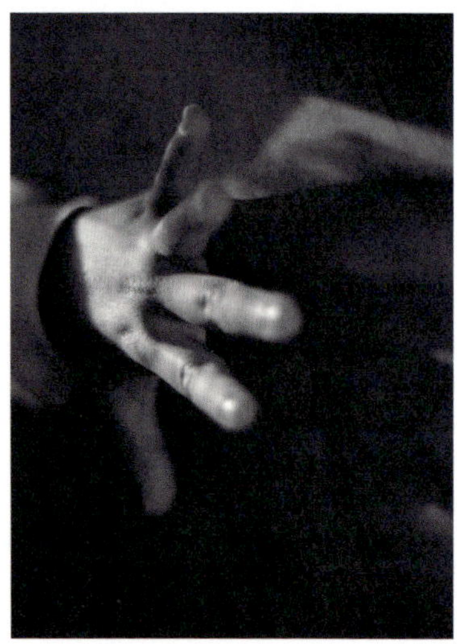
0970P

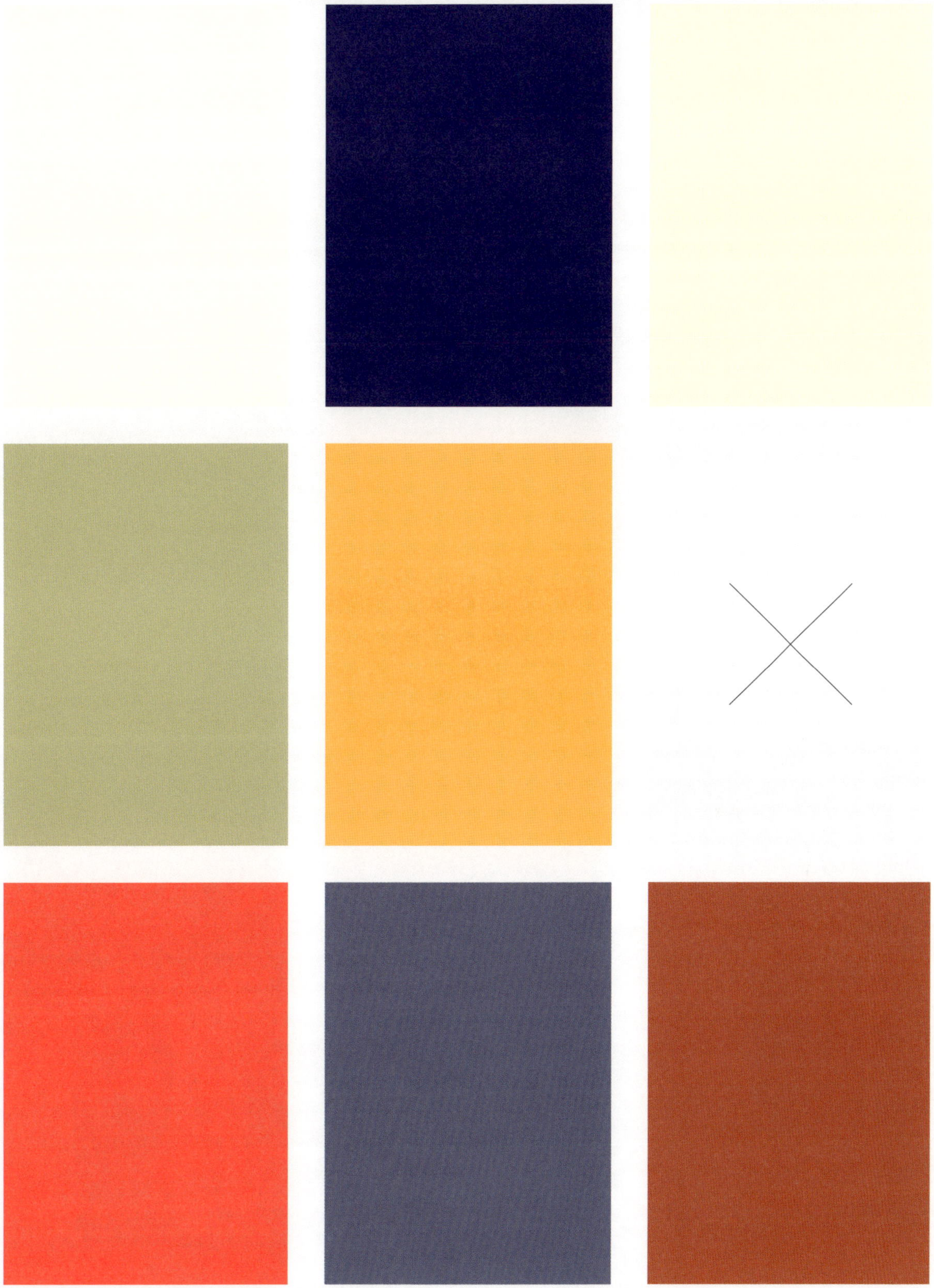

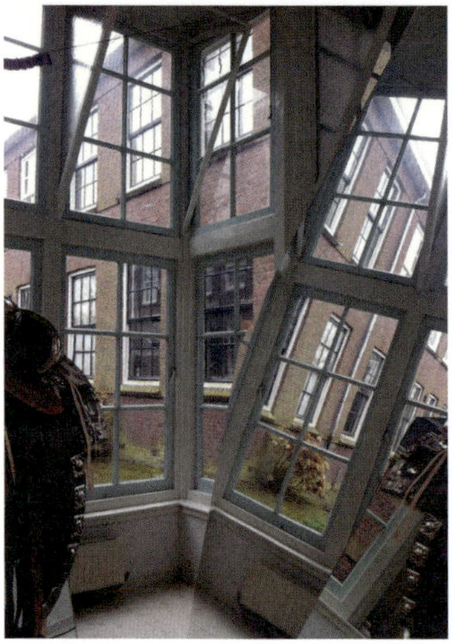
0156P

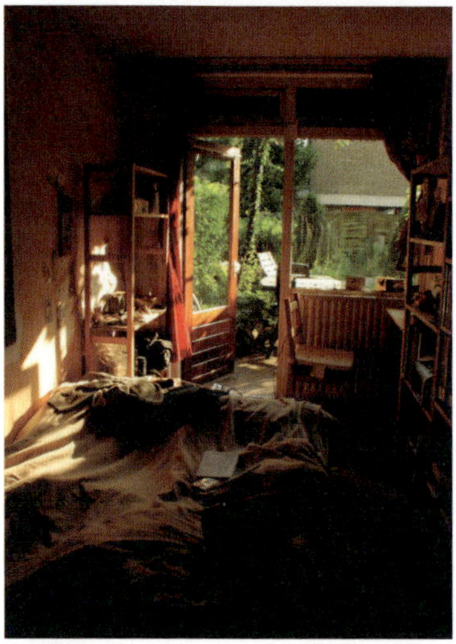
0366P

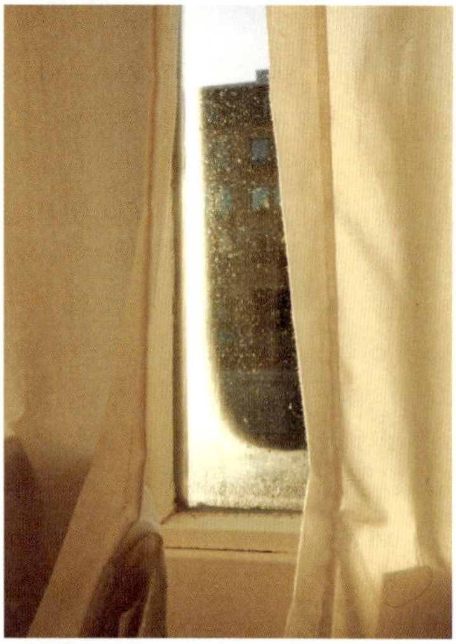
0718P

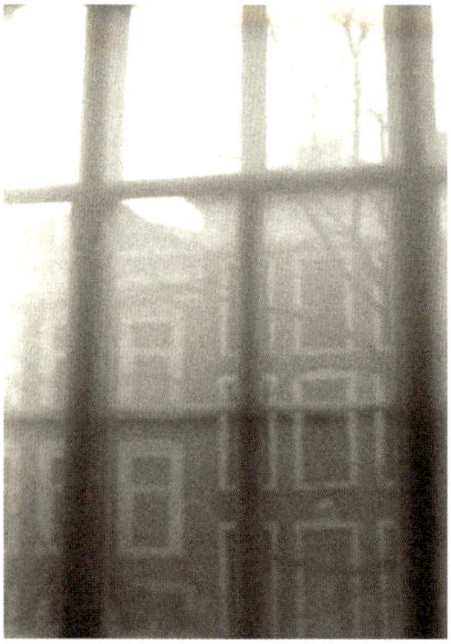
1230P

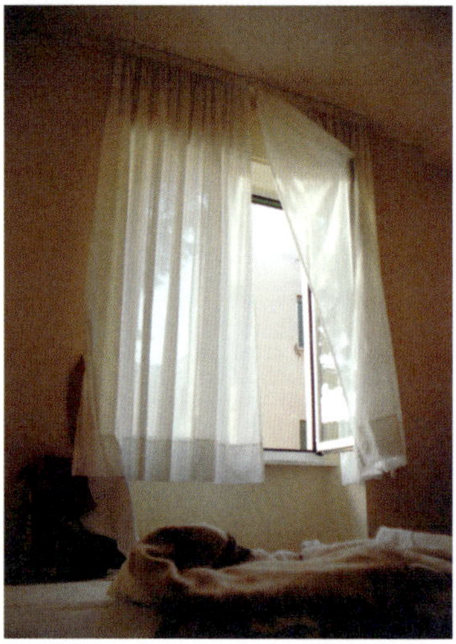
1256P

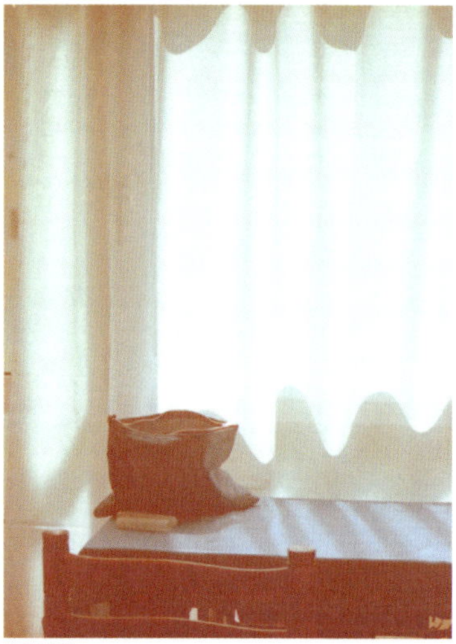
1258P

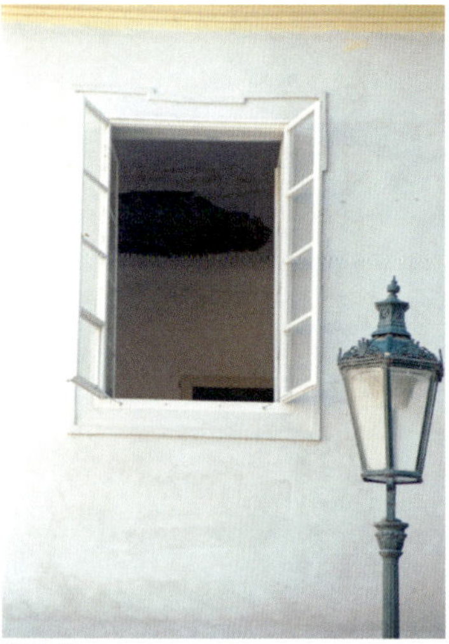
1260P

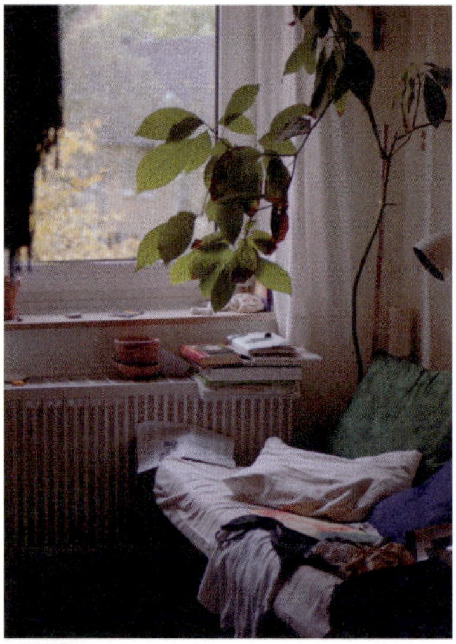
1261P

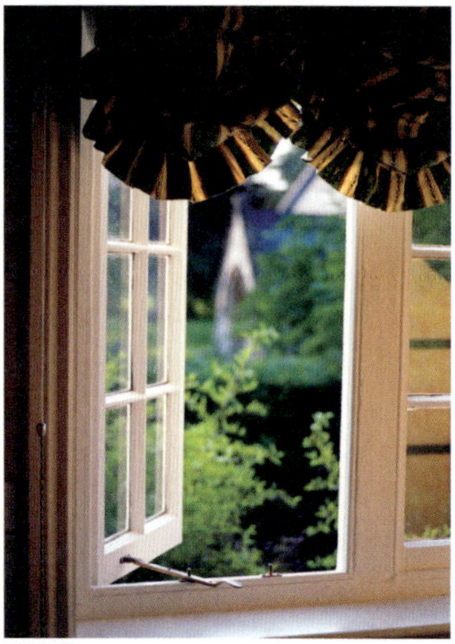
1277P

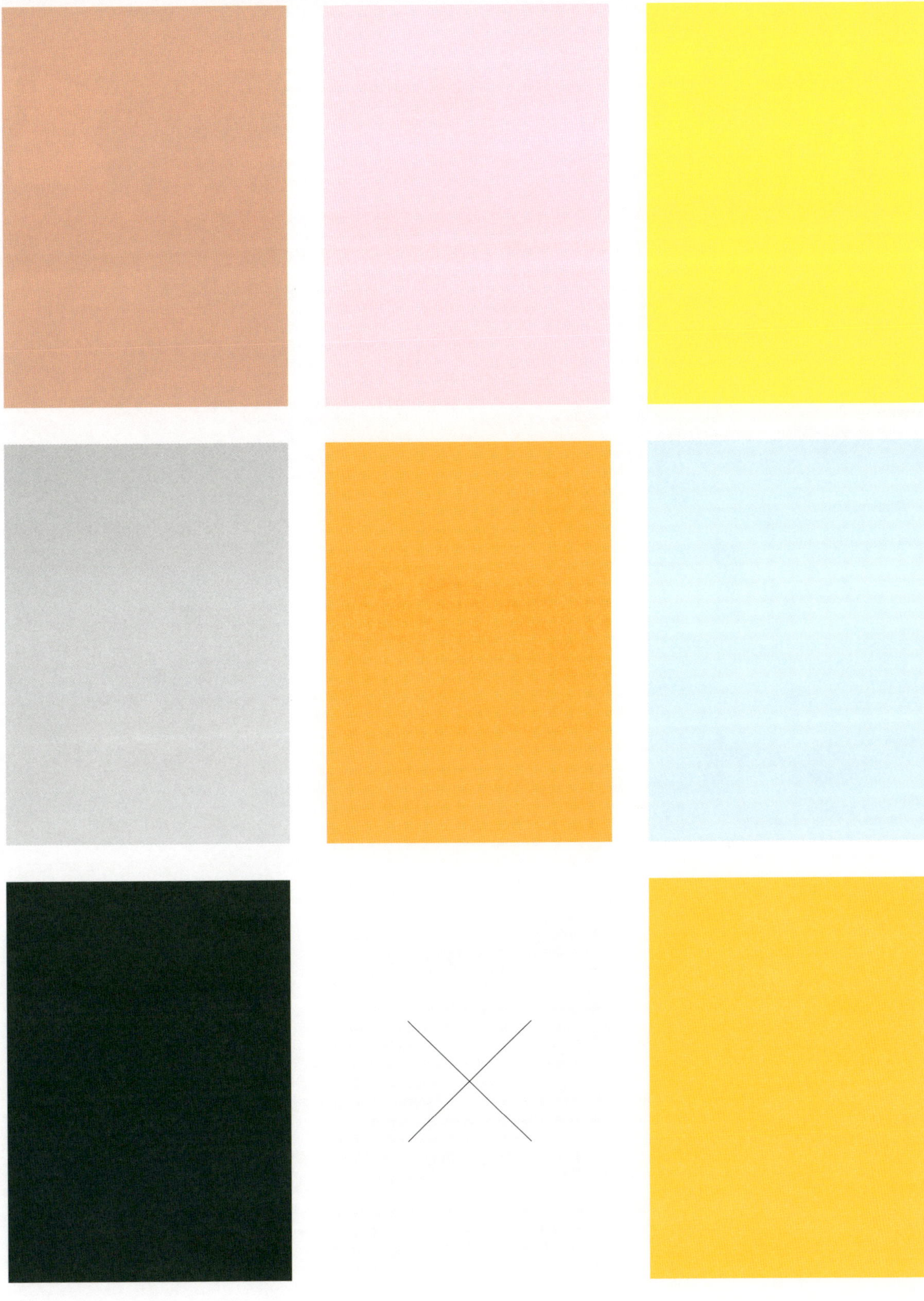

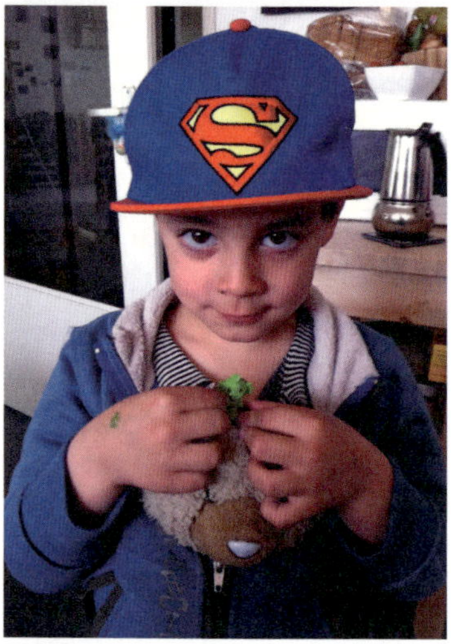
0199P

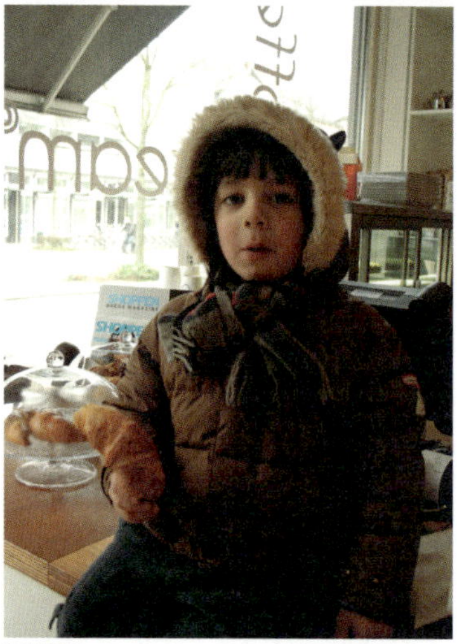
0353P

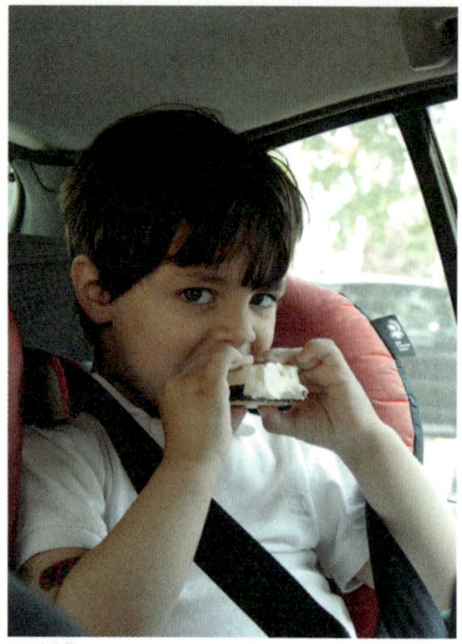
0439P

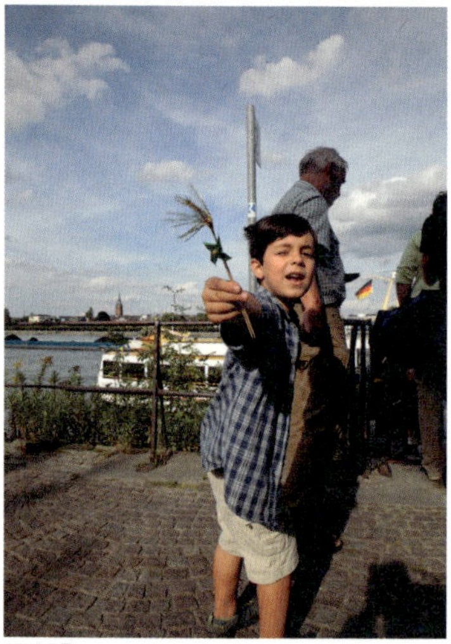
0723P

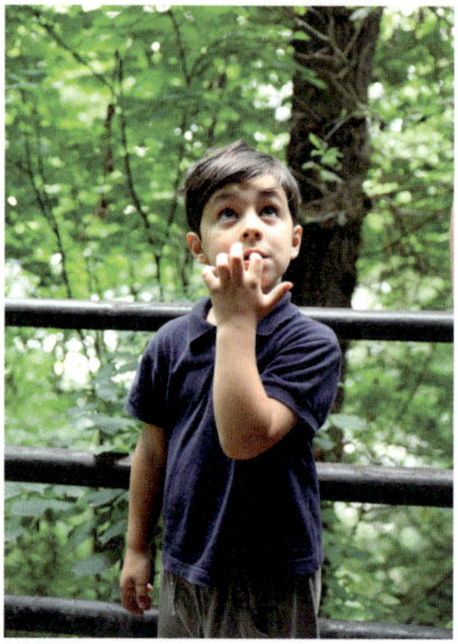
0732P

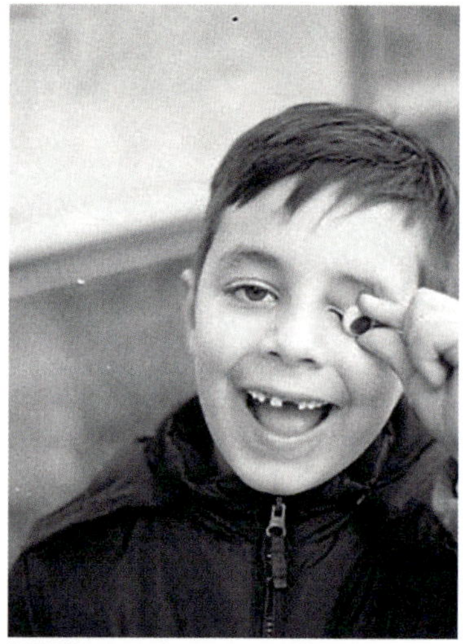
0897P

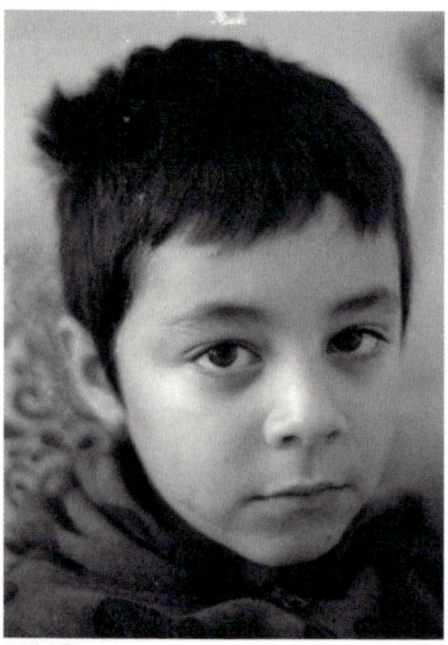
1001P

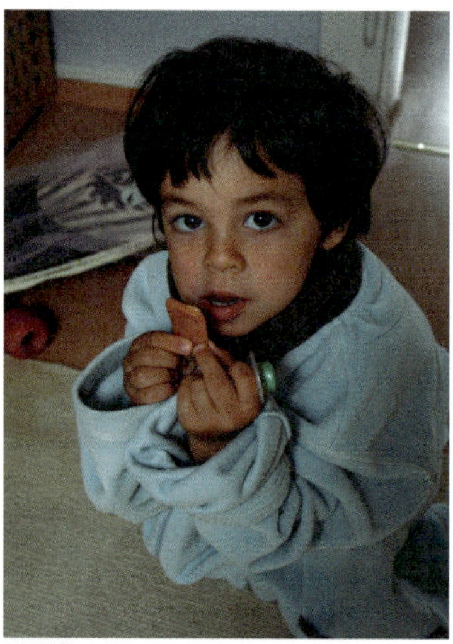
1176P

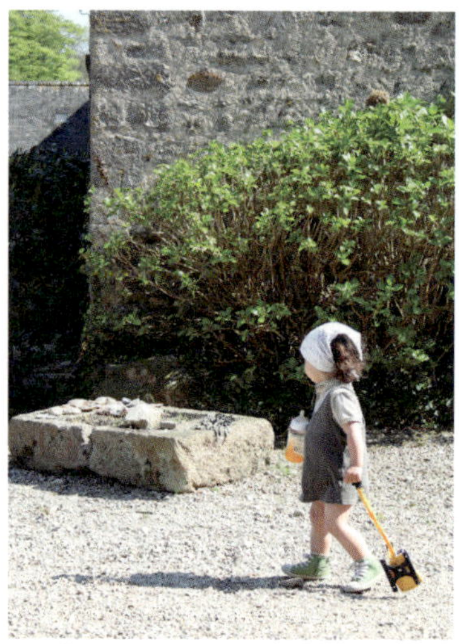
0632P

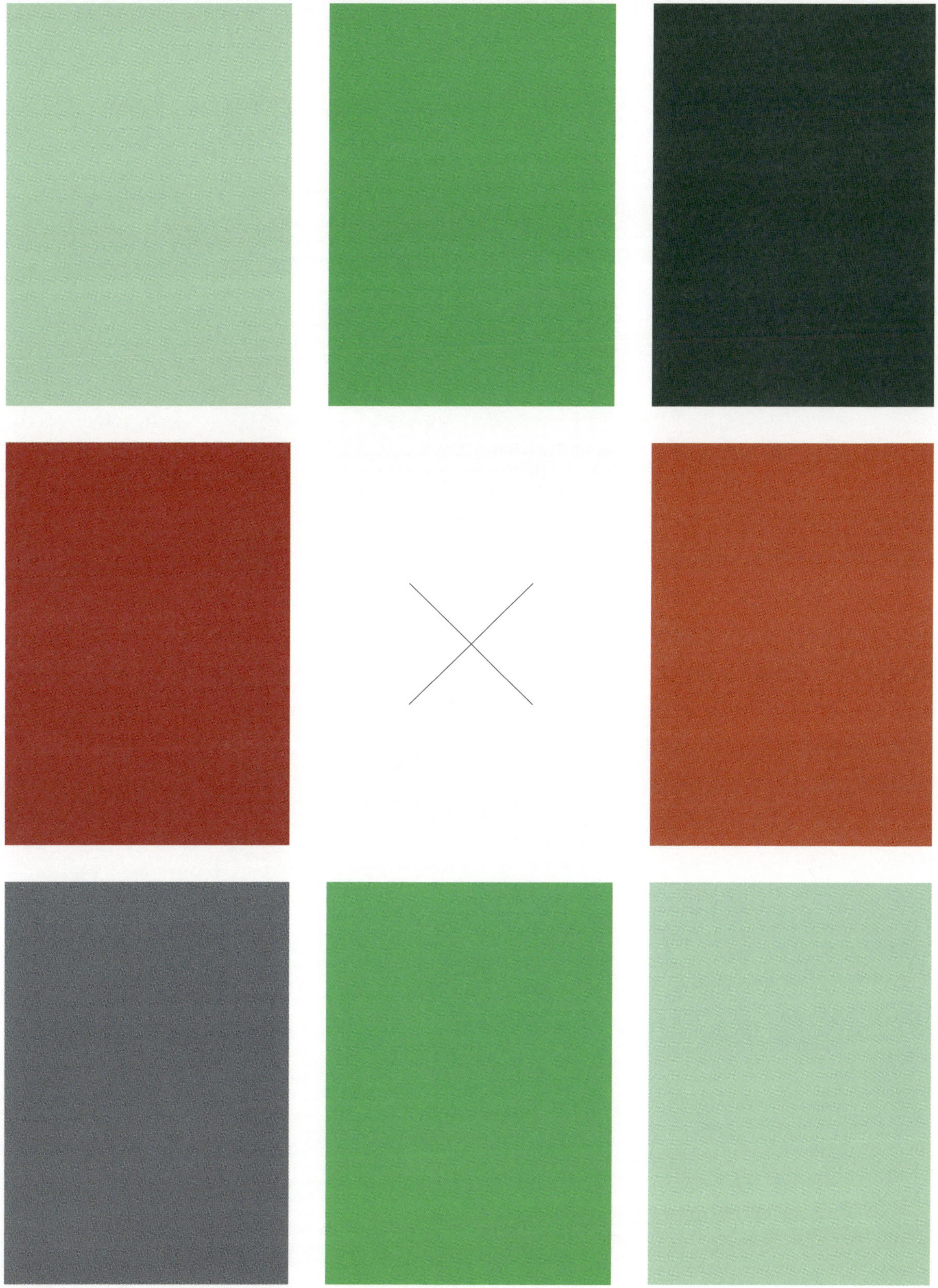

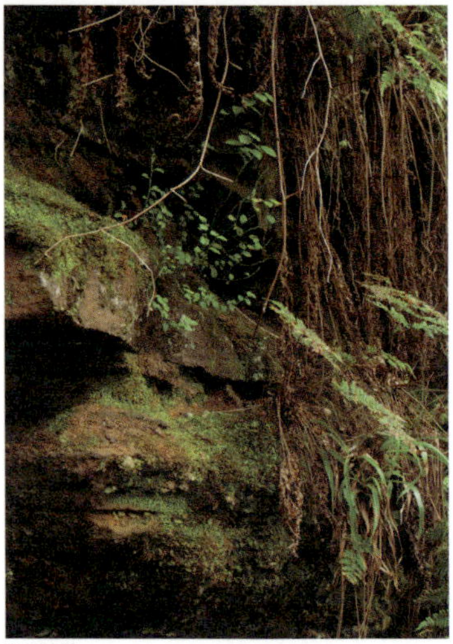
0232P

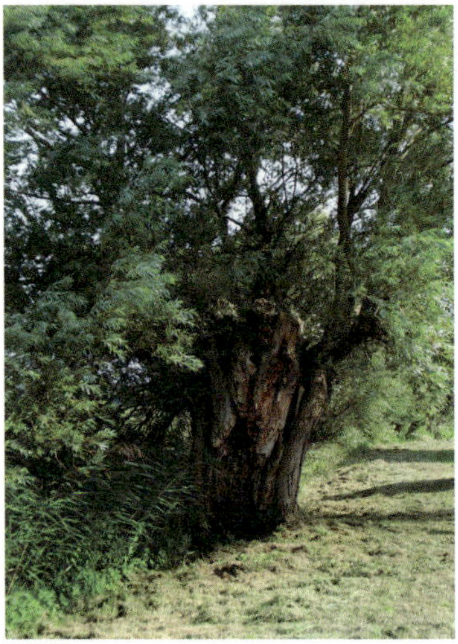
0635P

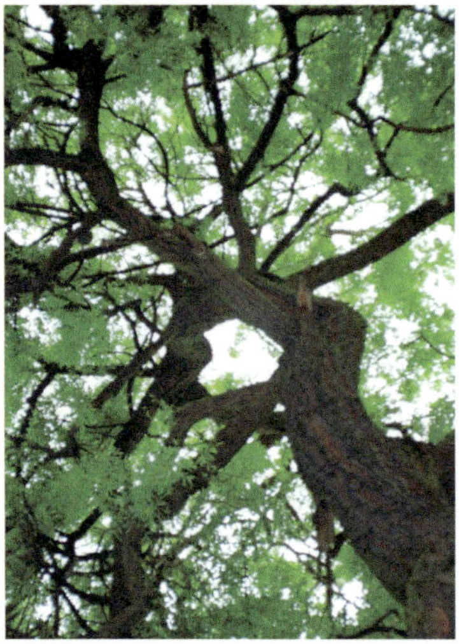
0820P

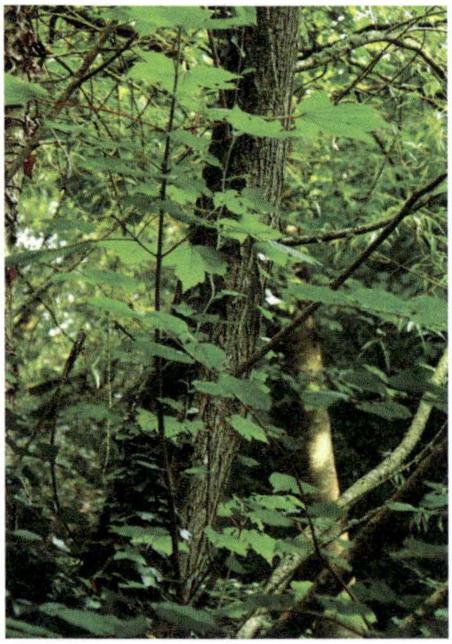
1222P

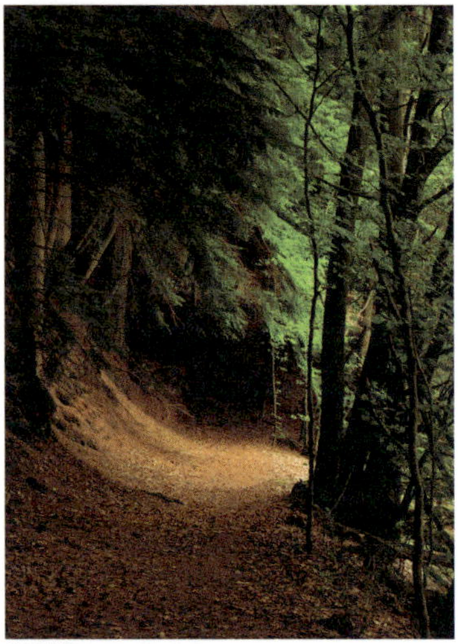
1227P

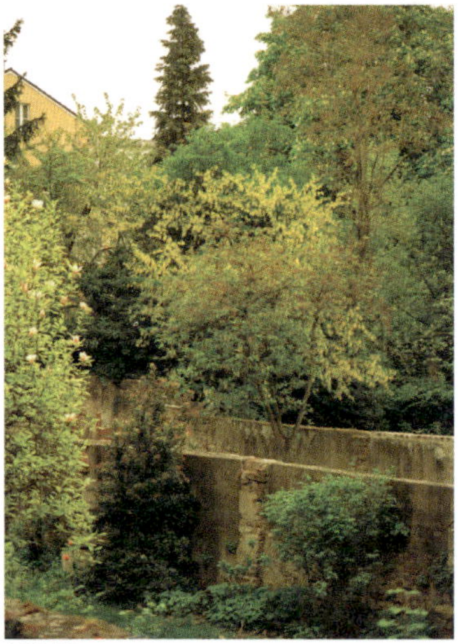
1234P

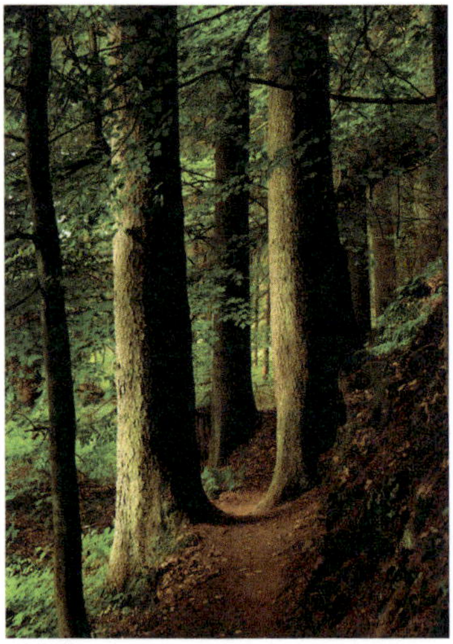
1238P

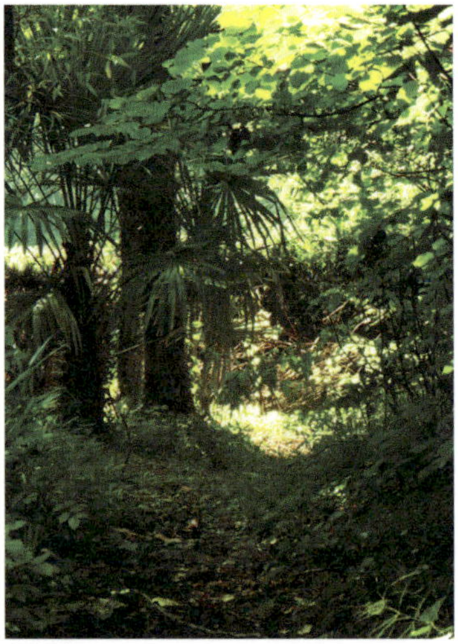
1243P

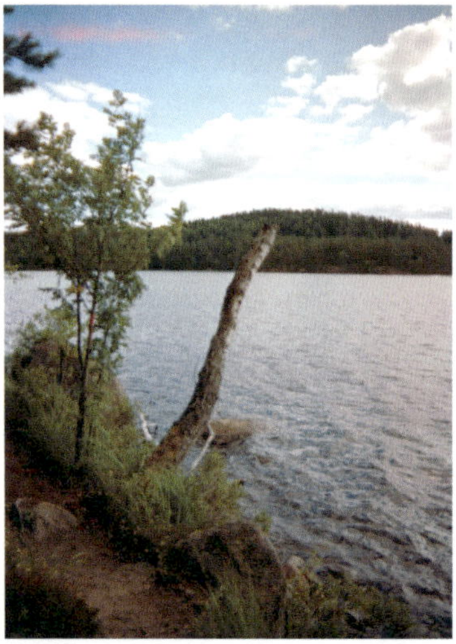
1263P

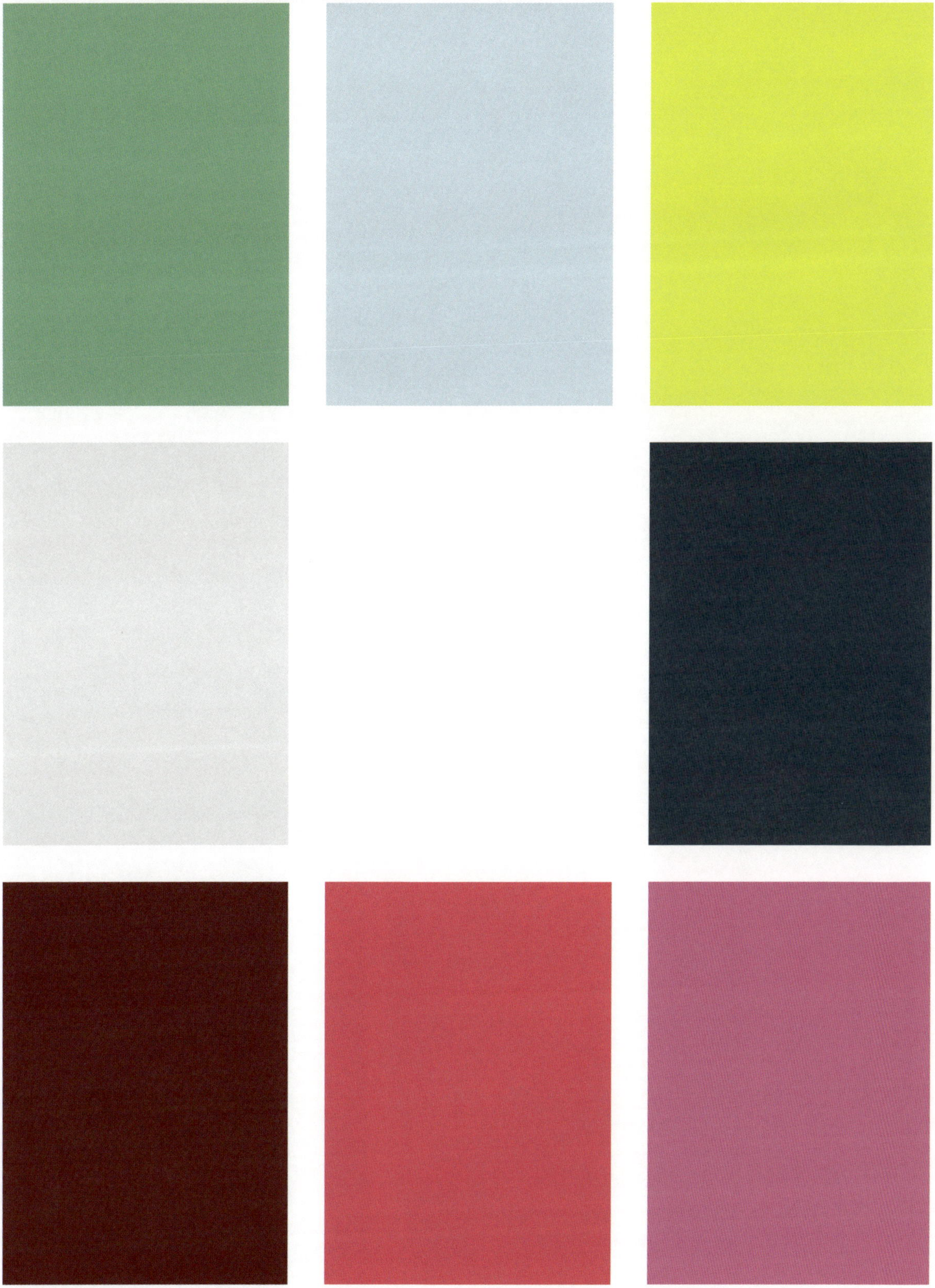

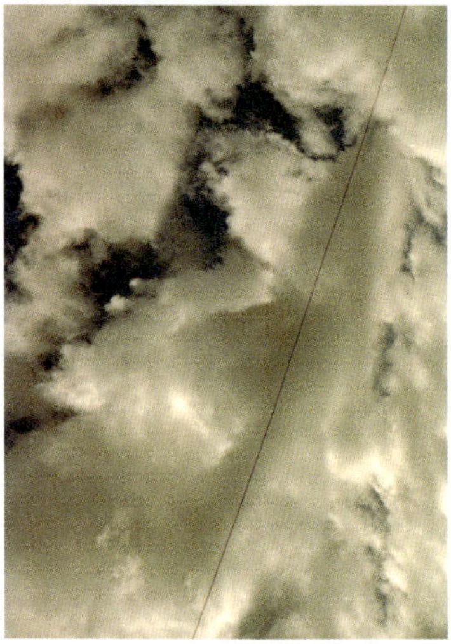
0365P

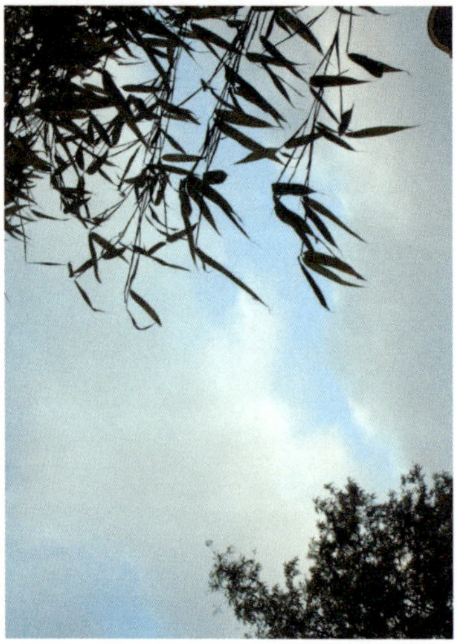
0466P

0624P

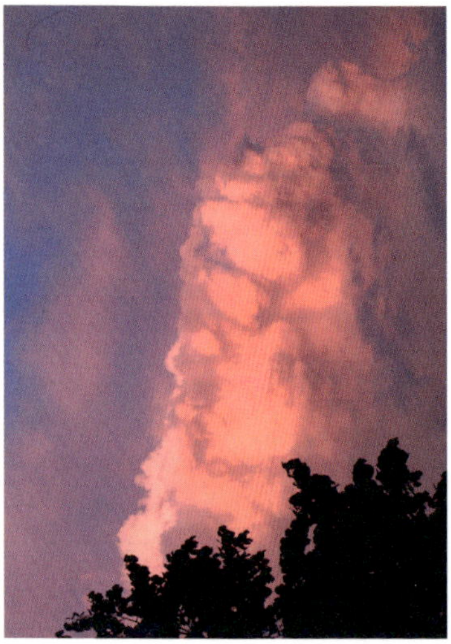
0654P

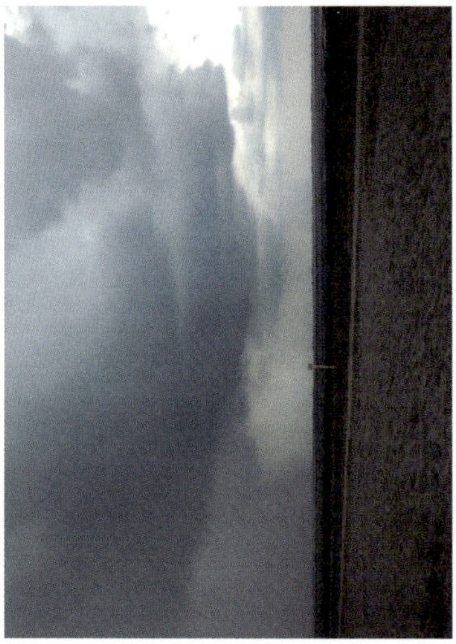
0764P

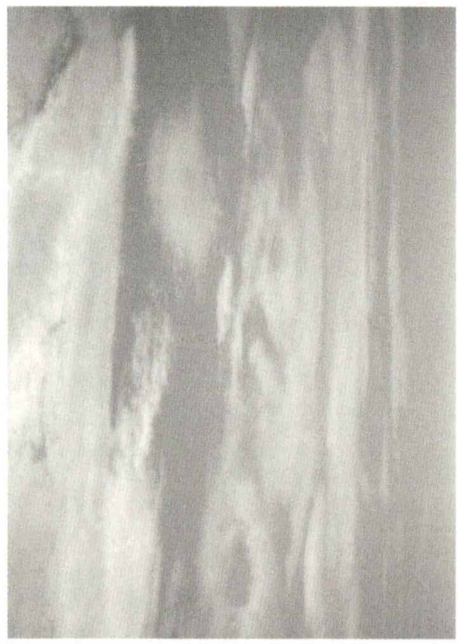
0841P

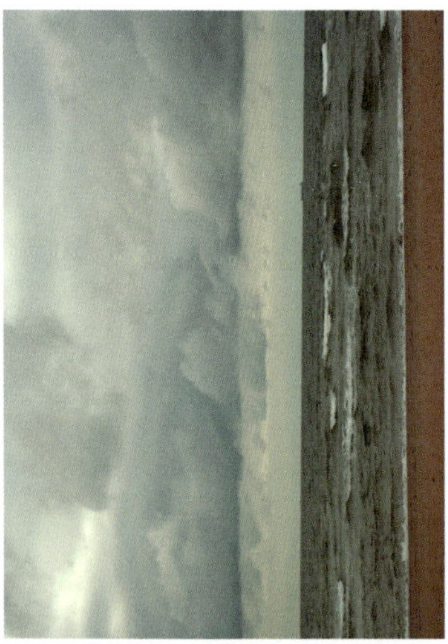
0851P

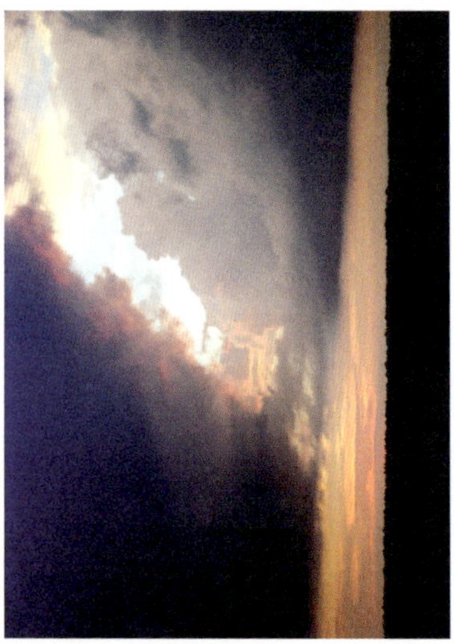
1373P

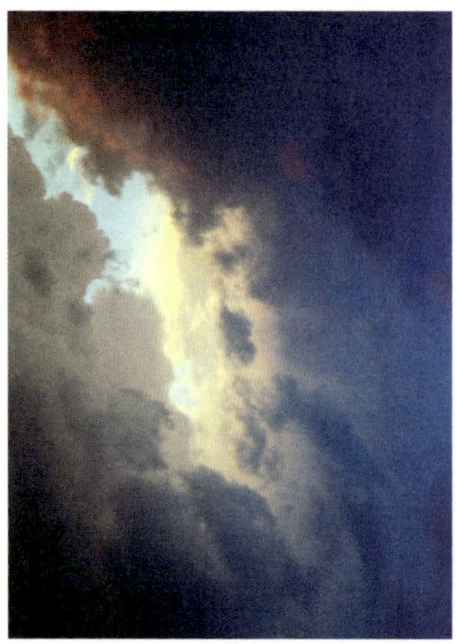
1391P

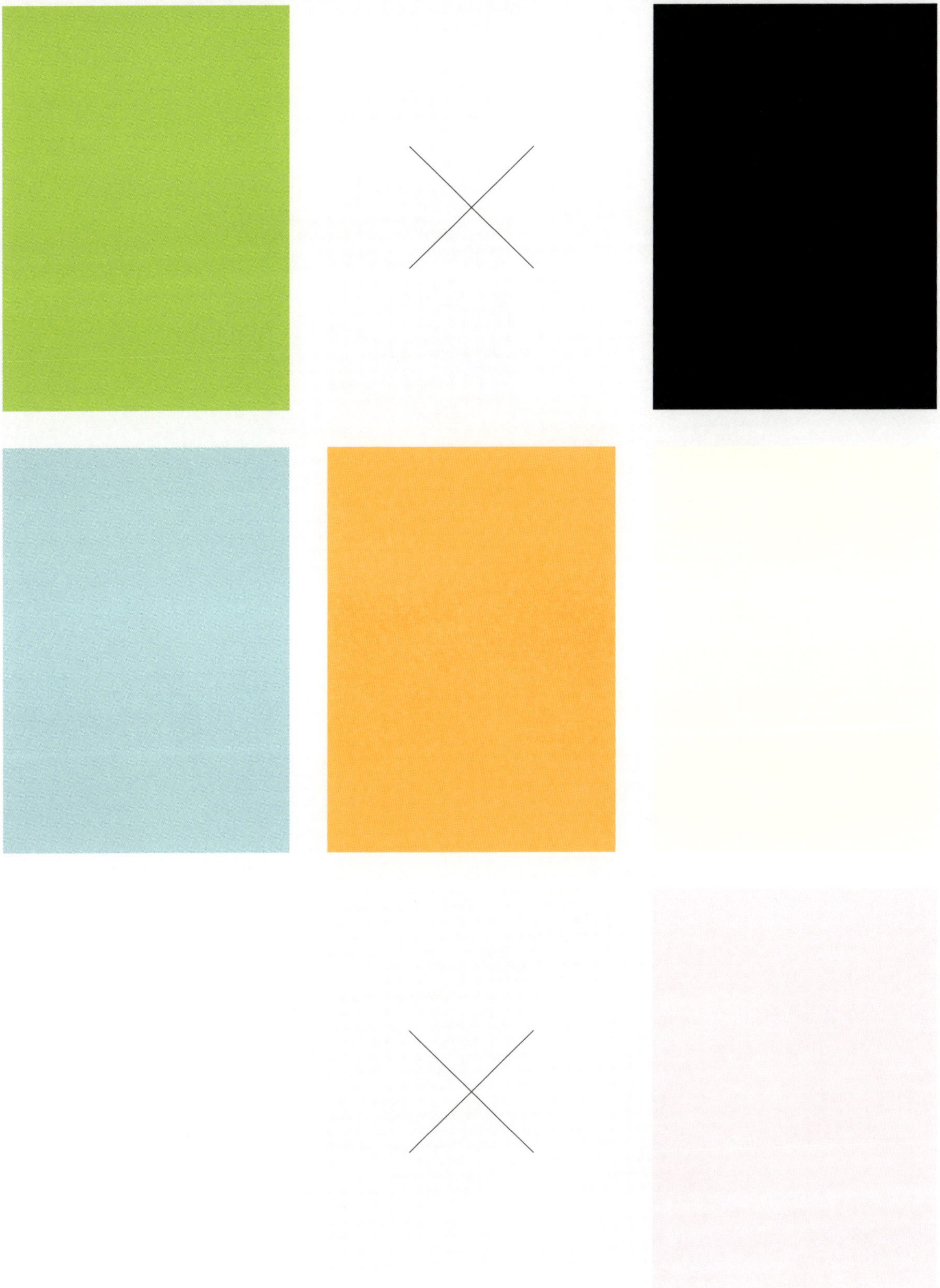

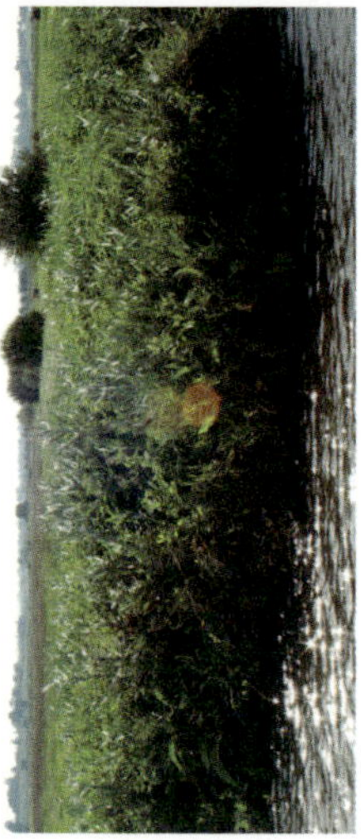
0410P

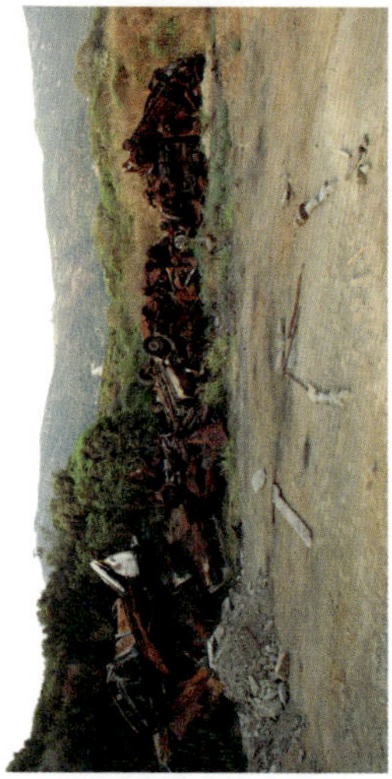
0542P

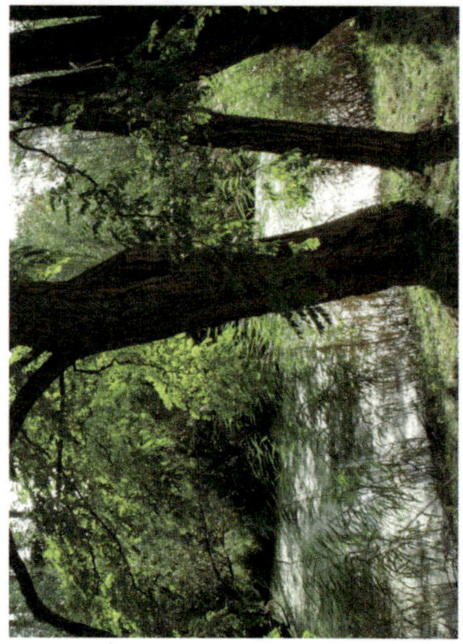
0557P

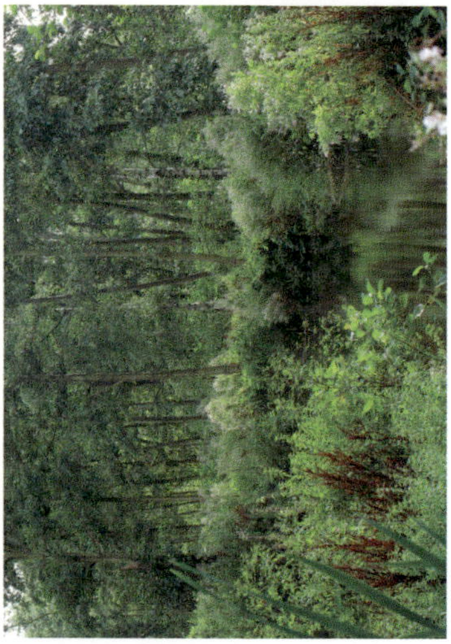
0629P

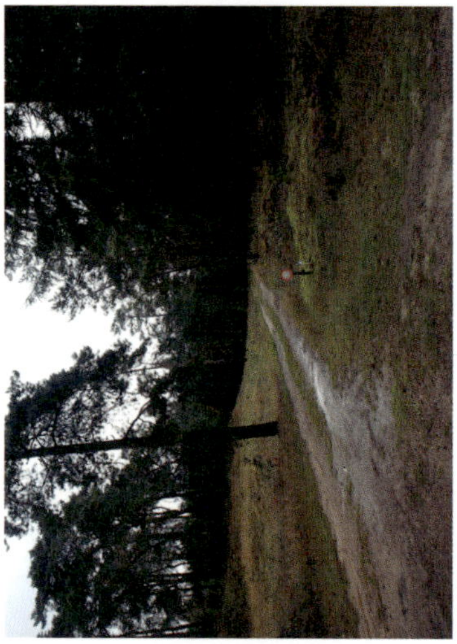
0797P

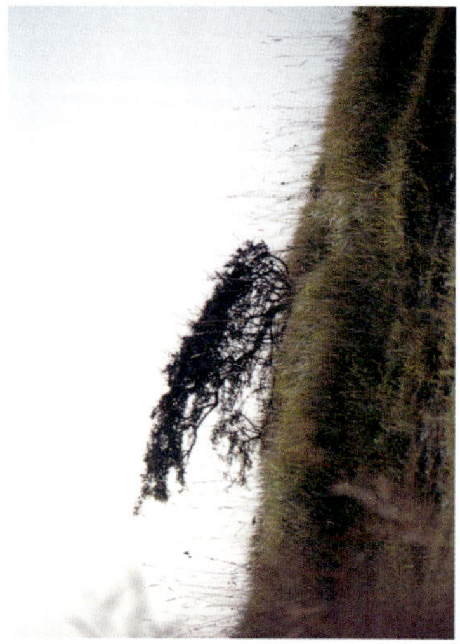
0810P

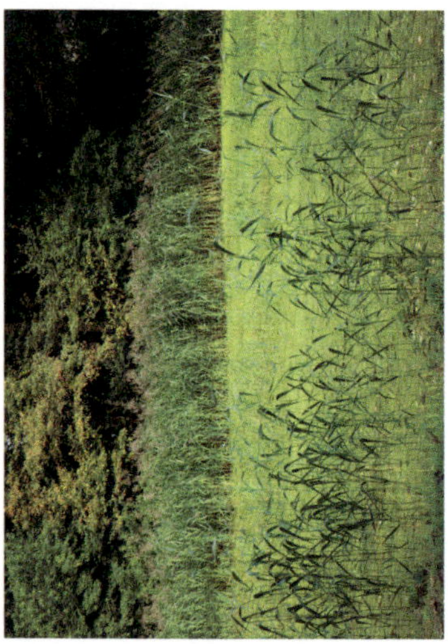
0839P

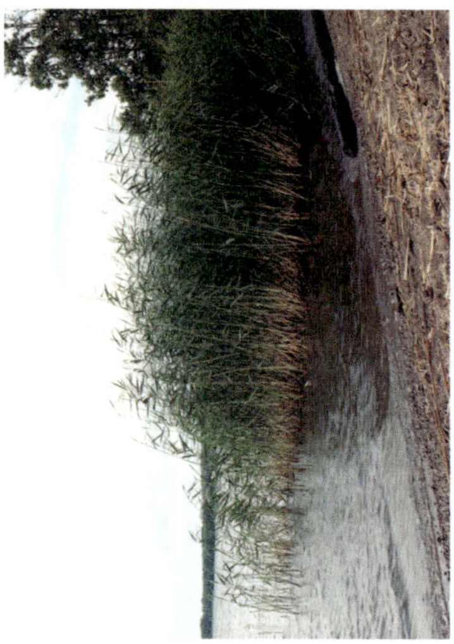
1375P

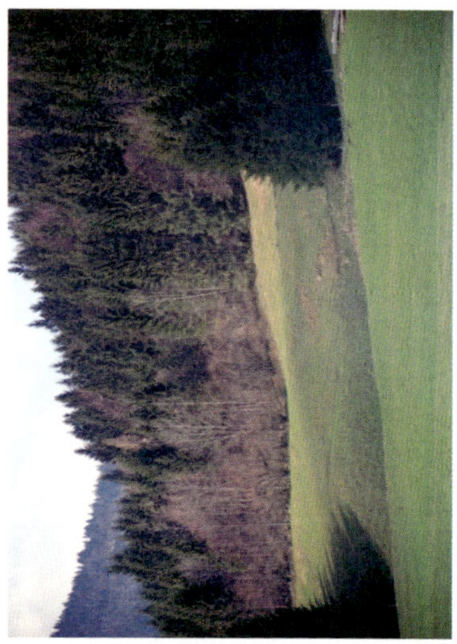
1381P

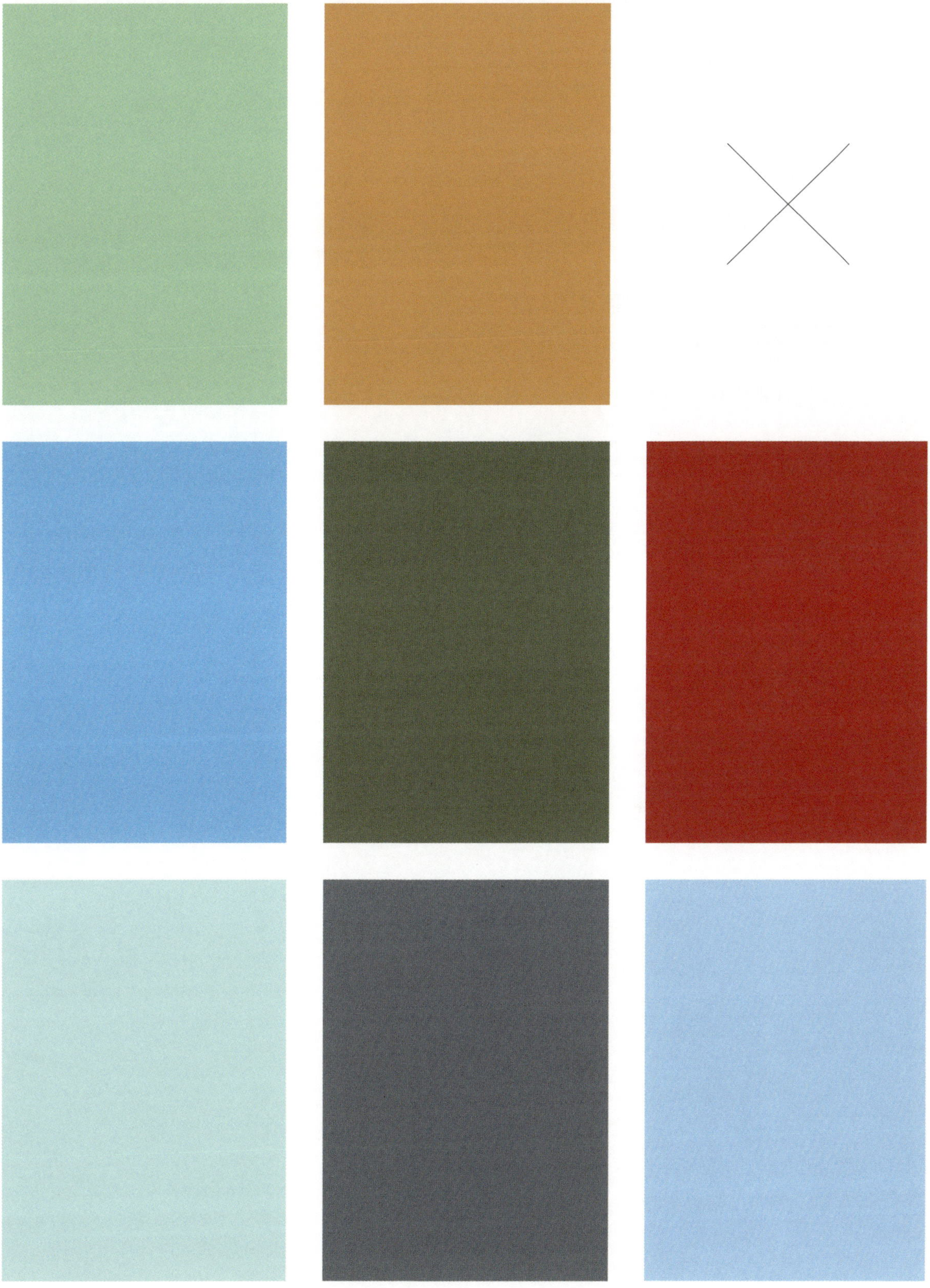

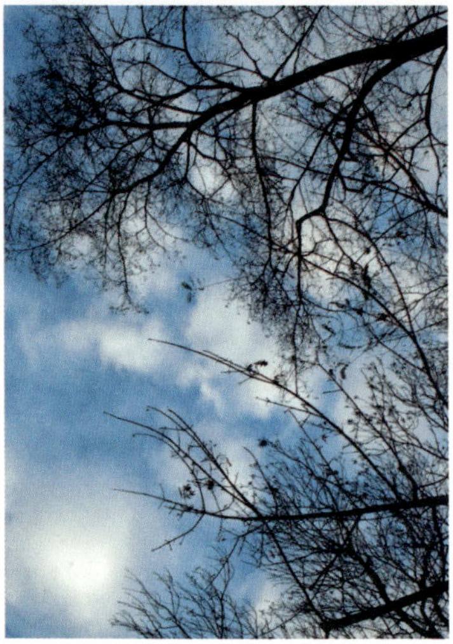
0729P

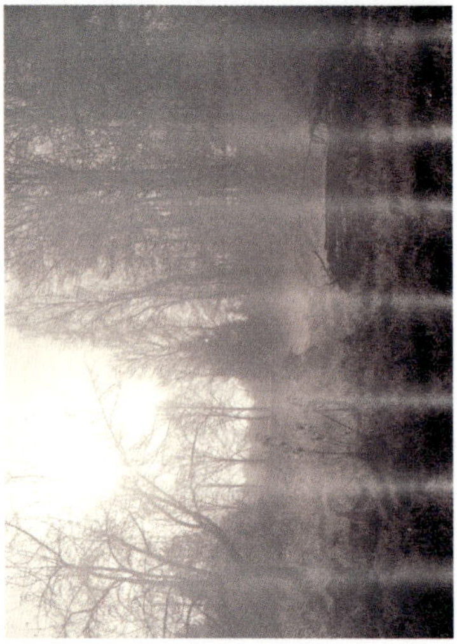
0769P

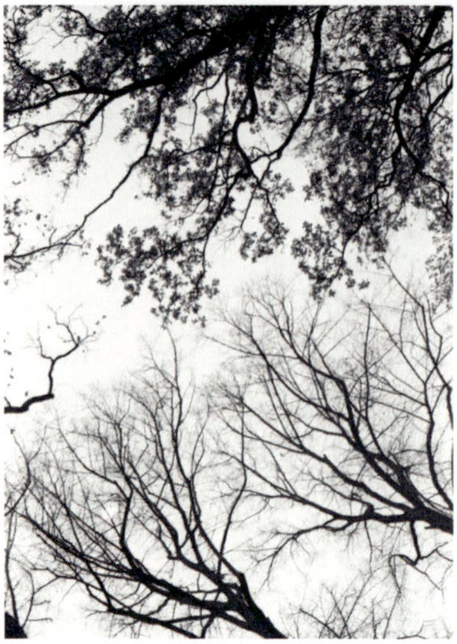
0816P

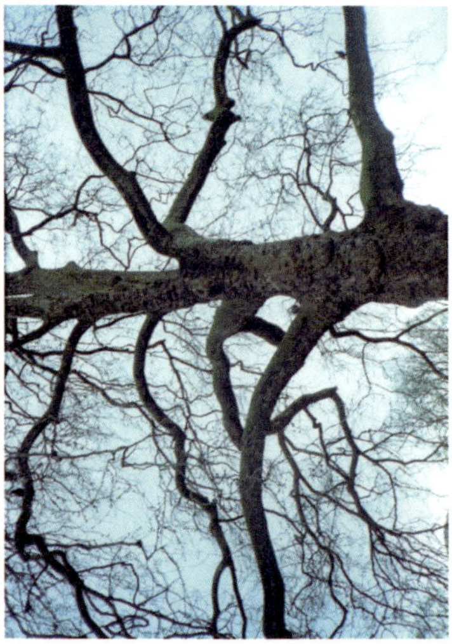
1332P

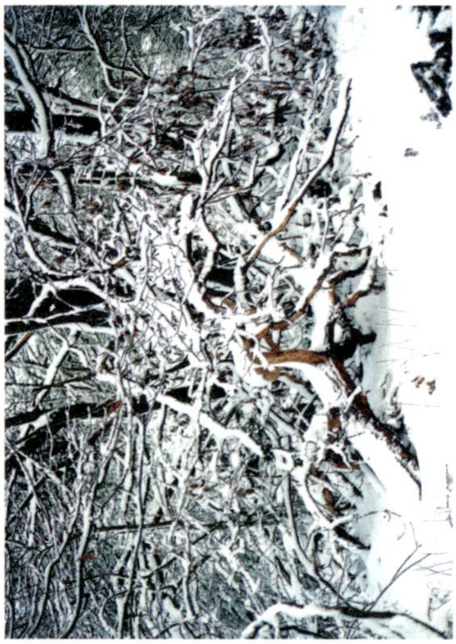
1359P

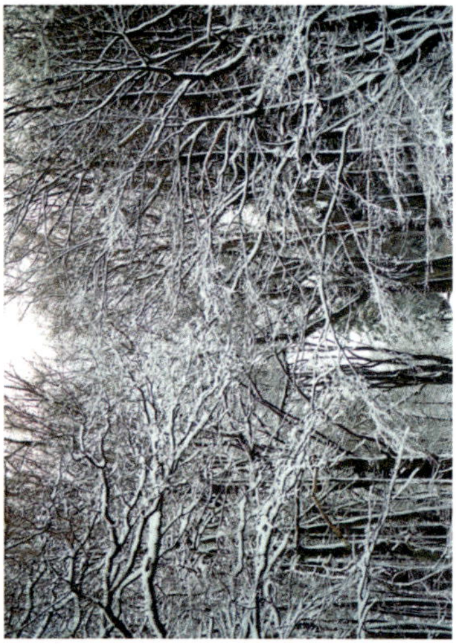
1406P

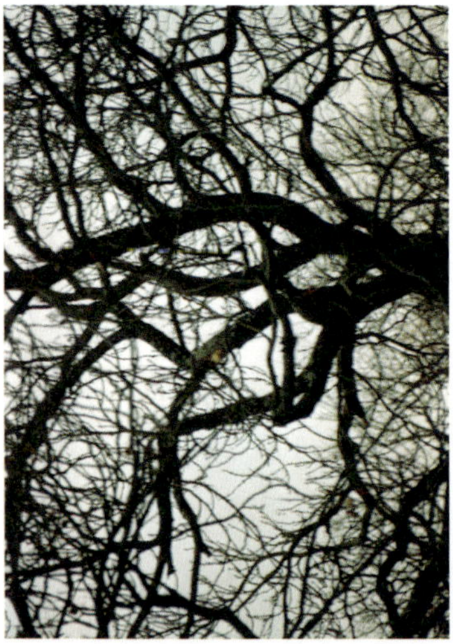
1418P

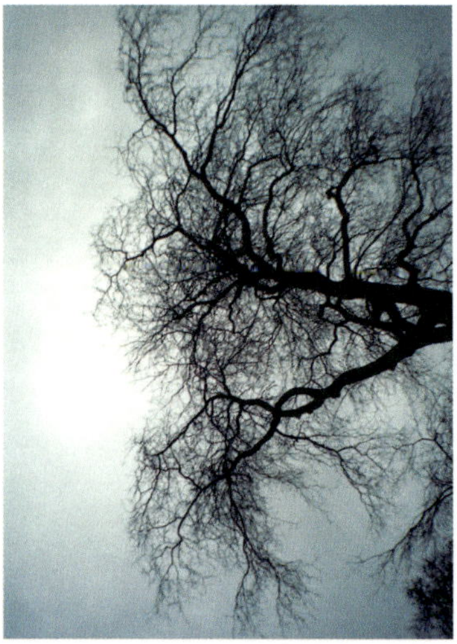
1432P

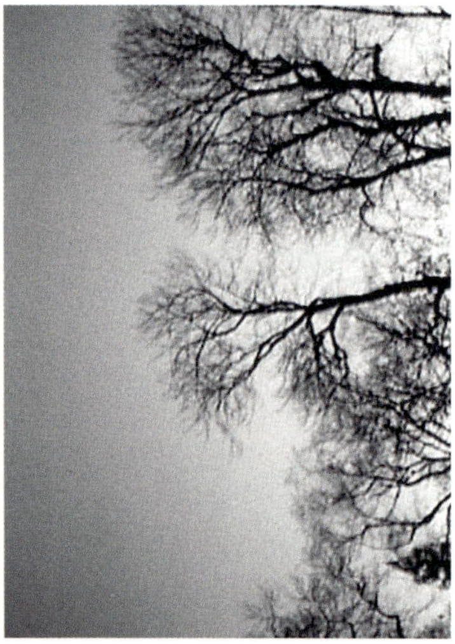
1455P

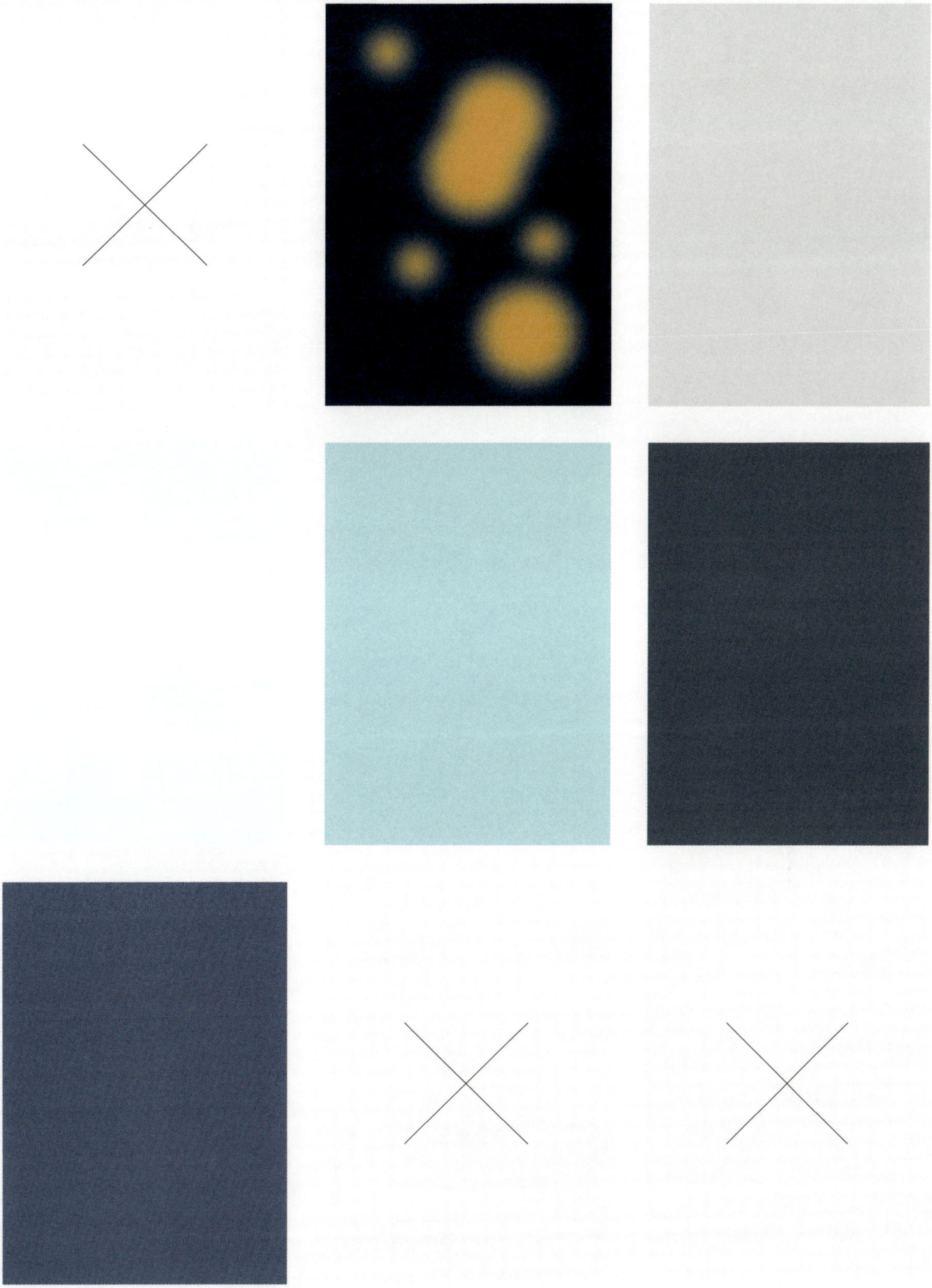

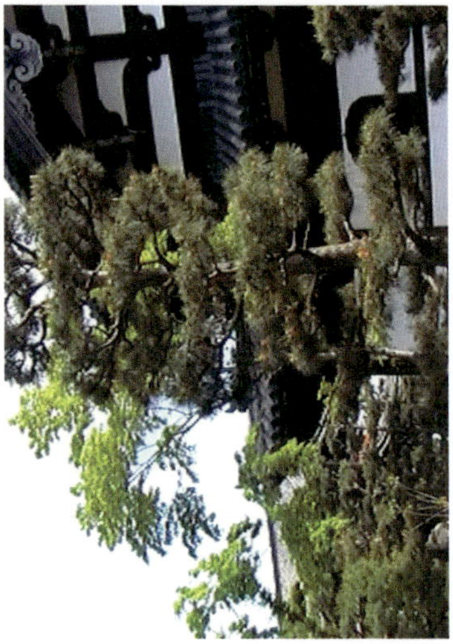
0850P

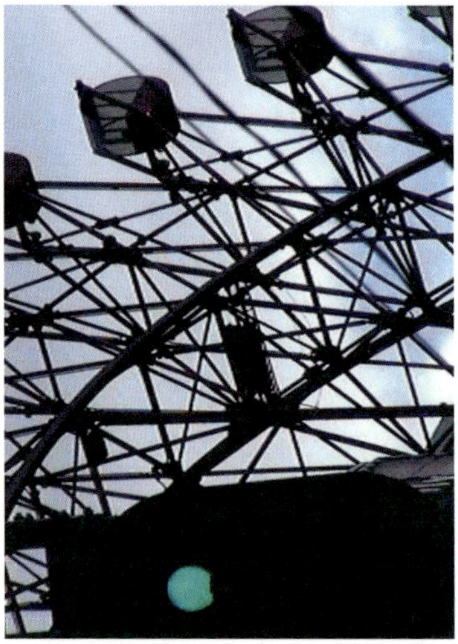
0908P

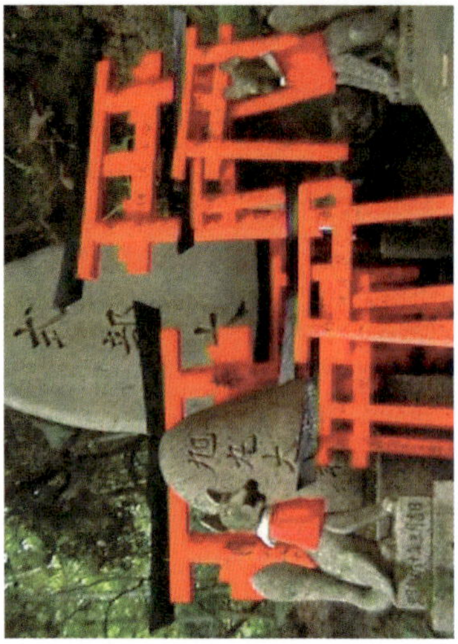
0909P

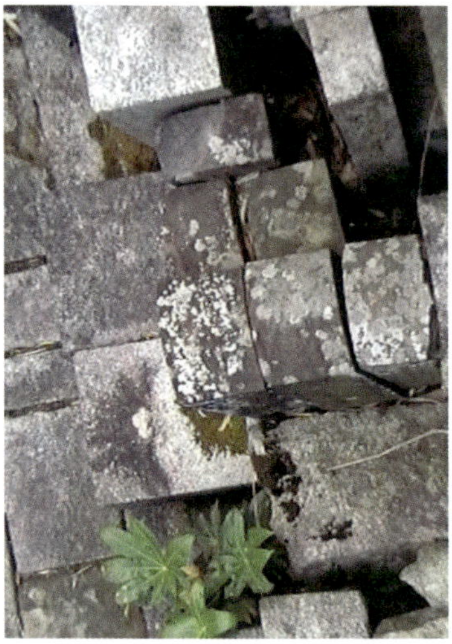
0910P

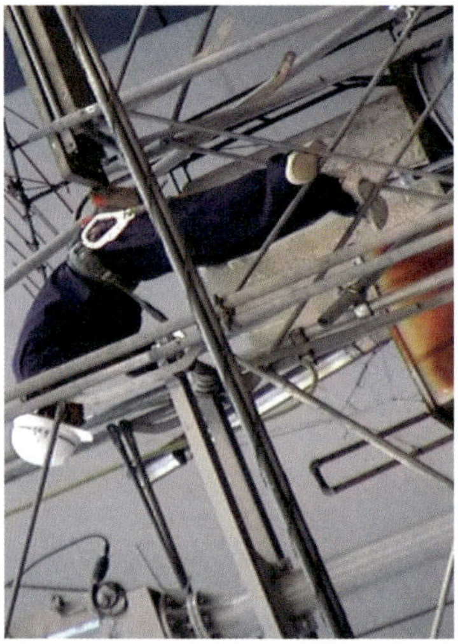
0969P

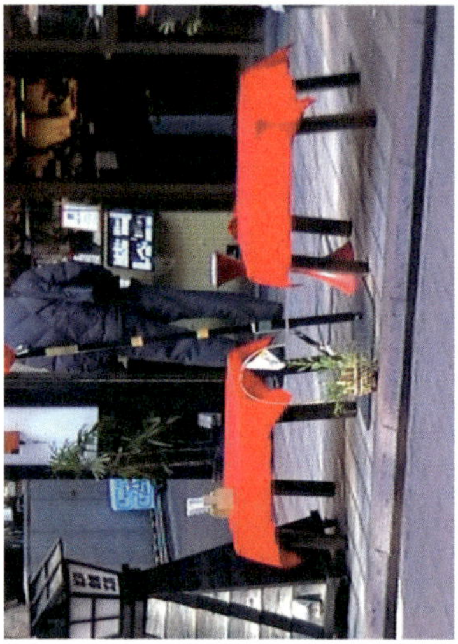
0972P

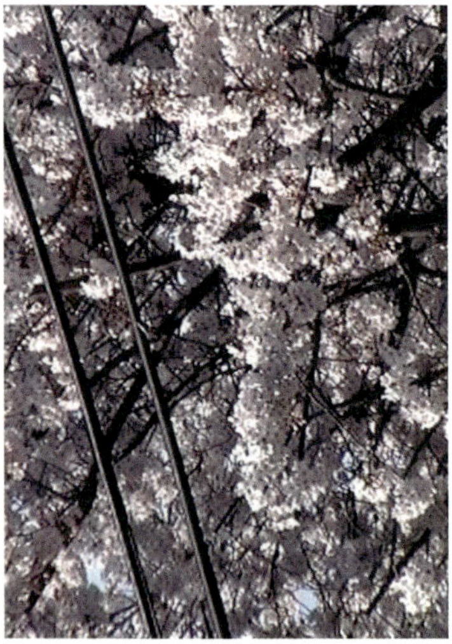
0992P

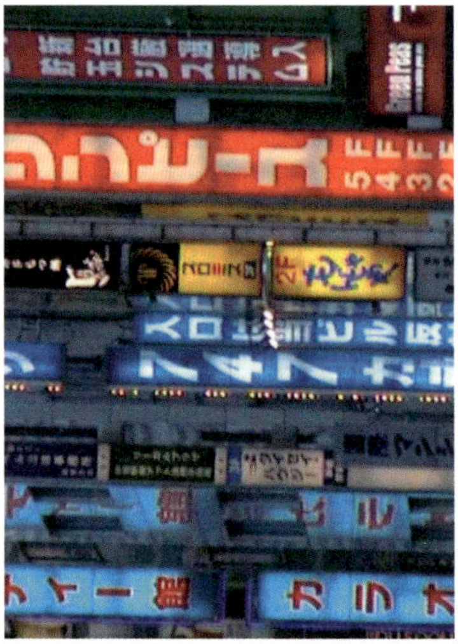
1011P

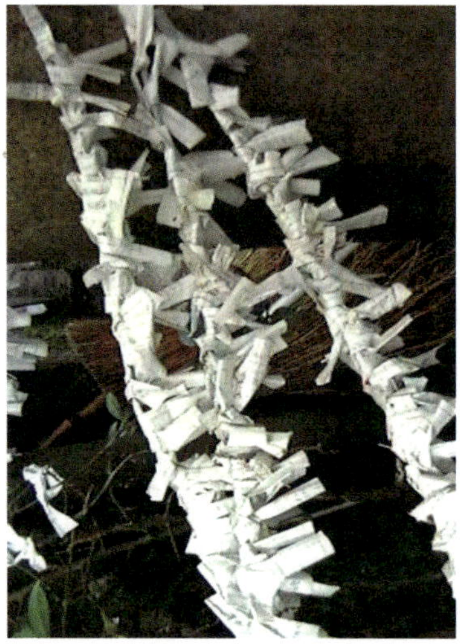
1487P

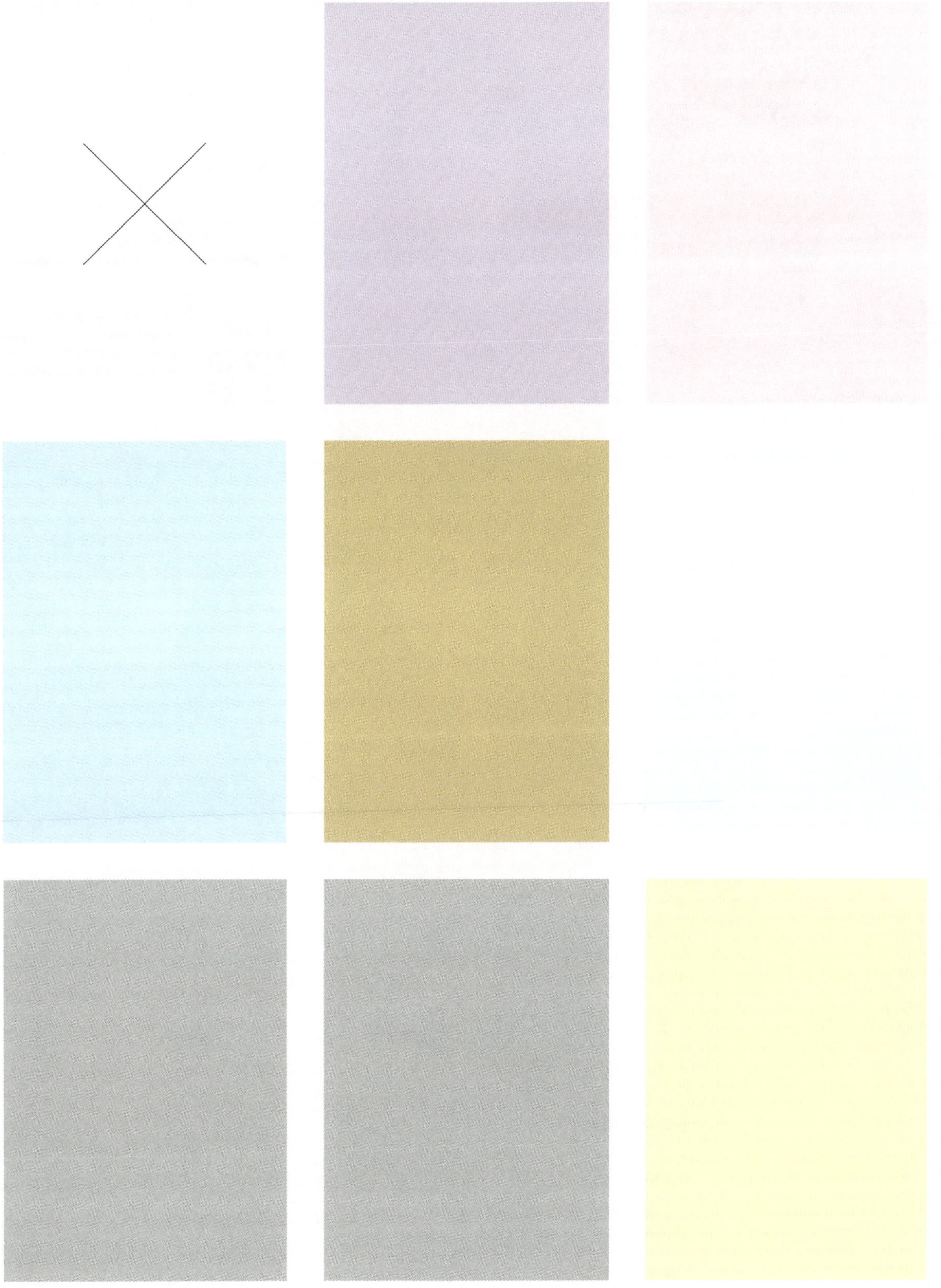

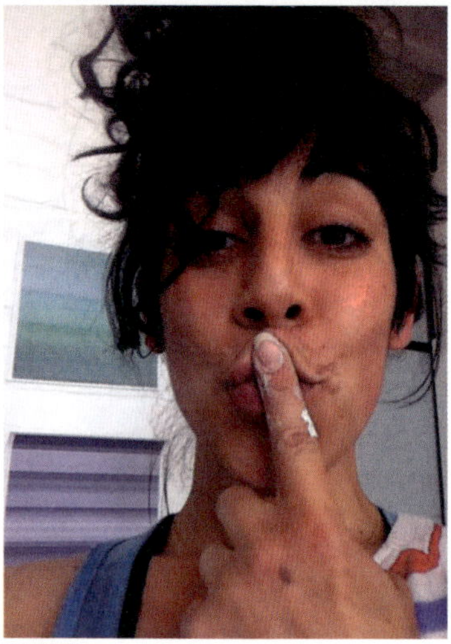
0903P

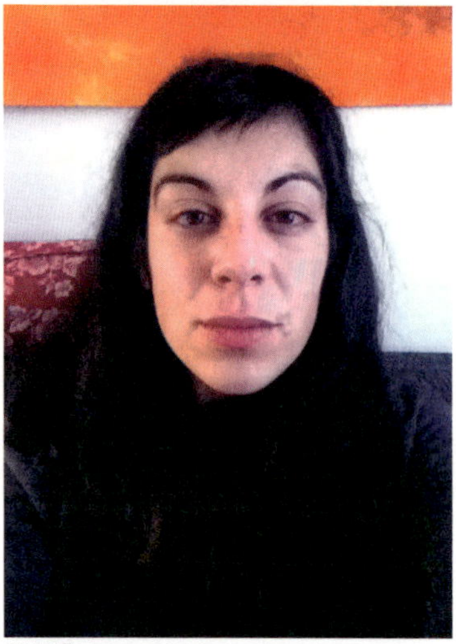
0920P

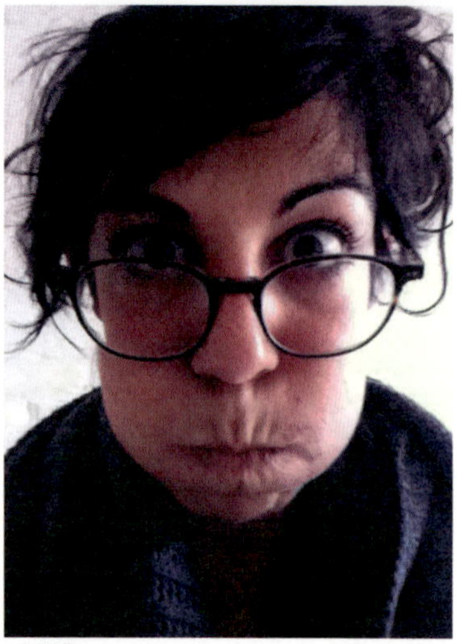
0922P

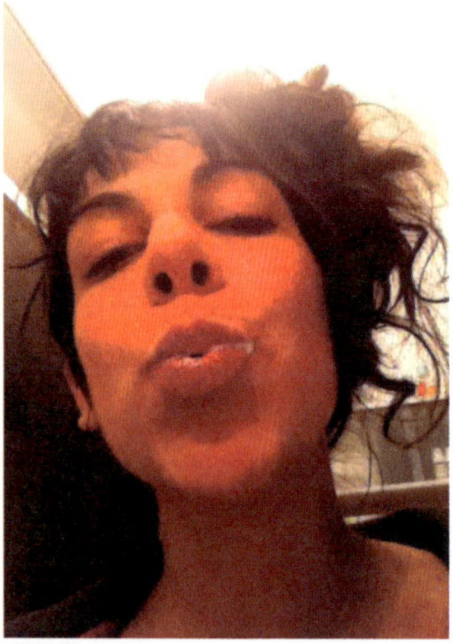
0923P

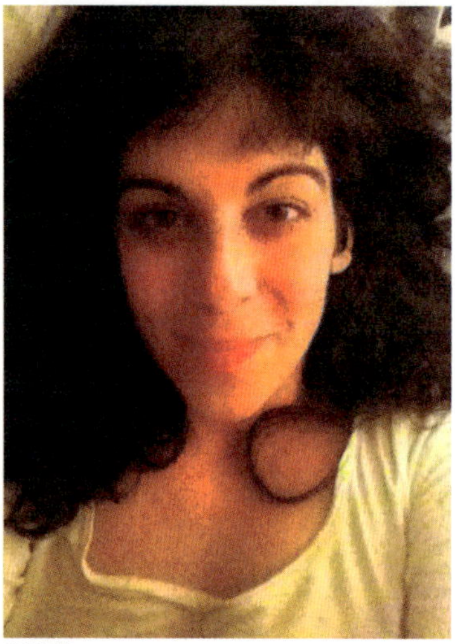
0924P

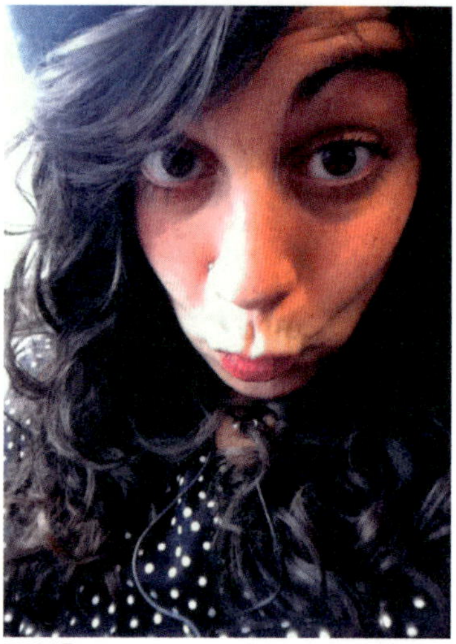
0926P

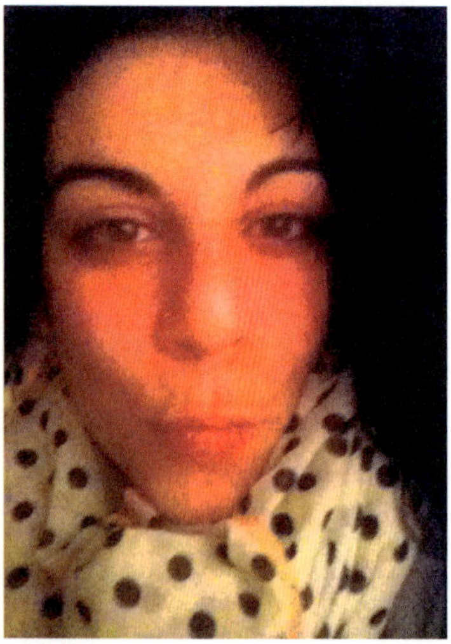
0927P

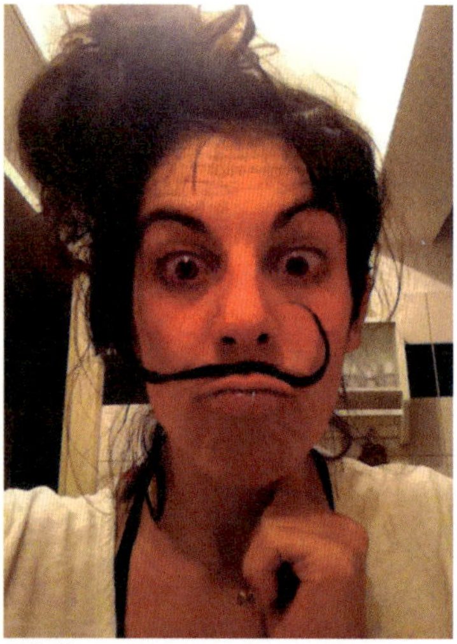
0936P

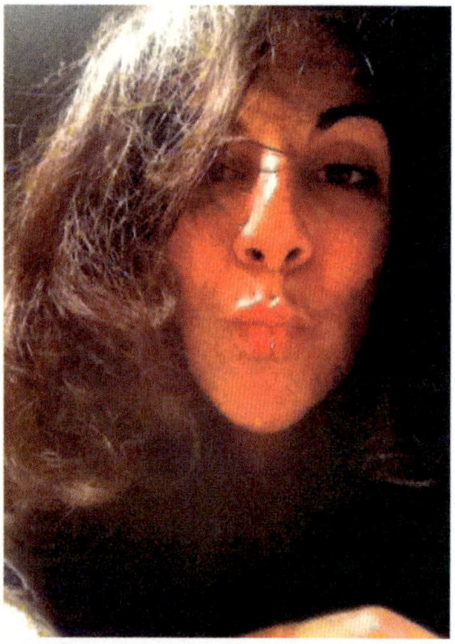
0941P

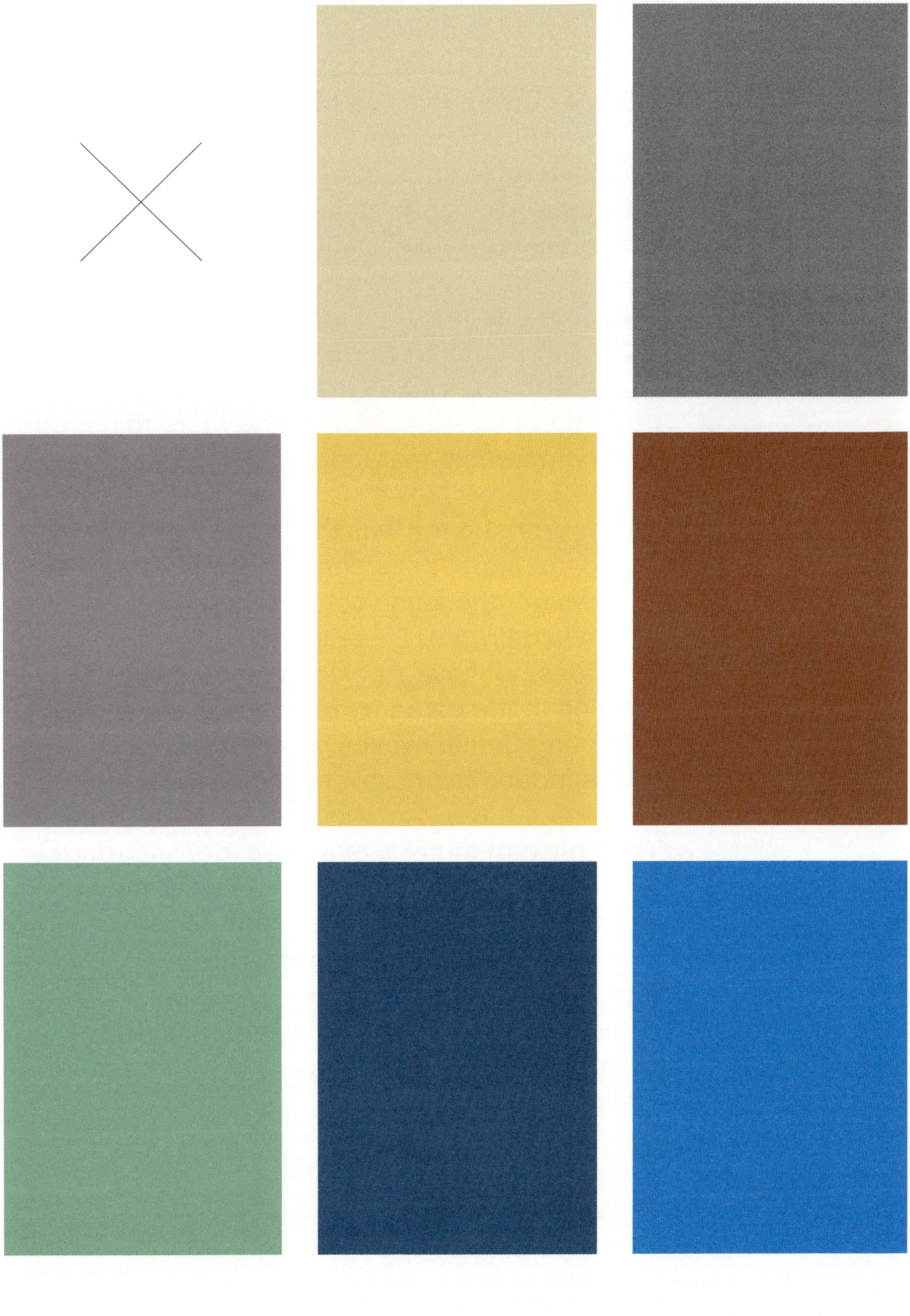

Some subjects do clearly evoke similar colours, but not enough to speak of a strong and reliable correlation. Out of twelve categories, I consider that no more than five provoke a considerable amount of similar colour reactions—the only subjects for which most of the images evoke a similar colour are trees in winter (blue tones)[1] and Japan (pastels).[2] I observe that these do indeed seem to induce a similar emotional colour response in many cases, but not in every instance. This could be further investigated by doing more research with comparable material.

What surprised me was that my own child would not appear in a consistent colour, as I always get the same colour whenever I think of him in daily life. There even is a photograph that does not have a colour association at all. Looking at the category for longer, I see that the young Emilio evokes green tones,[3] and that is indeed how I saw him in the first years.

The Emilio of today is of a deep, rich, saturated reddish earth colour, heavy, with a powdery structure. I can see that the more recent images do create that colour.[4] This shows that the emotional response is indeed linked to the very situation (here we touch on the essence of photography) and not to the person that is depicted.

As to the image in the middle,[5] why is there no synaesthetic response to someone so dear to me?

1

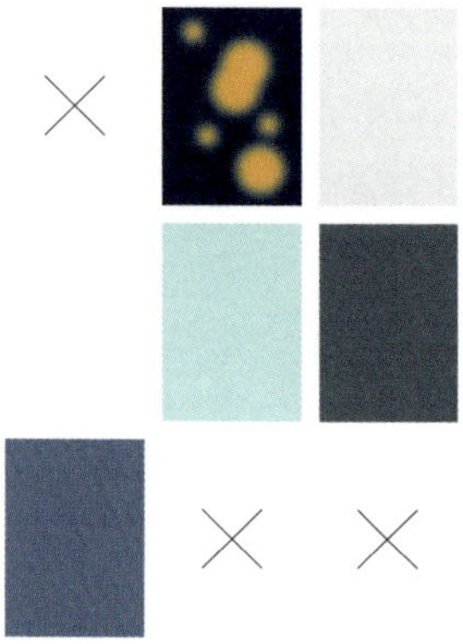

Trees in winter (p 96)

2

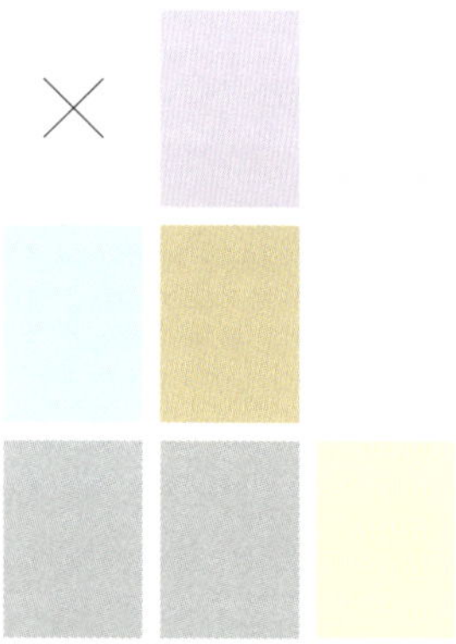

Japan (p 98)

3

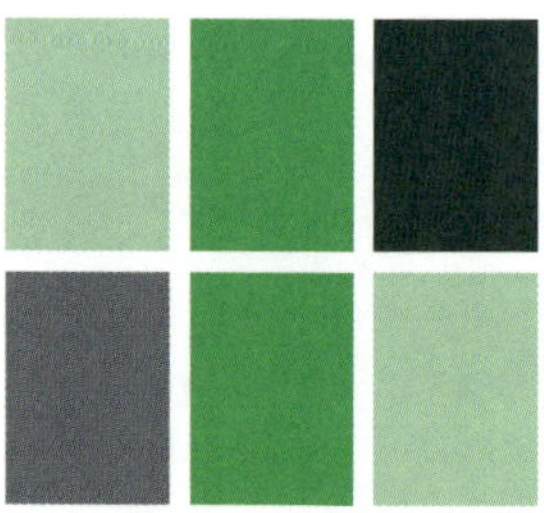

Emilio (p 88)

I remember shooting that picture in a forest near my hometown a few days after my boyfriend of many years left me in the summer of 2014. It was unexpected, degrading and extremely painful.

I was overwhelmed by it and, being hardly able to cope with the situation with my son around all day during the holidays, I decided to do many activities to keep us busy. We went to a forest nearby with a dear friend; that is where this picture was taken, with Emilio looking at a bird.

I am guessing it is because I felt so much pain and did not have any time to process what had happened that there is no colour association. In difficult times as much as in situations with sensory overload, my synaesthesia seems to fade or even disappear completely. That could be an explanation as to why this image has no 'feeling'—the situation I shot did not have any—or rather: not a tangible one.

I find the same phenomenon in the category of self-portraits.[6] I am not a selfie person but all of these, which were on my phone since I sent them to my boyfriend when he was far away, were part of my photographic archive and so some of them were chosen to complete the database.

To my surprise, the first selfie has no colour. I remember what that situation felt like: it was a few days before my graduation in June 2015. I was under immense pressure to get my book

4

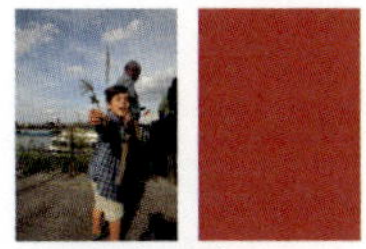

Emilio (p 88)

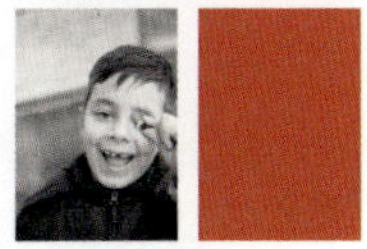

Emilio (p 88)

5

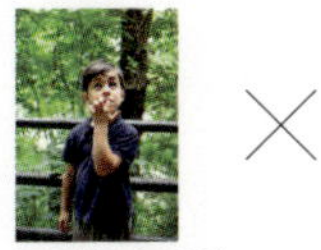

Emilio (p 88)

6

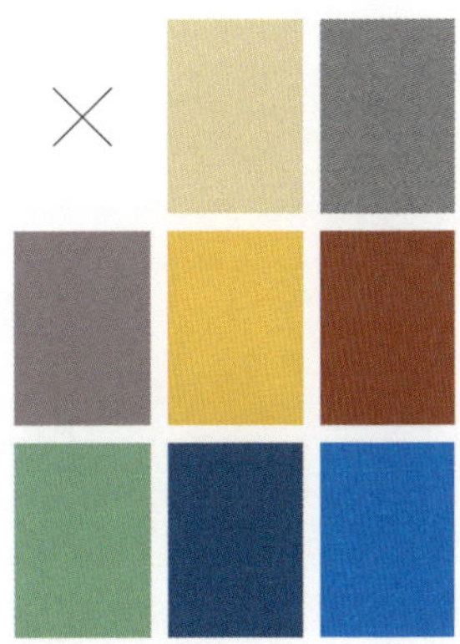

Self-portraits (p 100)

7

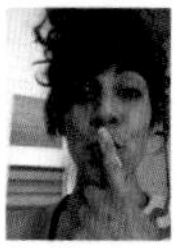

Self-portraits (p 100)

and big print ready on time—you can see the paint on my fingers.[7] I was under such stress that I slept literally 3 hours a night, trying to set everything straight for the graduation exam while dealing with single parenthood and an exhausting court procedure. I assume that this picture—like the one I describe above—does not have a synaesthetic response because the situation was so overwhelming that the colours were absent. Lack of sleep could account for it as well, although I would not know whether the sleep deprivation itself or the dizzy and overwhelmed mindset it created are causing the absence of any colour. All other selfies appear in blue tones (the colour of my boyfriend Sam back then, I guess because I think of him when I see them) and yellow/brown/broken white, which is how I felt at that time—shades of comfort, happiness and confidence, a notion of home and hope, and the colour of trust.

I conclude that my synaesthesia cannot be categorised by isolated stimuli, but that there needs to be an emotional component to trigger a colour association. Apparently, the context is crucial instead of any single element in the image on its own. A tree does not always evoke the same colour, nor do different people in my life. I do not have consistent emotional relationships with hands, obviously. It seems to be the emotion aroused by the image which determines the shade of the synaesthetic reaction. So apparently a subject can only trigger a consistent colour response if my emotional connection to it is more or less consistent. This would, again, be interesting to observe in a long-term study: how, for example, the synaesthetic colour association for my son would evolve throughout the years.

In accordance with those findings, the absence of emotions in the context of a photograph's creation leads to the absence of a synaesthetic association when I look at it later. This corresponds with my observation that it is not the subject of the image that triggers a colour but rather the memory of the situation, the photograph reminding me of how I felt back then, that determines the synaesthetic response.
I conclude that my synaesthetic reactions are consistent only if my emotional reaction to them is consistent, too. Are there other determining factors? Can the colour of the emotion not be influenced by colours that appear in the photograph?

Chapter 3
Tonality

To find out whether or not the colour I feel is related to the colours that appear in the photograph, I generalised each image into colour ranges, displayed as colour bars. As there is a chance that the synaesthetic response might be triggered or influenced by colours in the image, they can now easily be compared to find similarities between the colours in the image and the colour of the emotion it evokes.

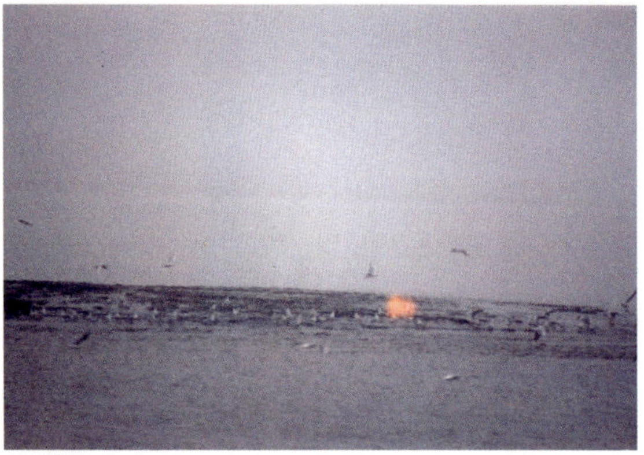
0015P

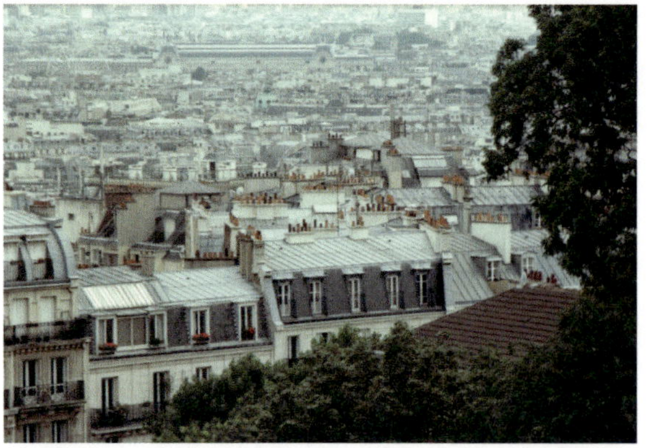
0019P

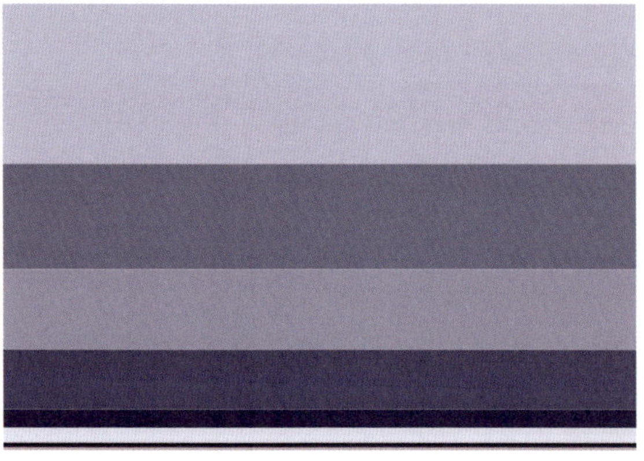

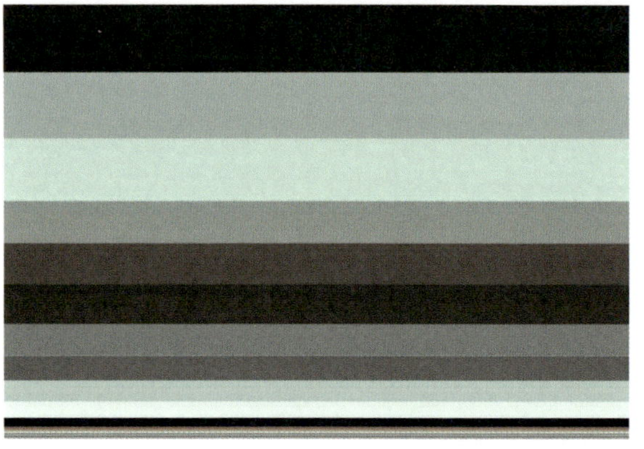

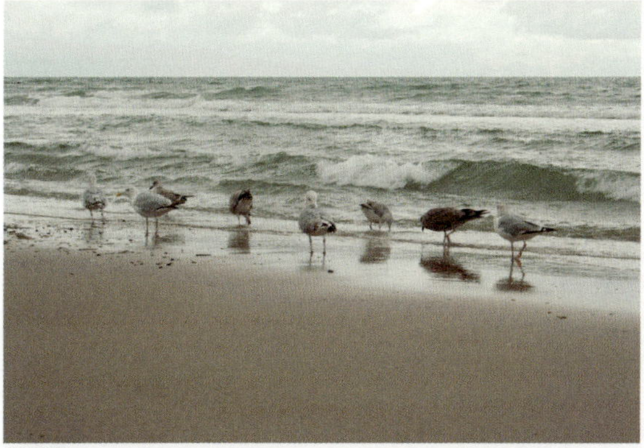

0048P

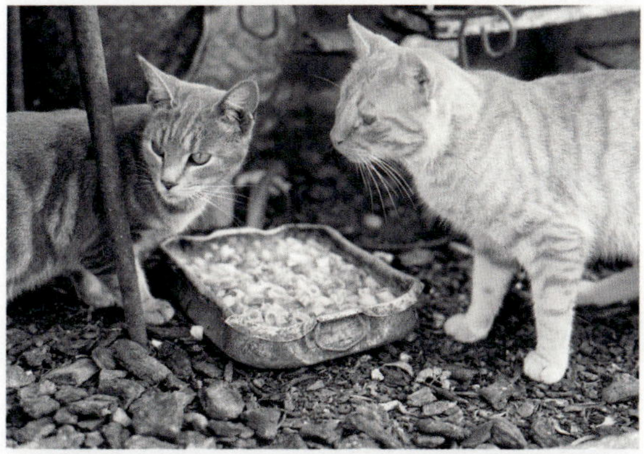

0114P

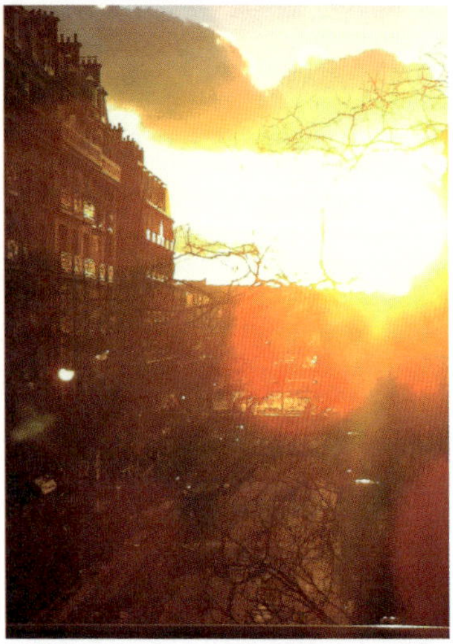
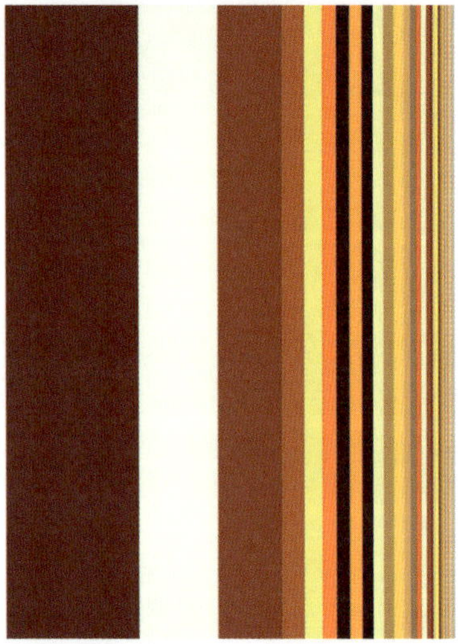

0028P

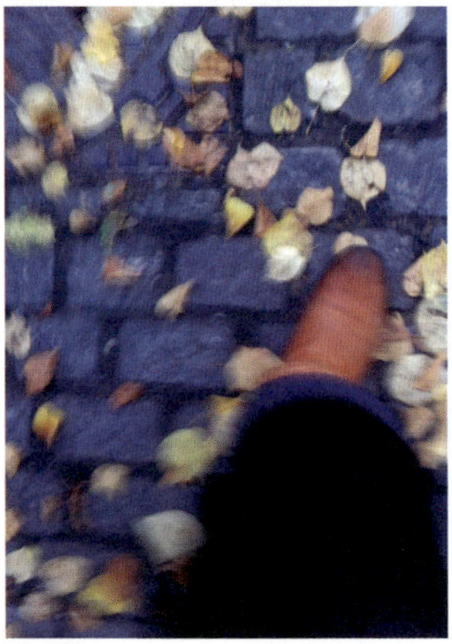

0111P

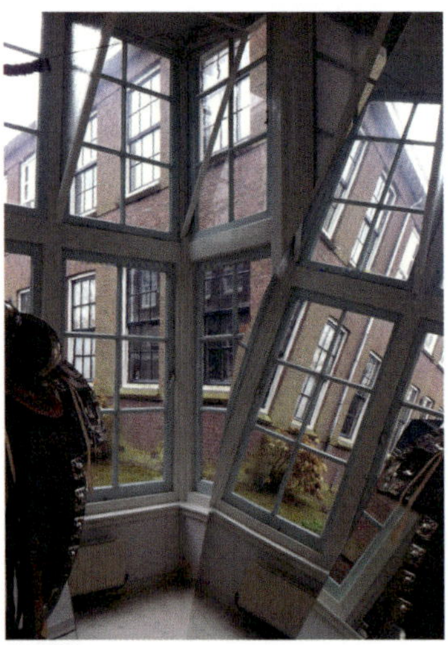

0156P

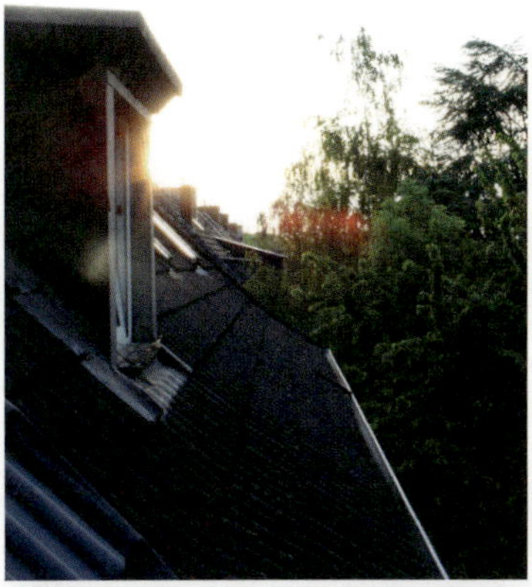

0240P

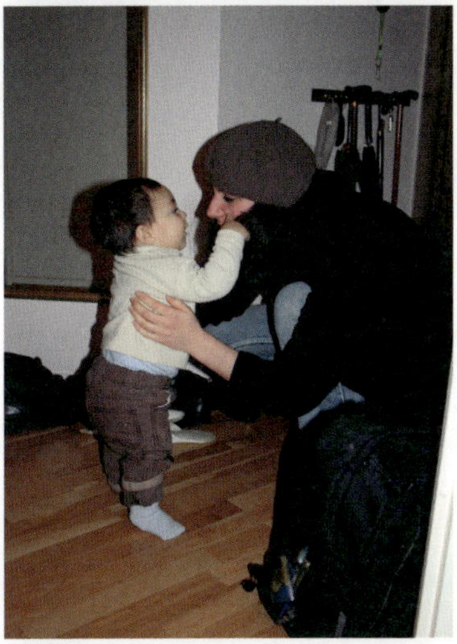

0403P

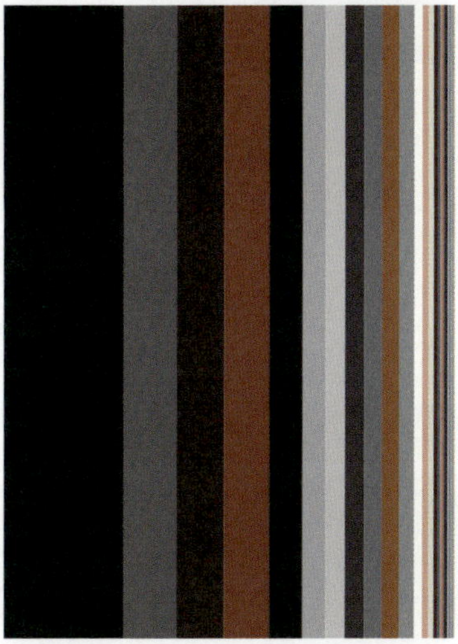

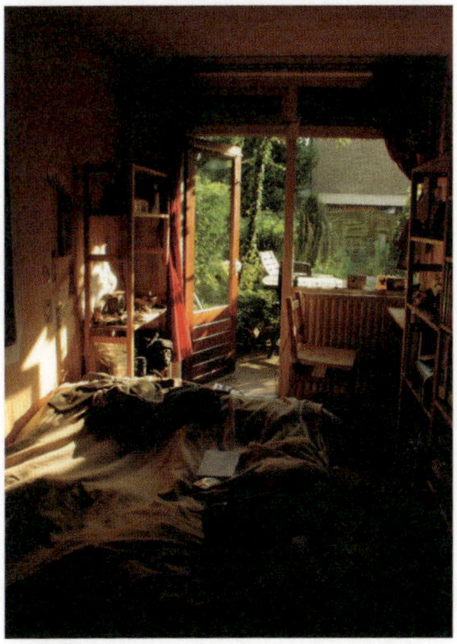

0366P

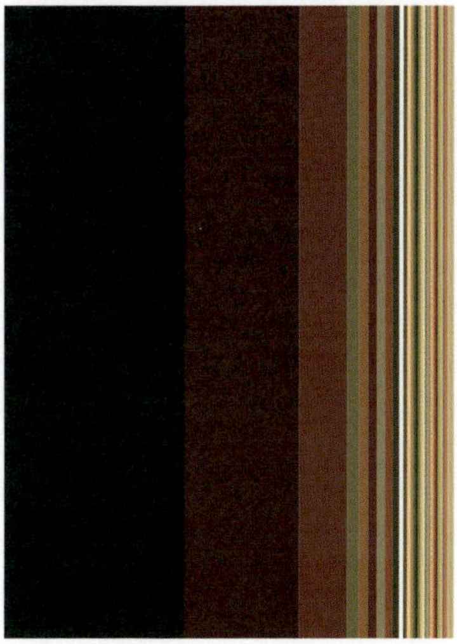

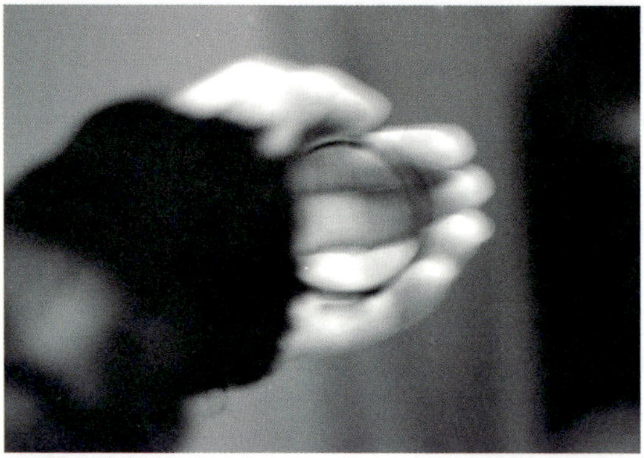

0116P

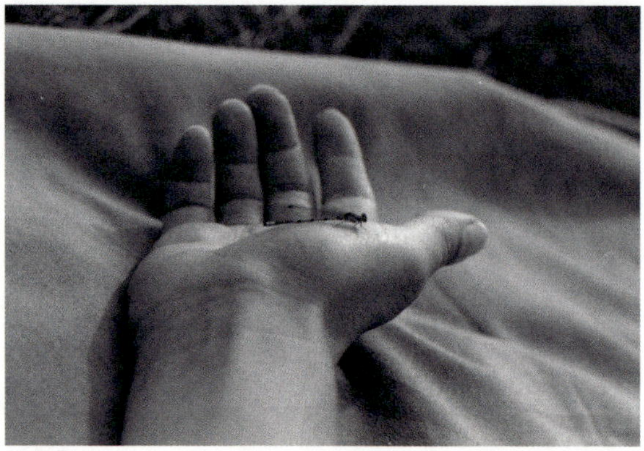

0179P

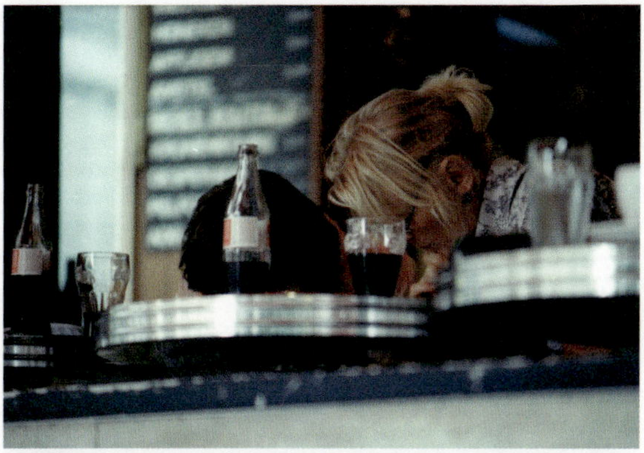
0220P

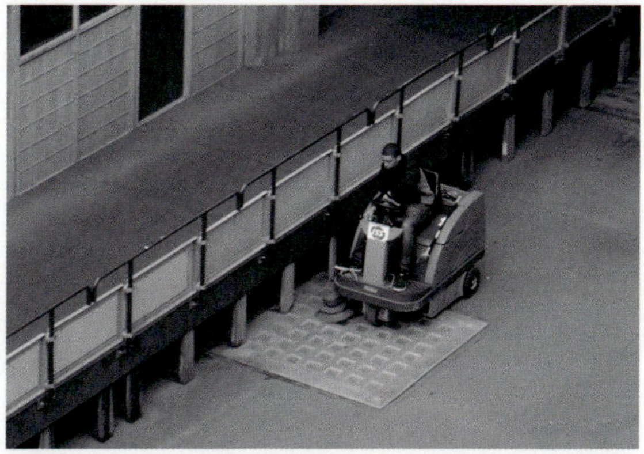
0260P

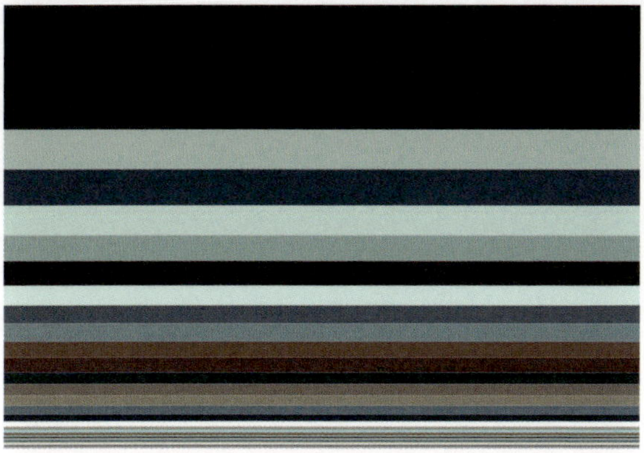

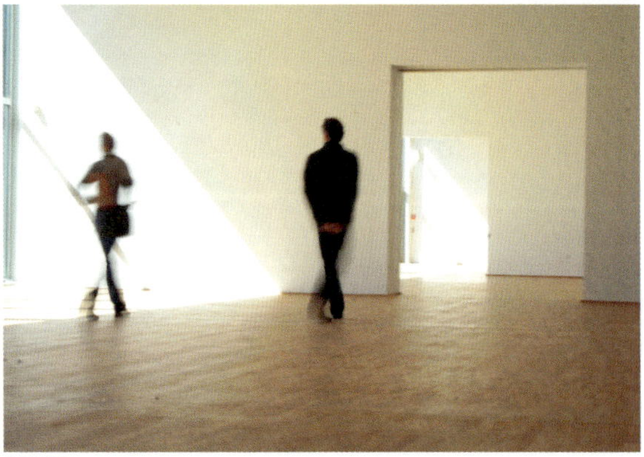

0334P

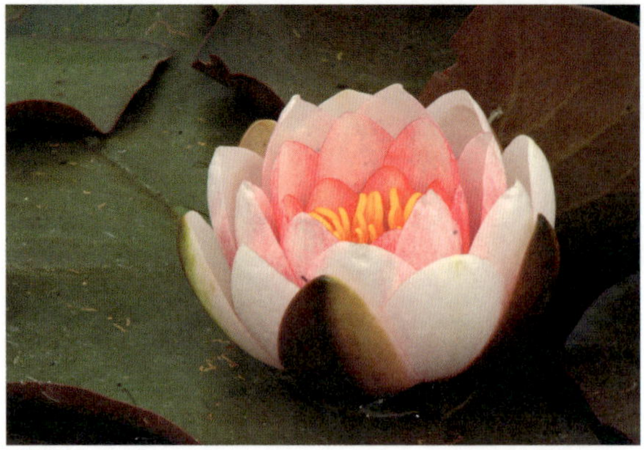

0335P

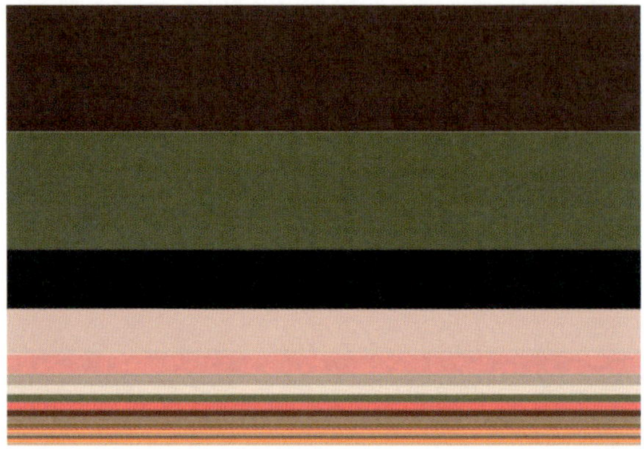

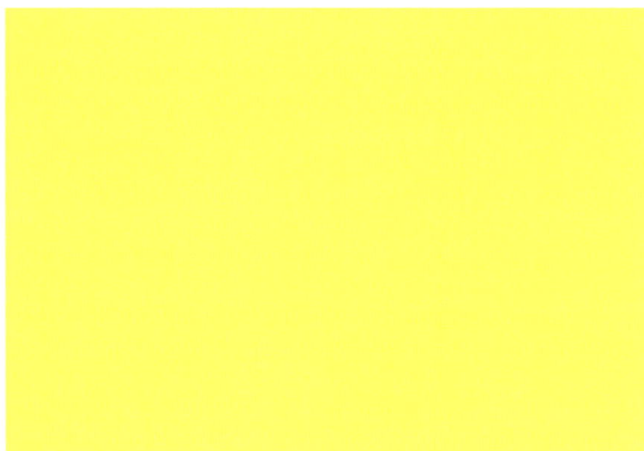

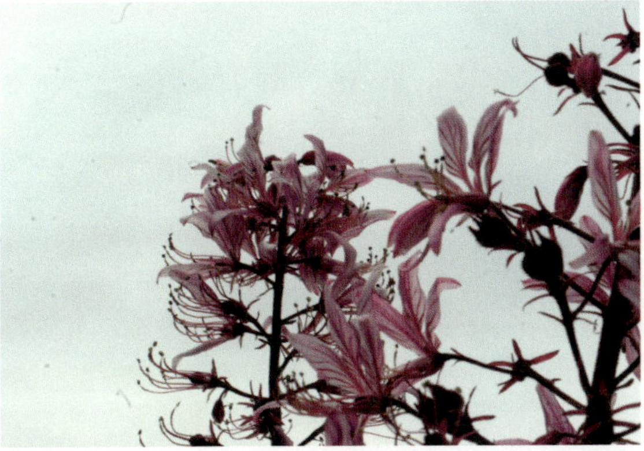

0337P

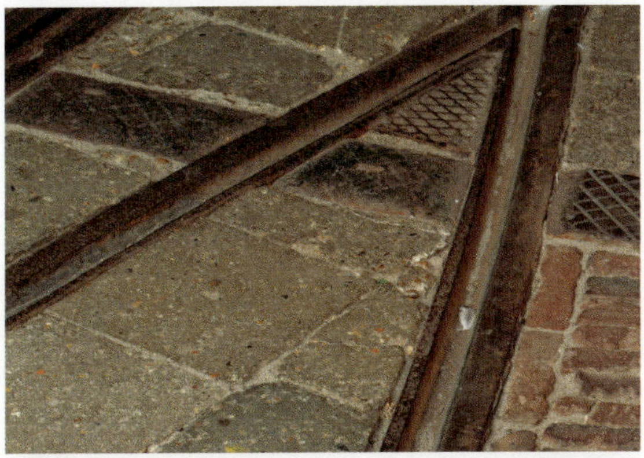

0486P

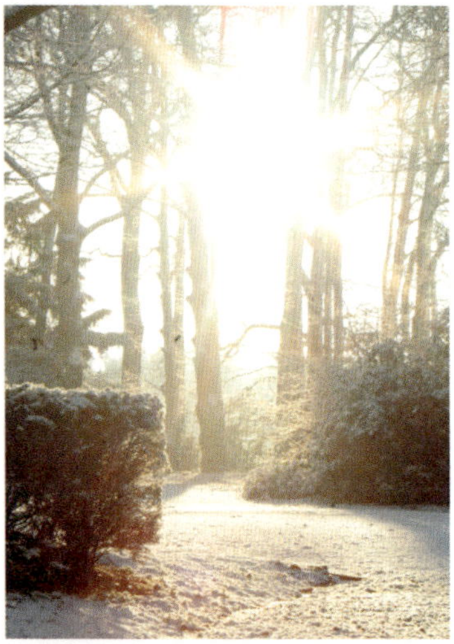

0472P

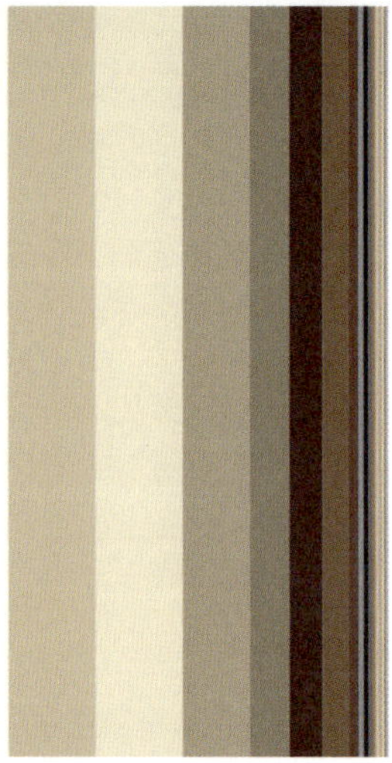

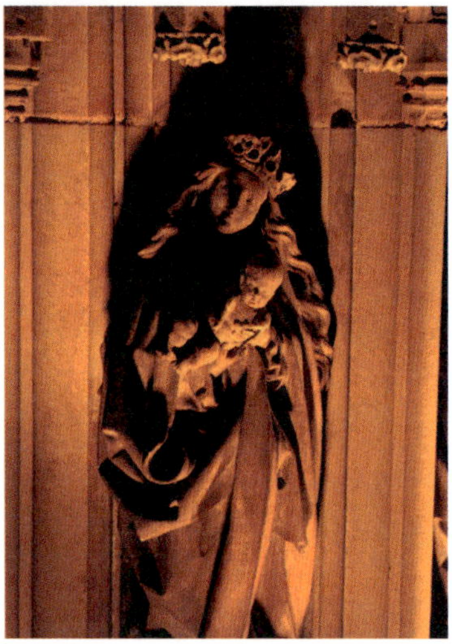

0474P

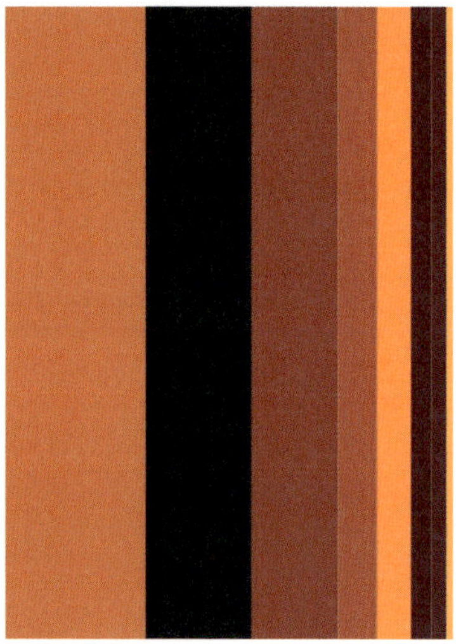

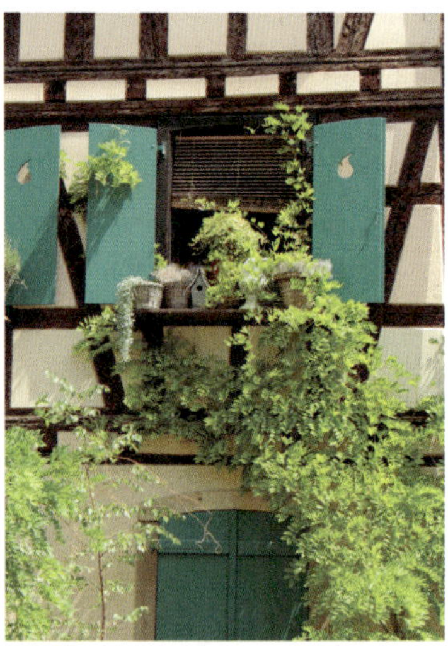

0530P

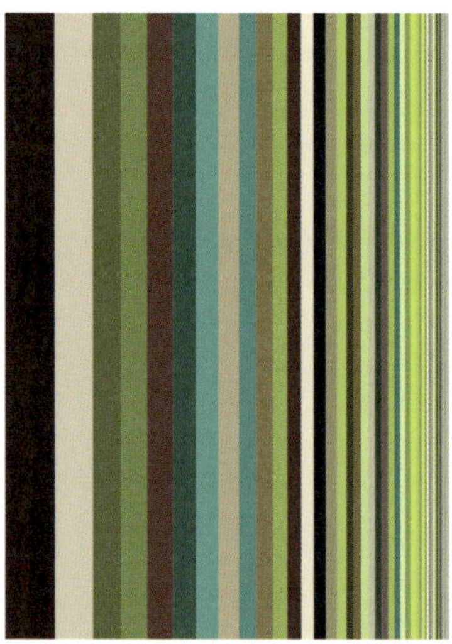

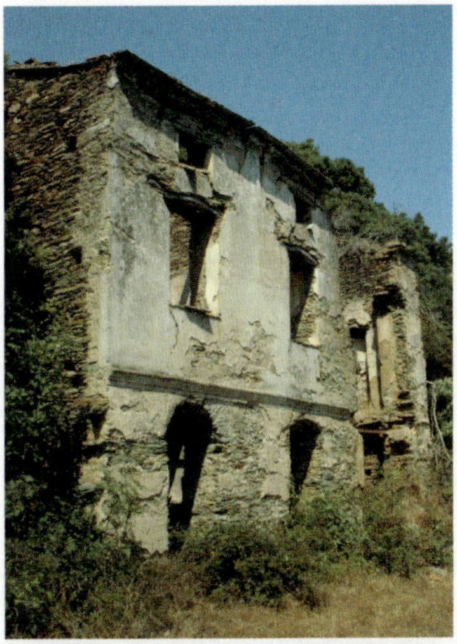

0540P

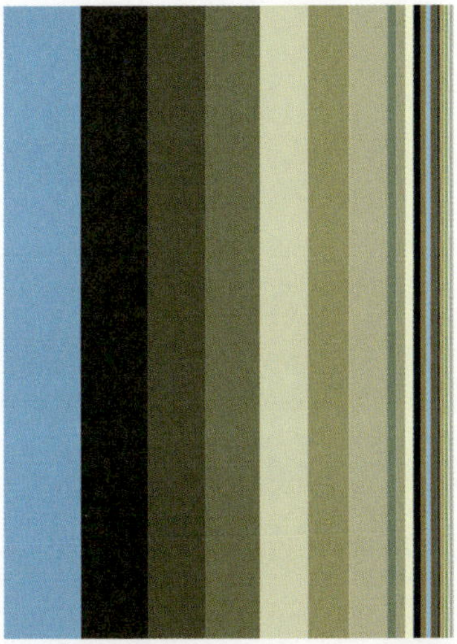

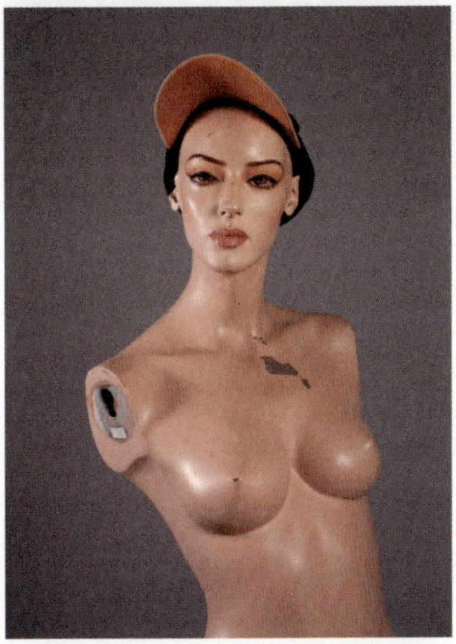

0541P

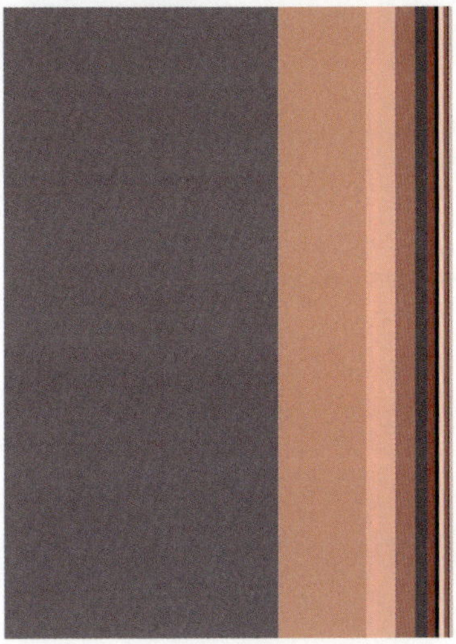

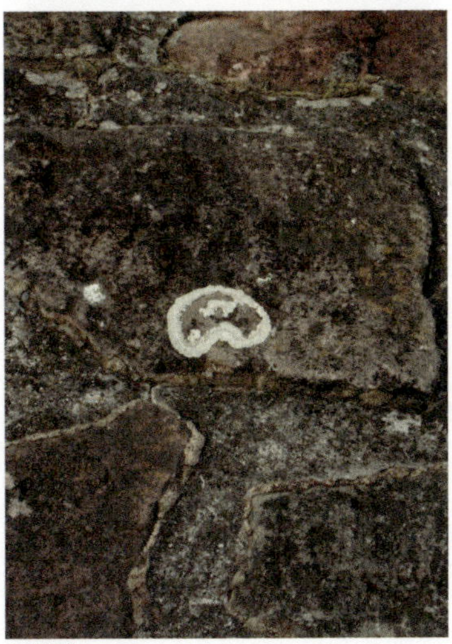

0547P

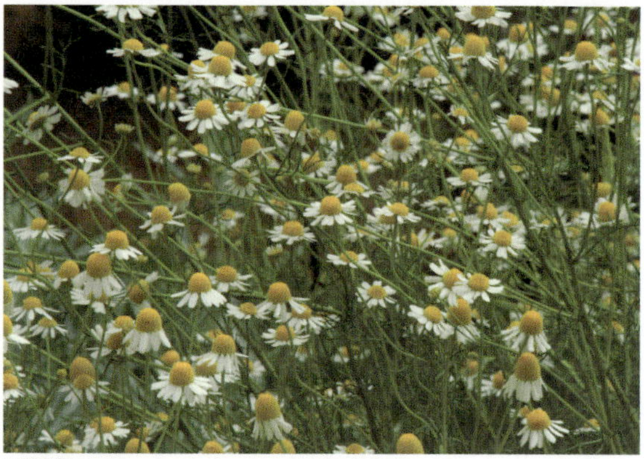
0496P

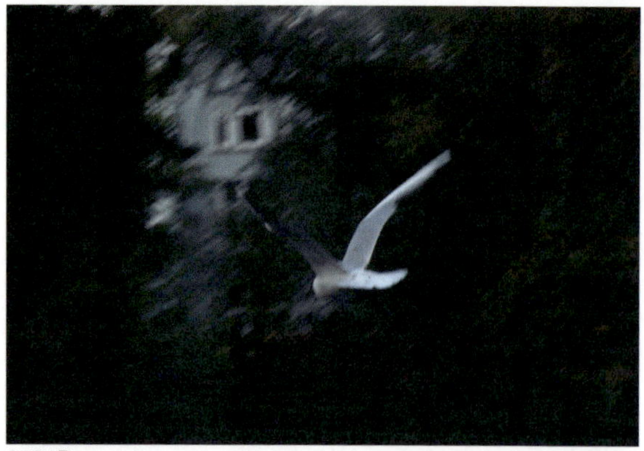
0534P

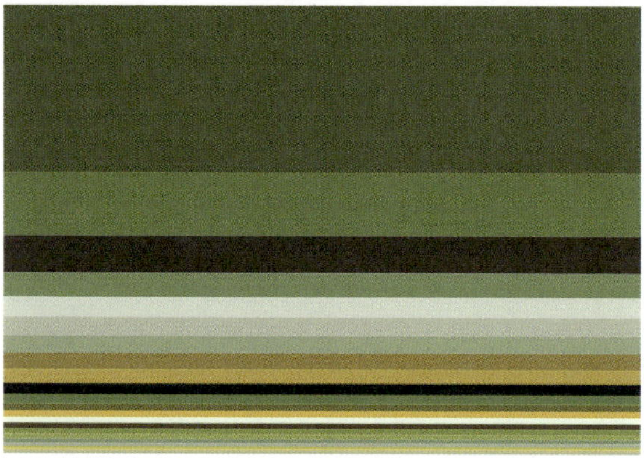

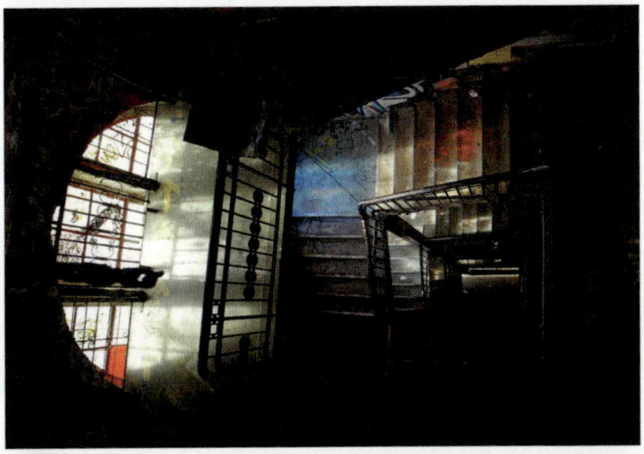
0552P

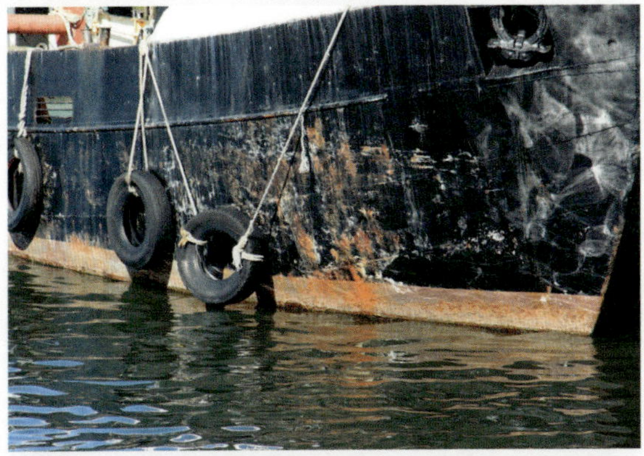
0562P

0564P

0624P

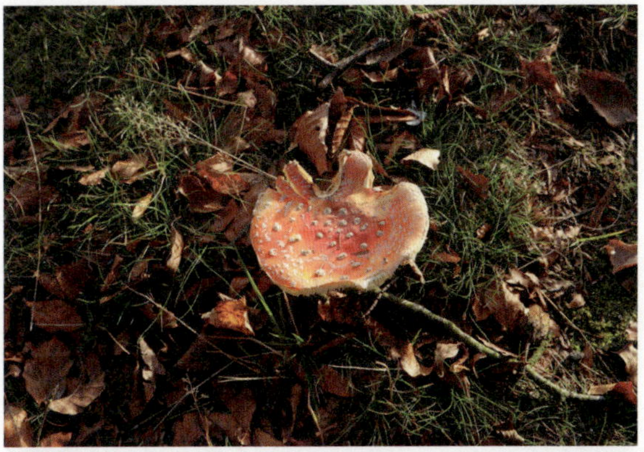

0771P

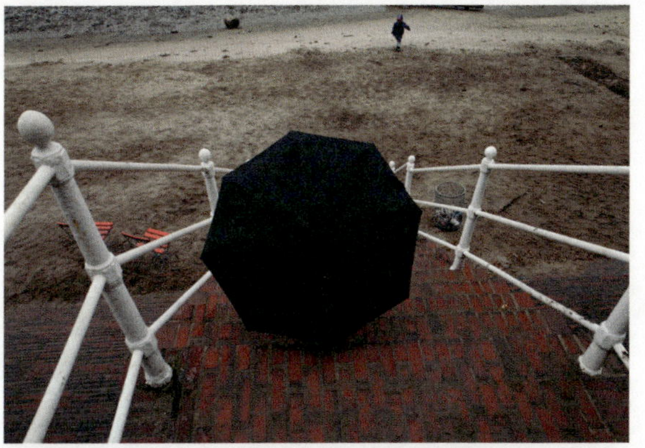

0775P

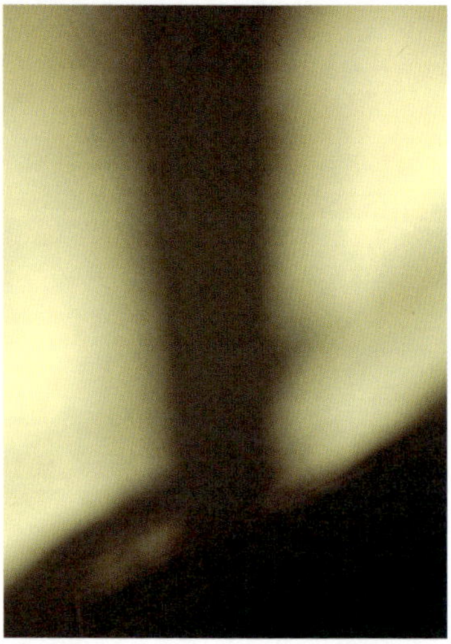
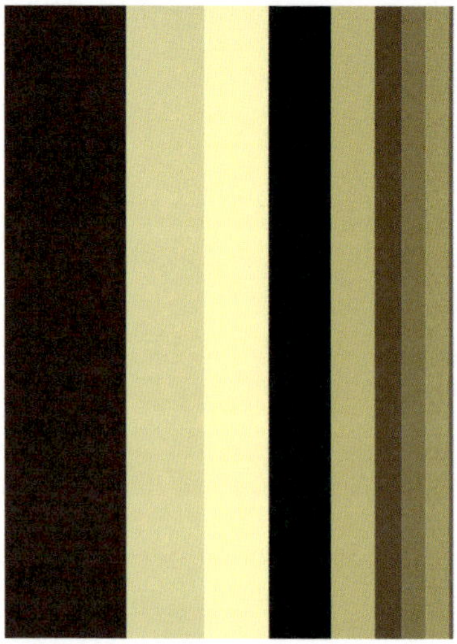

0565P

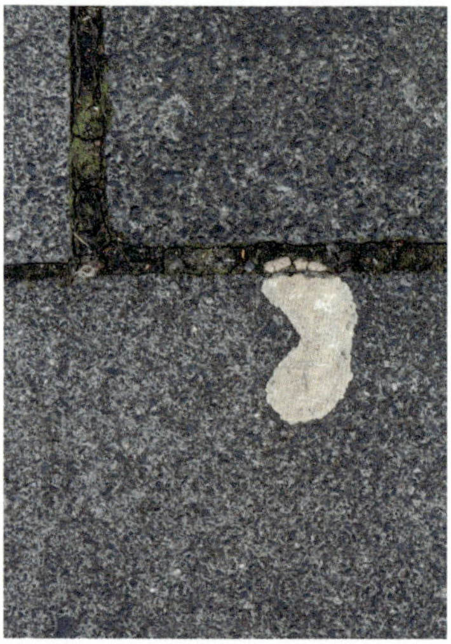

0604P

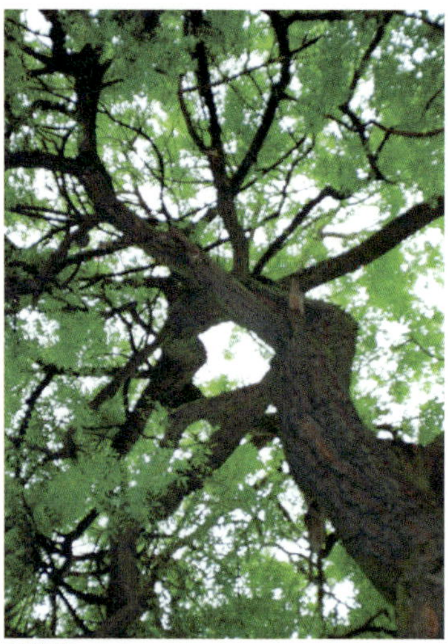
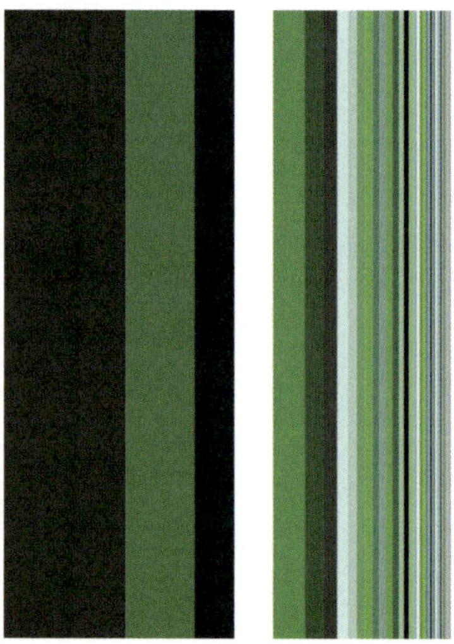

0820P

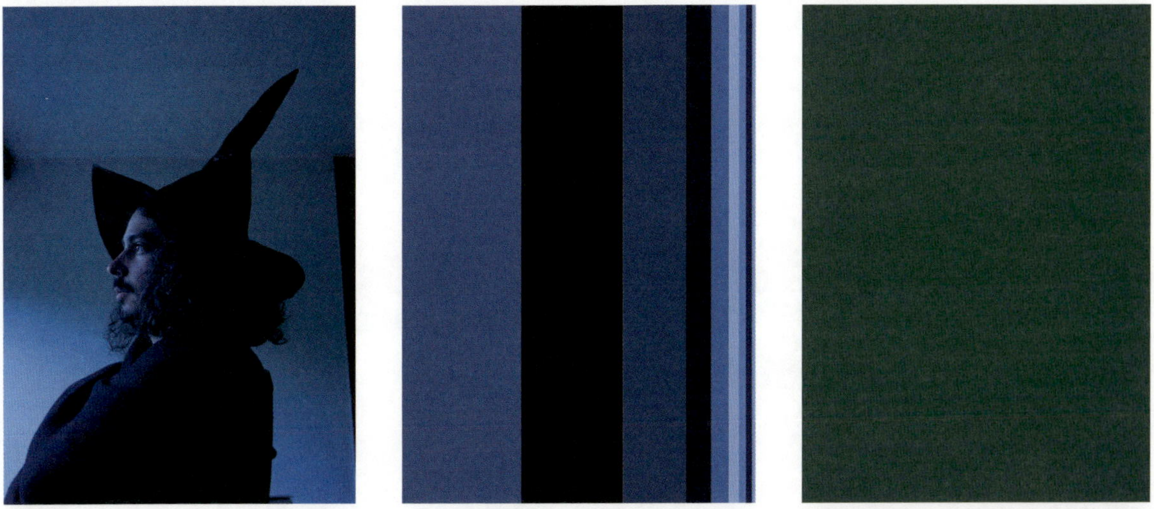
0868P

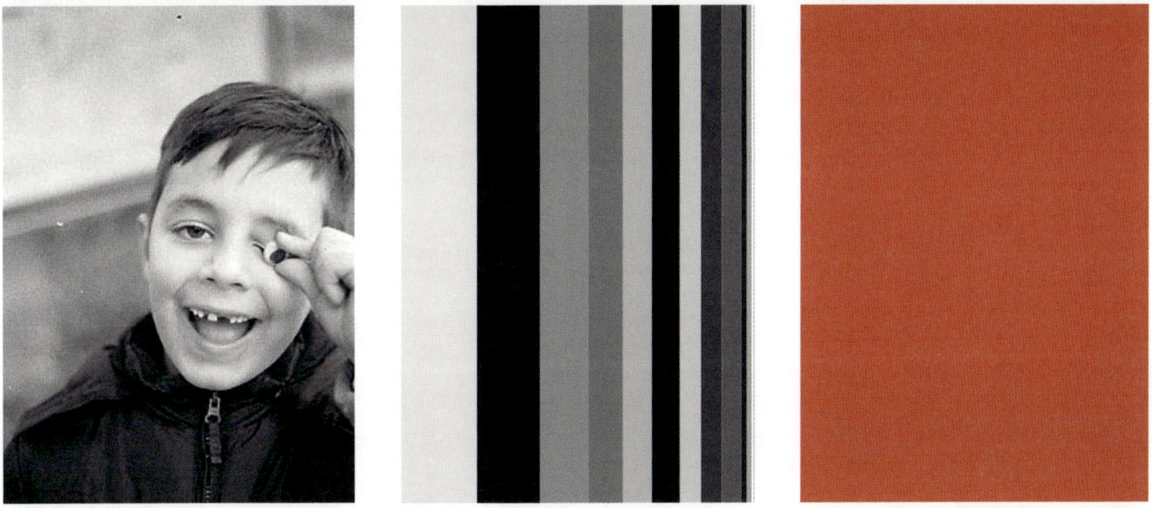
0897P

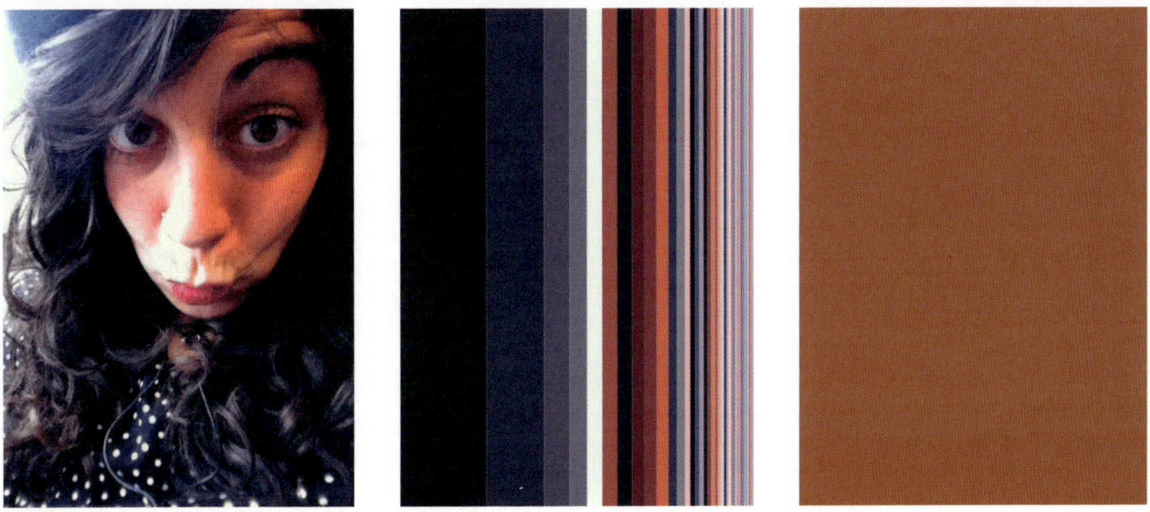
0926P

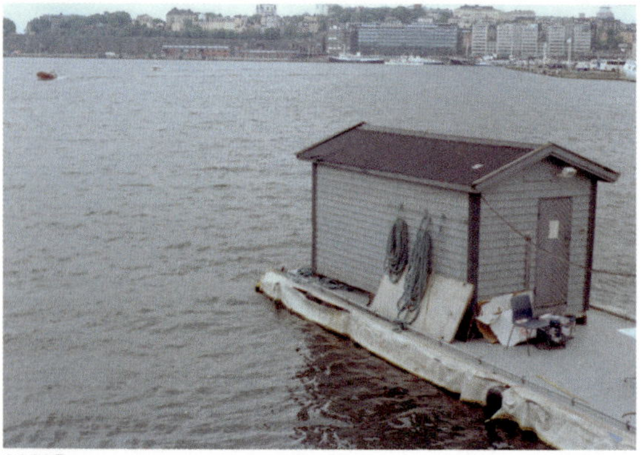

0803P

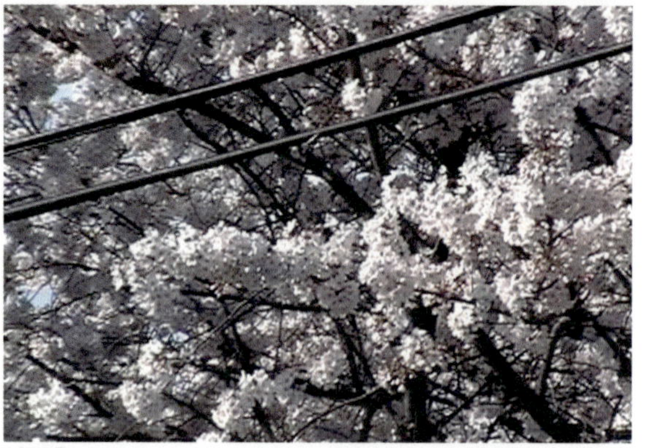

0992P

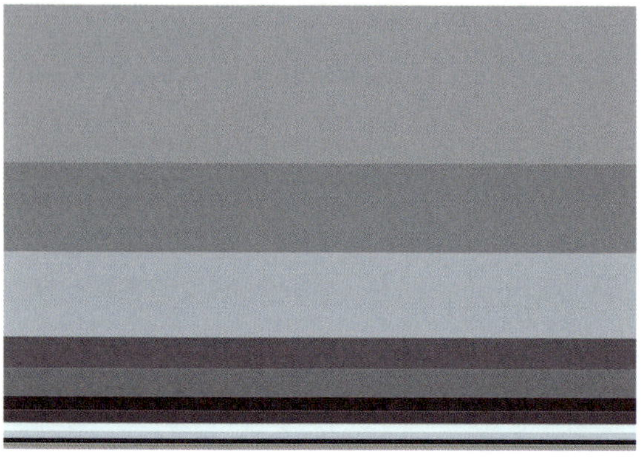

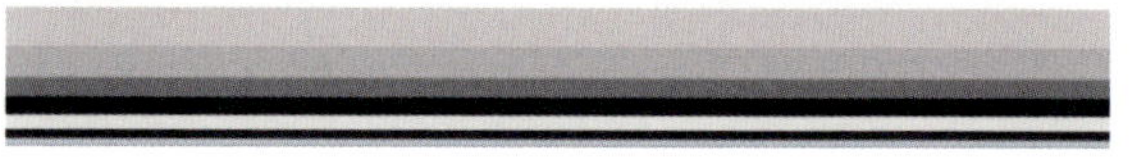

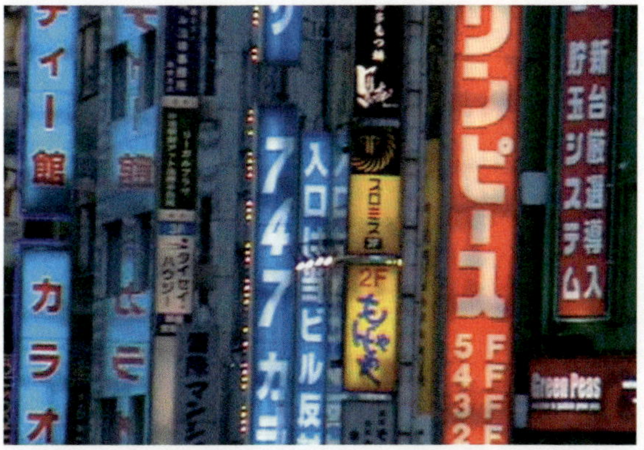

1011P

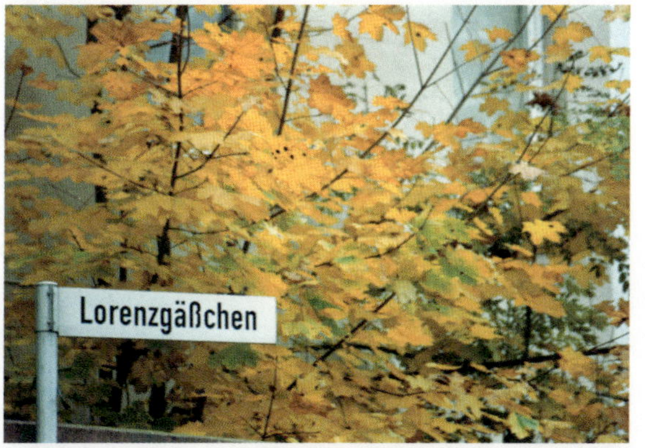

1369P

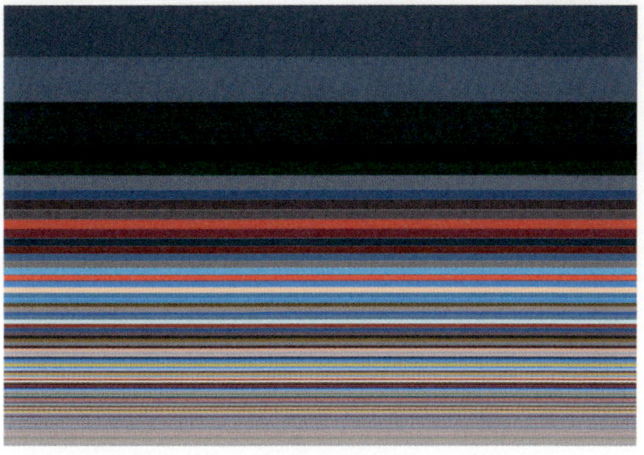

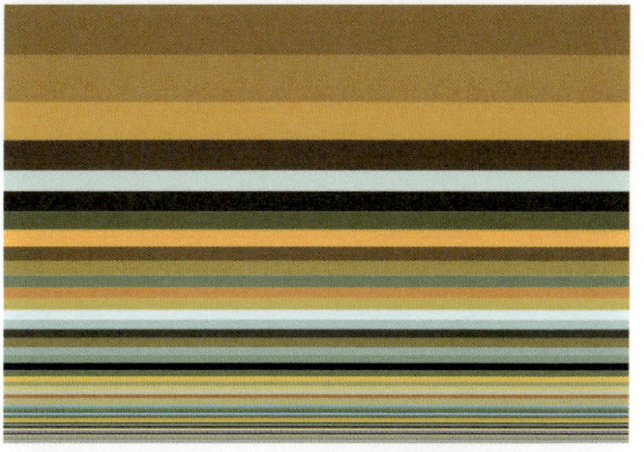

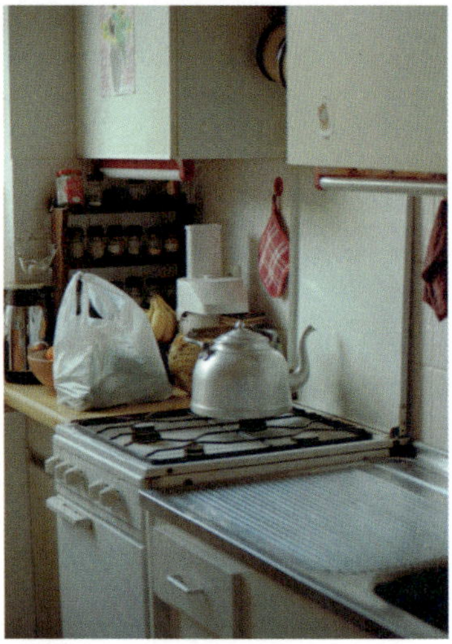

1057P

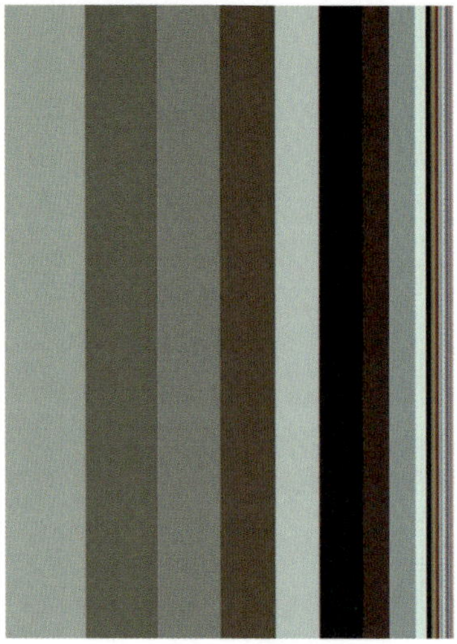

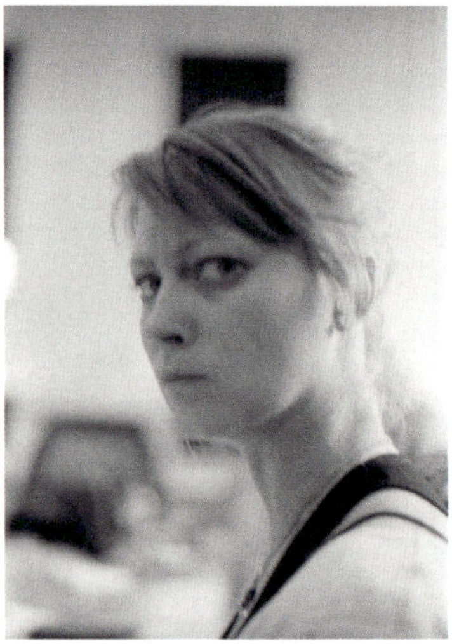

1068P

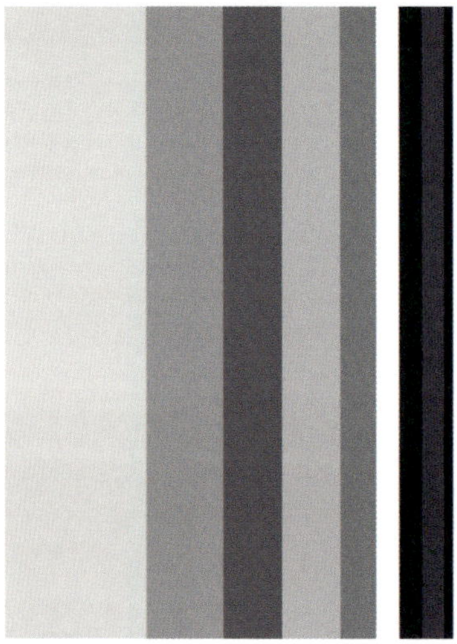

1085P

1123P

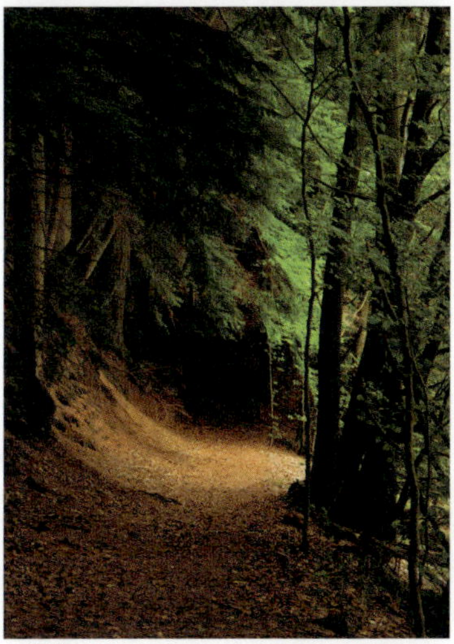

1227P

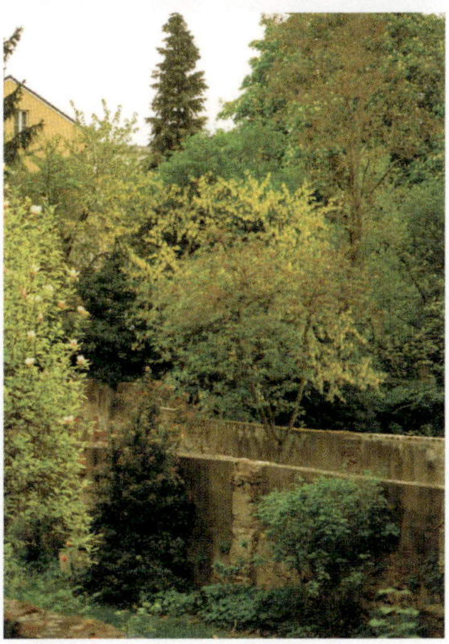
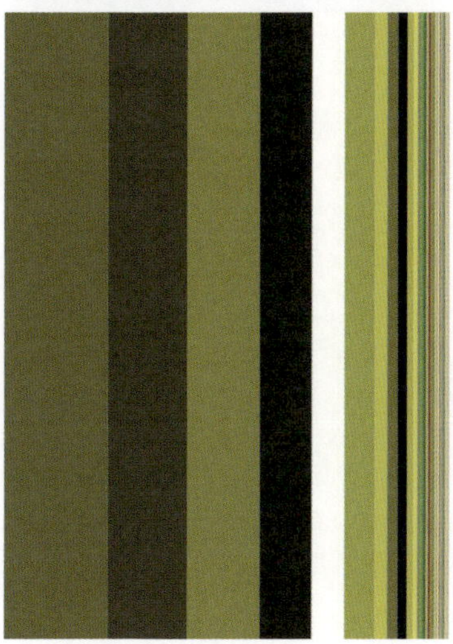

1234P

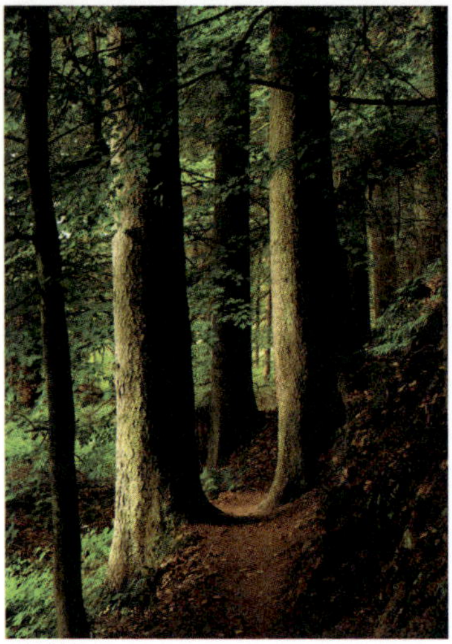
1238P

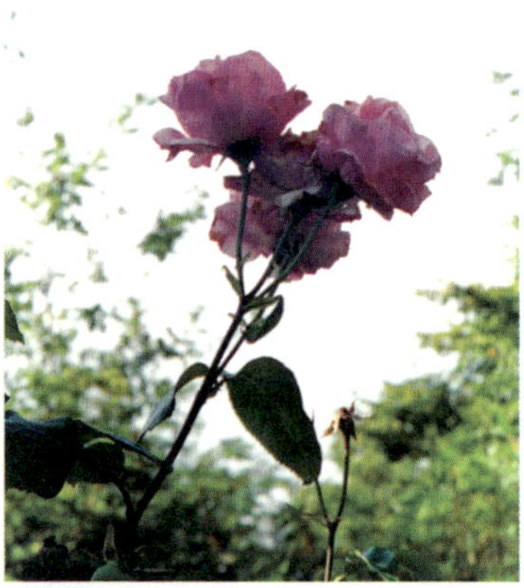
1244P

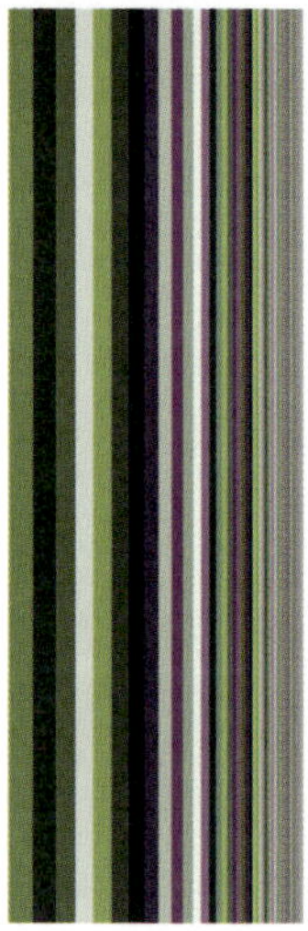

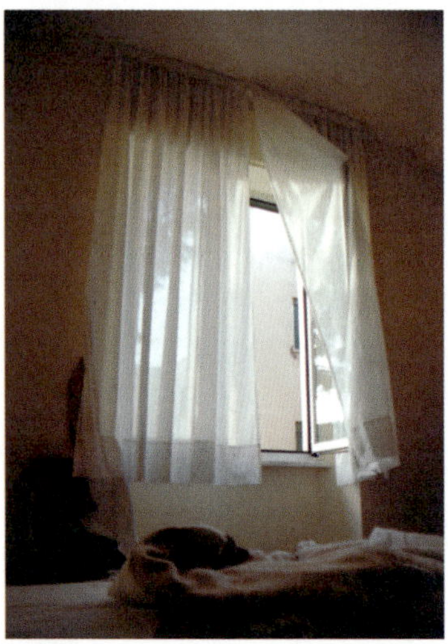
1256P

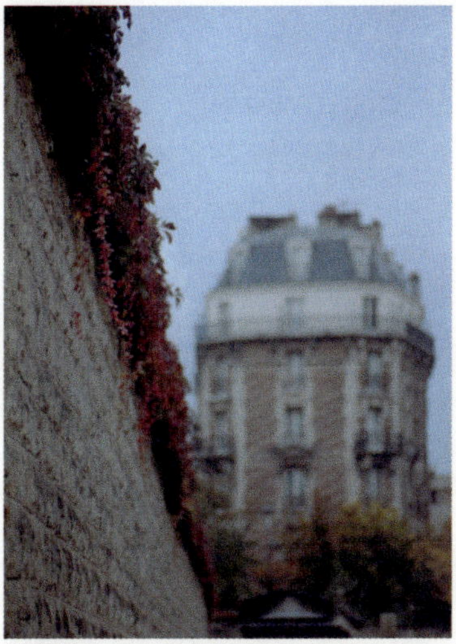

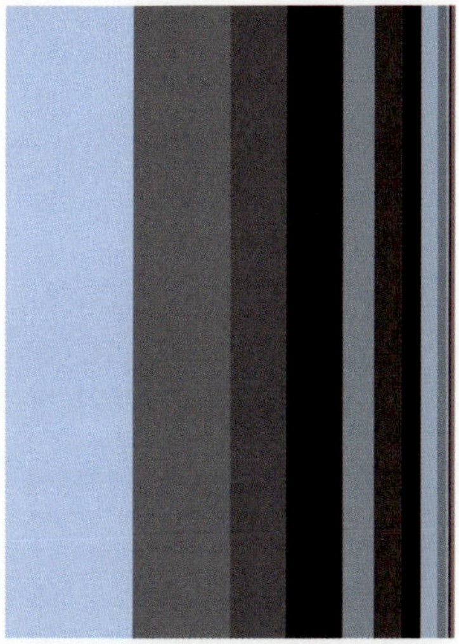

1271P

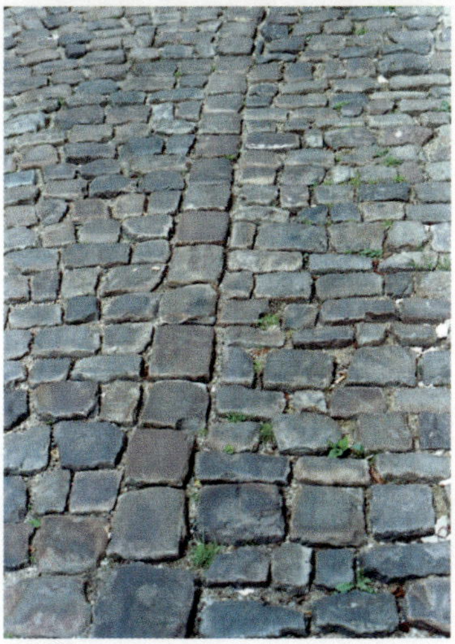

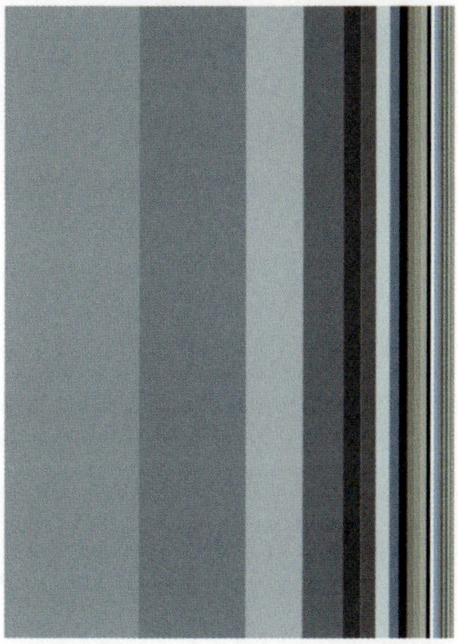

1295P

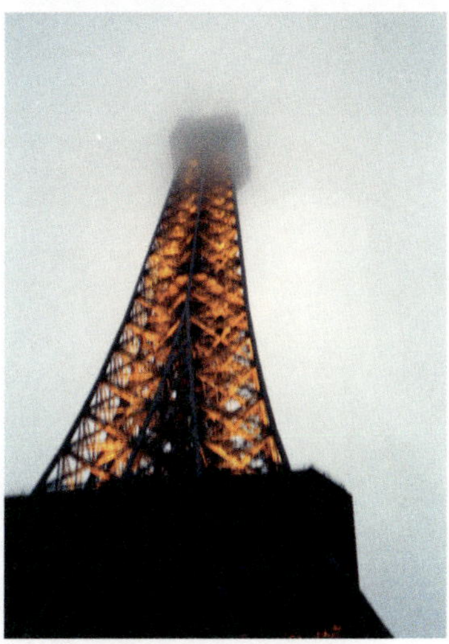

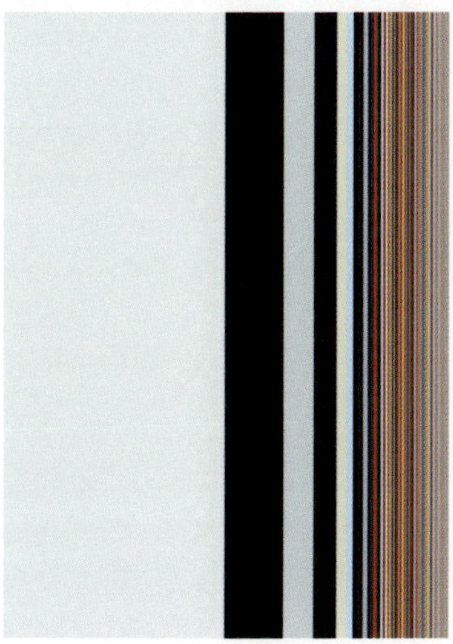

1311P

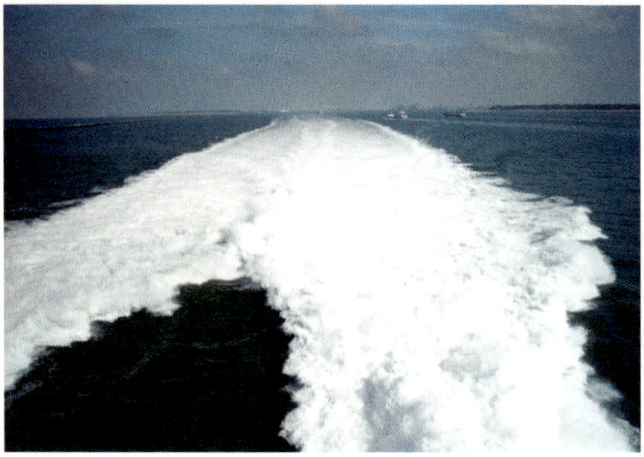

1328P

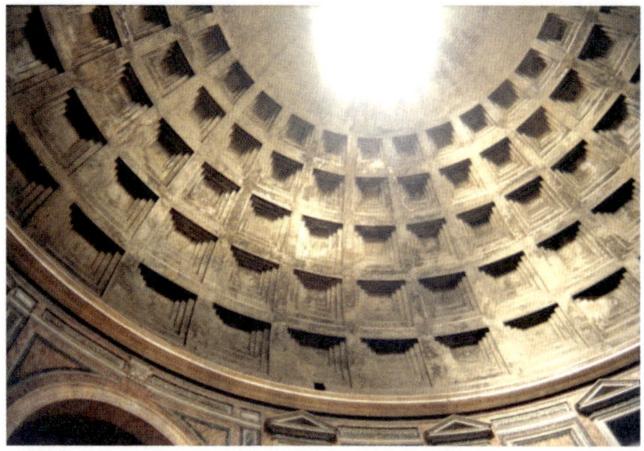

1377P

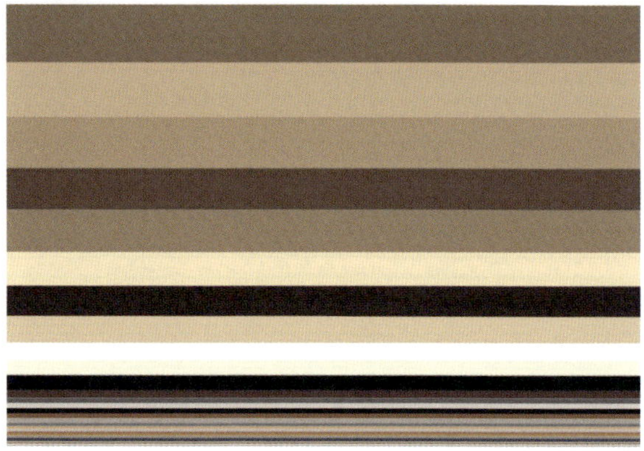

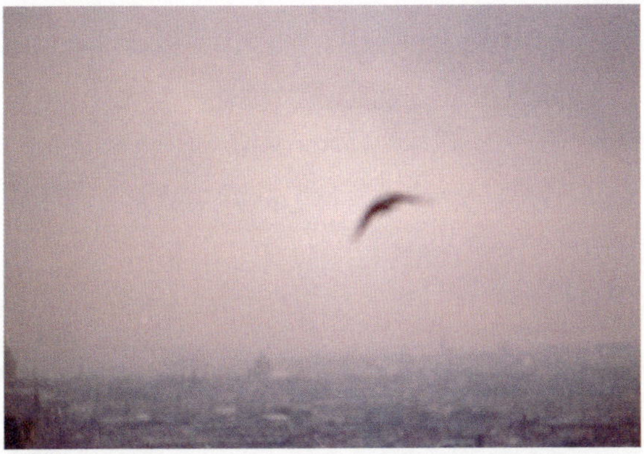

1396P

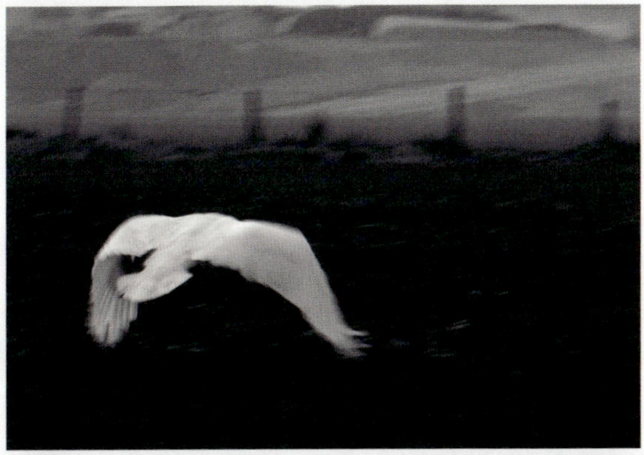

1482P

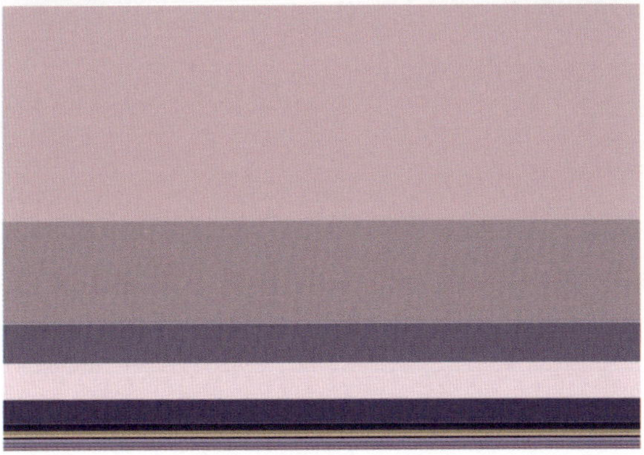

1

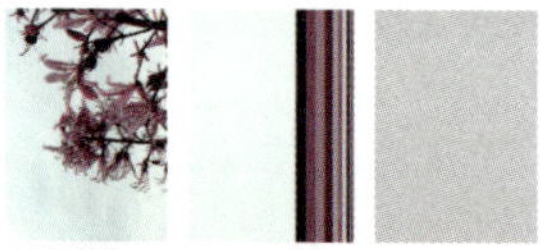

0337P

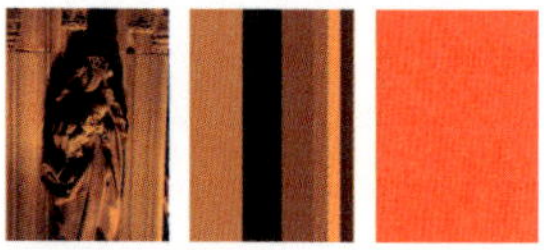

0474P

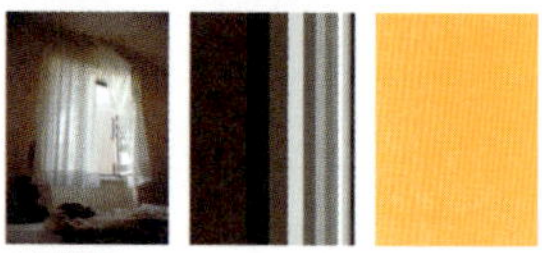

1256P

2

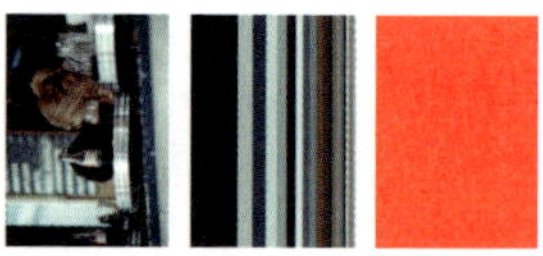

0220P

3

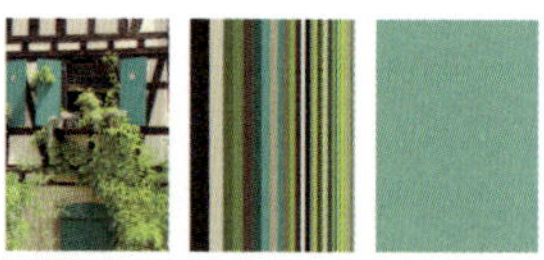

0530P

4

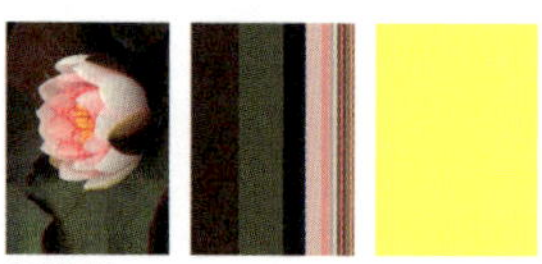

0335P

A clear minority of the images show similarities in colour tone and emotional response—we see that out of 61 photographs, 11 display some sort of resemblance, roughly one-sixth. For those images, I would diagnose a (slight) similarity between the colours of the photograph and the colour of the feeling I get when I look at it. In three of these images[1] the colour of the emotion seems to resemble the photograph, but looking more closely at the isolated colour bars, the colour of the emotion does not really appear in the image.

However, it remains hard to say whether there might not be other factors in the image. The coke bottle[2] mentioned before might be defining the red that I feel when I see that photograph. In the picture of a window in Strasbourg, the wooden panels stand out quite starkly and the colour of the emotion has a very similar tone.[3] As to a more abstract example, there is a hidden but vivid yellow at the very centre of an image that naturally draws our visual attention:[4] it is at the core of an unfolding flower. The yellow can barely be seen amongst the colour bars but is still the visual centre of the image, and might by virtue of that just as easily create a shade of yellow in my mind when I look at it. The same happens in three other images:[5] a central element draws our attention and has the same colour as the synaesthetic reaction—in these cases, there might be an influence despite that colour barely appearing within the colour bars that measure the amount of a colour in the photograph.

In other photographs, a colour might be majorly present and still not be related to the synaesthetic response at all. The best demonstration of this might be the black and white images that respectively trigger shades of mint green, bright blue, orange, yellow, off-white and a deep dark blue.[6]

It remains very difficult to interpret the resemblance between the colours of a photograph and the synaesthetic reaction that I see, simply because my judgment on which photographs are possibly related to their synaesthetic reaction remains purely subjective as well. In some cases, I might have seen a resemblance where others would not. Other images display some similarities that might attract somebody else's attention, but which I deemed insignificant.

All in all, I can identify elements of a photograph that might have influenced my synaesthetic reaction, be it clearly visible colours or rather hidden elements. I do not, however, conclude that those really influence the colour of the emotion evoked when I look at them. There is simply not enough overlay to assume that the synaesthetic reaction is clearly influenced by the colours of the photograph. So even if I acknowledge a number of possible correlations between the colour of the image and the colour of the synaesthetic concurrent, I do not see a clear influence.

Only approximately 18% of the images that were used for this test show similarities. Even

5

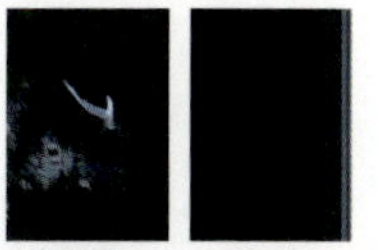

0534P

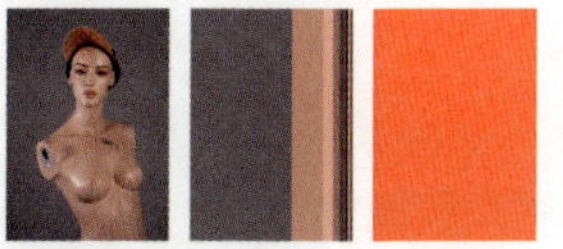

0541P

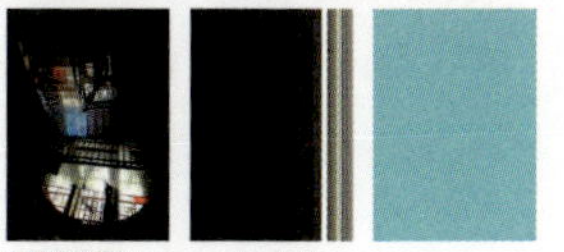

0552P

6

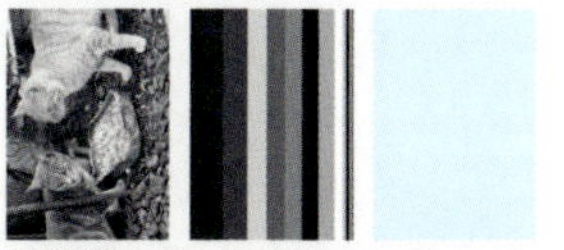

0114P

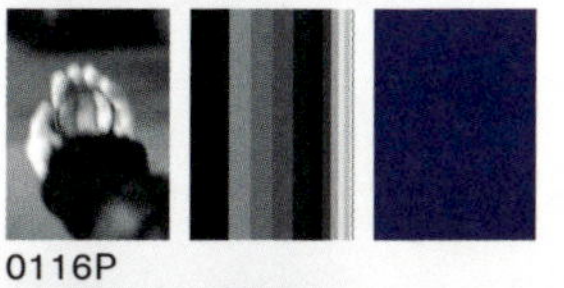

0116P

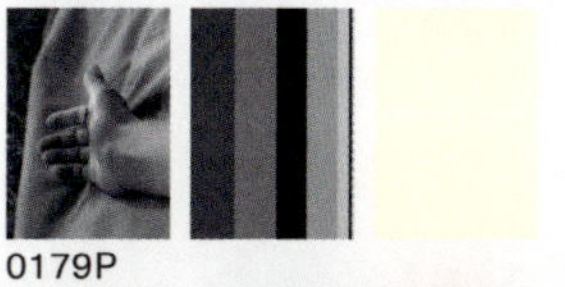

0179P

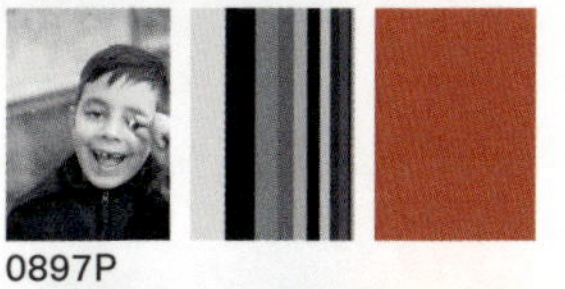

0897P

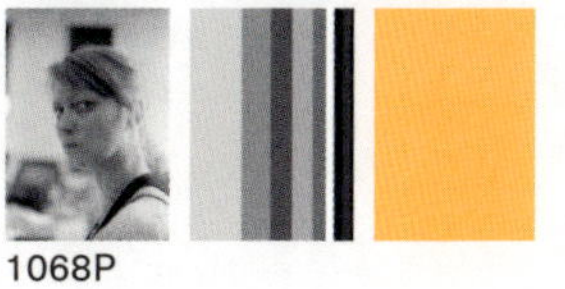

1068P

if I cannot exclude the possibility of a connection, this is not a solid enough basis for assuming that the colours in a photograph have a clearly noticeable influence on the colours of the emotions linked to it. If the colours of the image play no significant role in the generation of the coloured emotional response, what other factors influence my synaesthesia? What exactly makes a photograph feel yellow to me? I noticed that the colour reactions are not always consistent, so I decided to examine this more closely.

Chapter 4
Consistency

Because I was wondering if an image would always create the same sensory reaction, I tracked my emotional response to the photographs during a period of 36 days. For this part of the research, I used a selection of 24 of my photographs. During this research, I wrote down the synaesthetic reaction I had when looking at each particular picture.

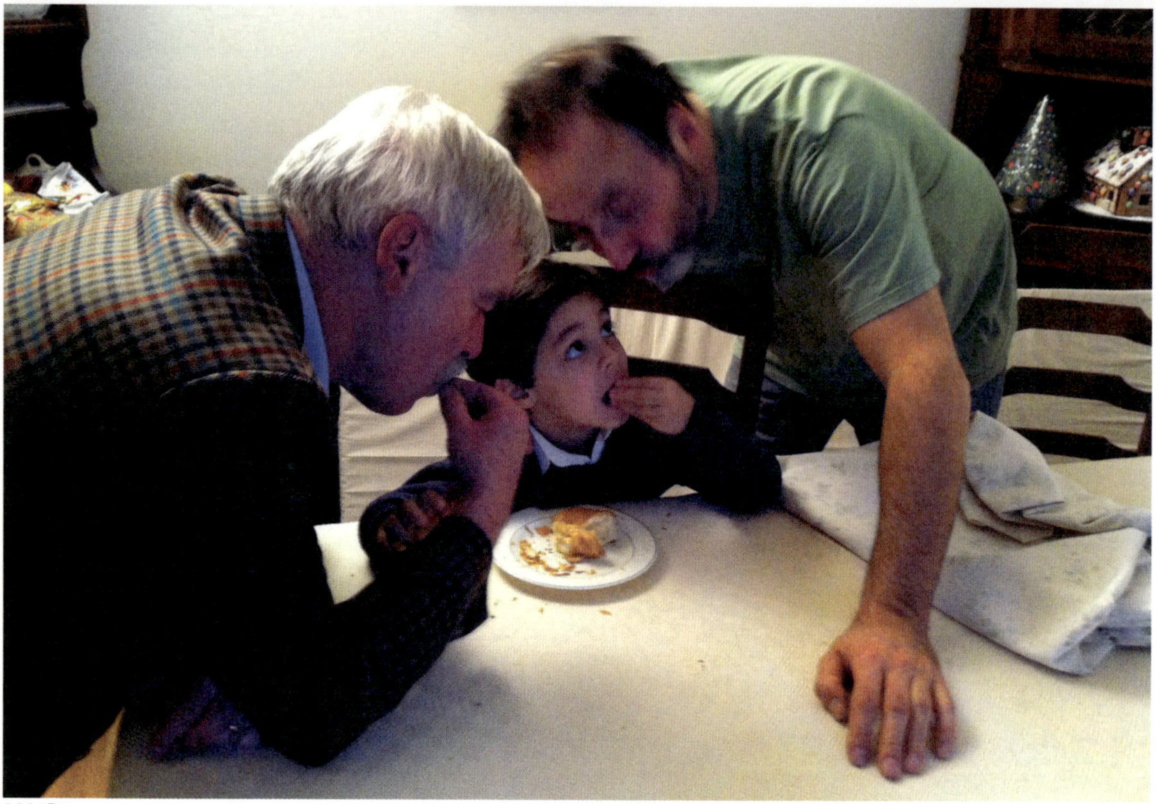

0004P

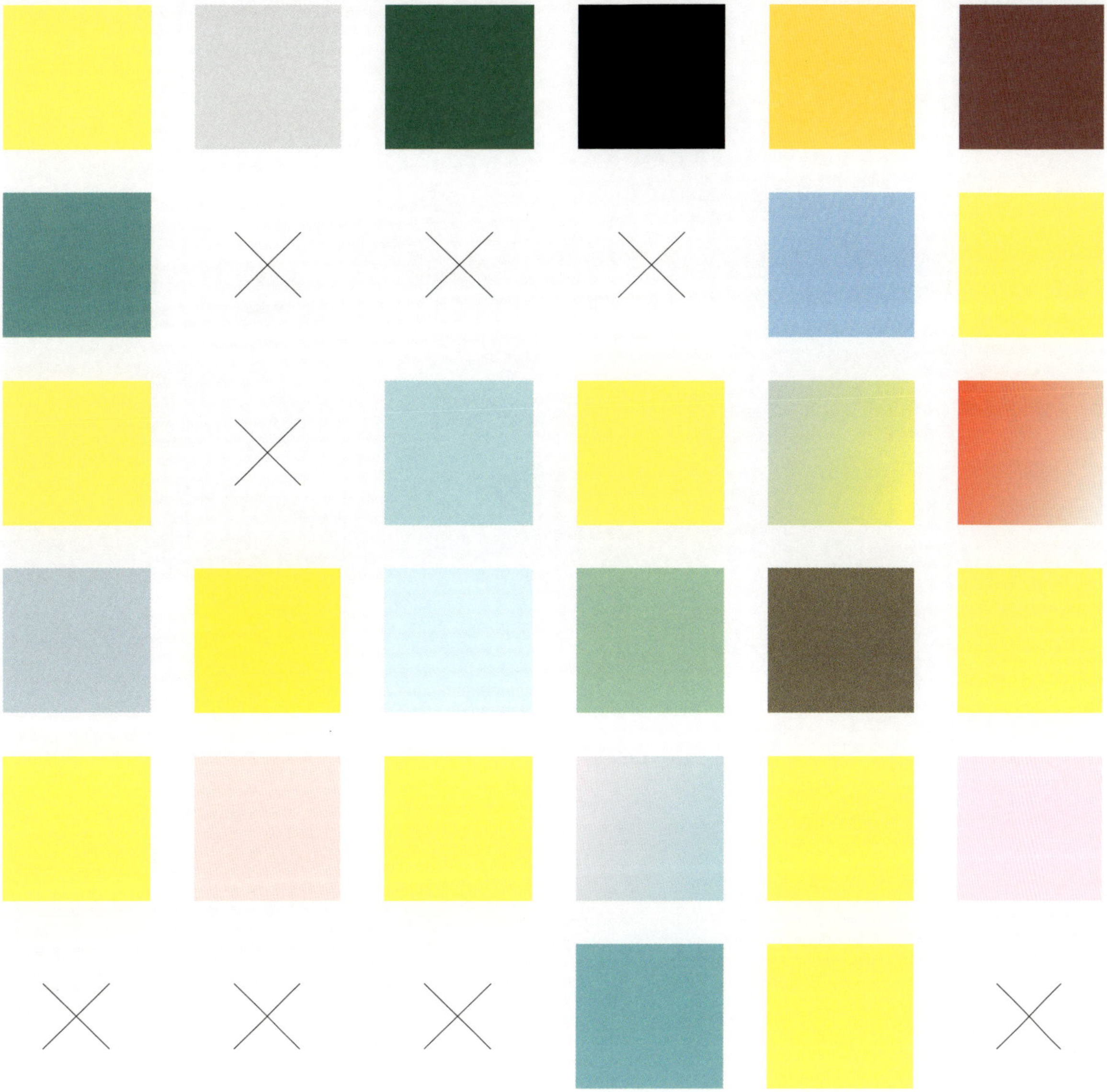

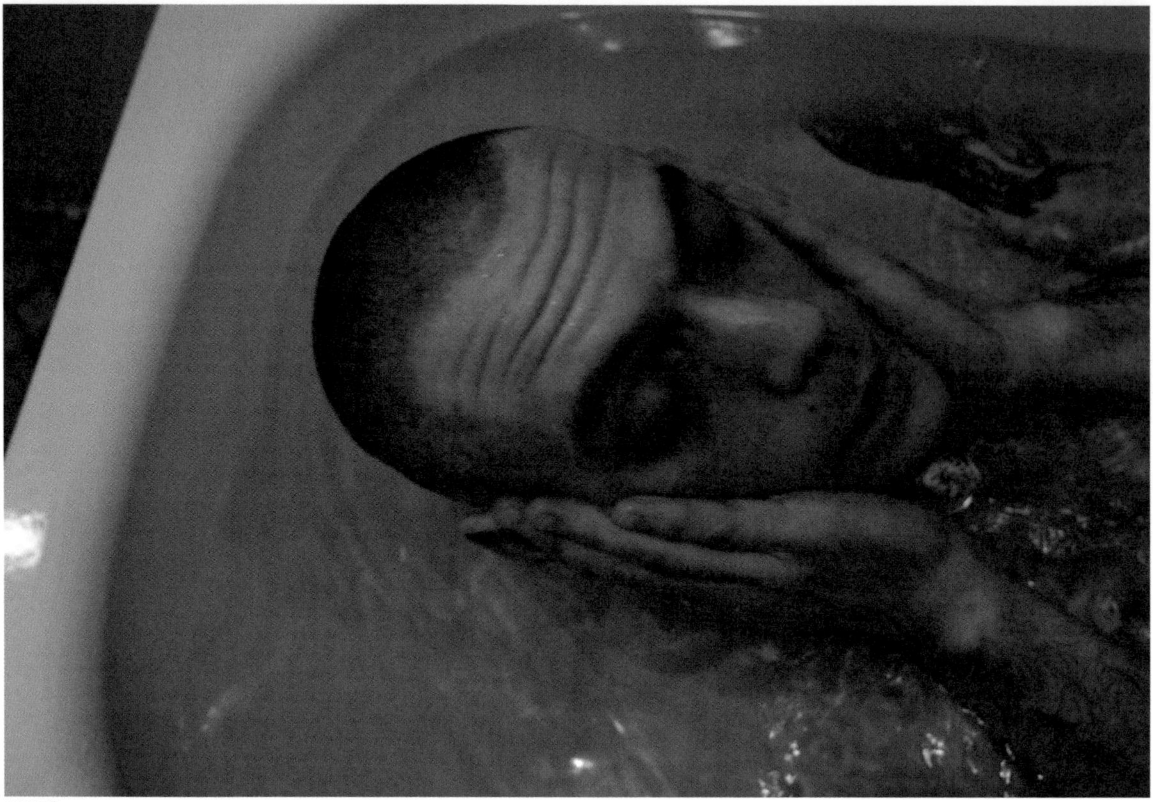

0036P

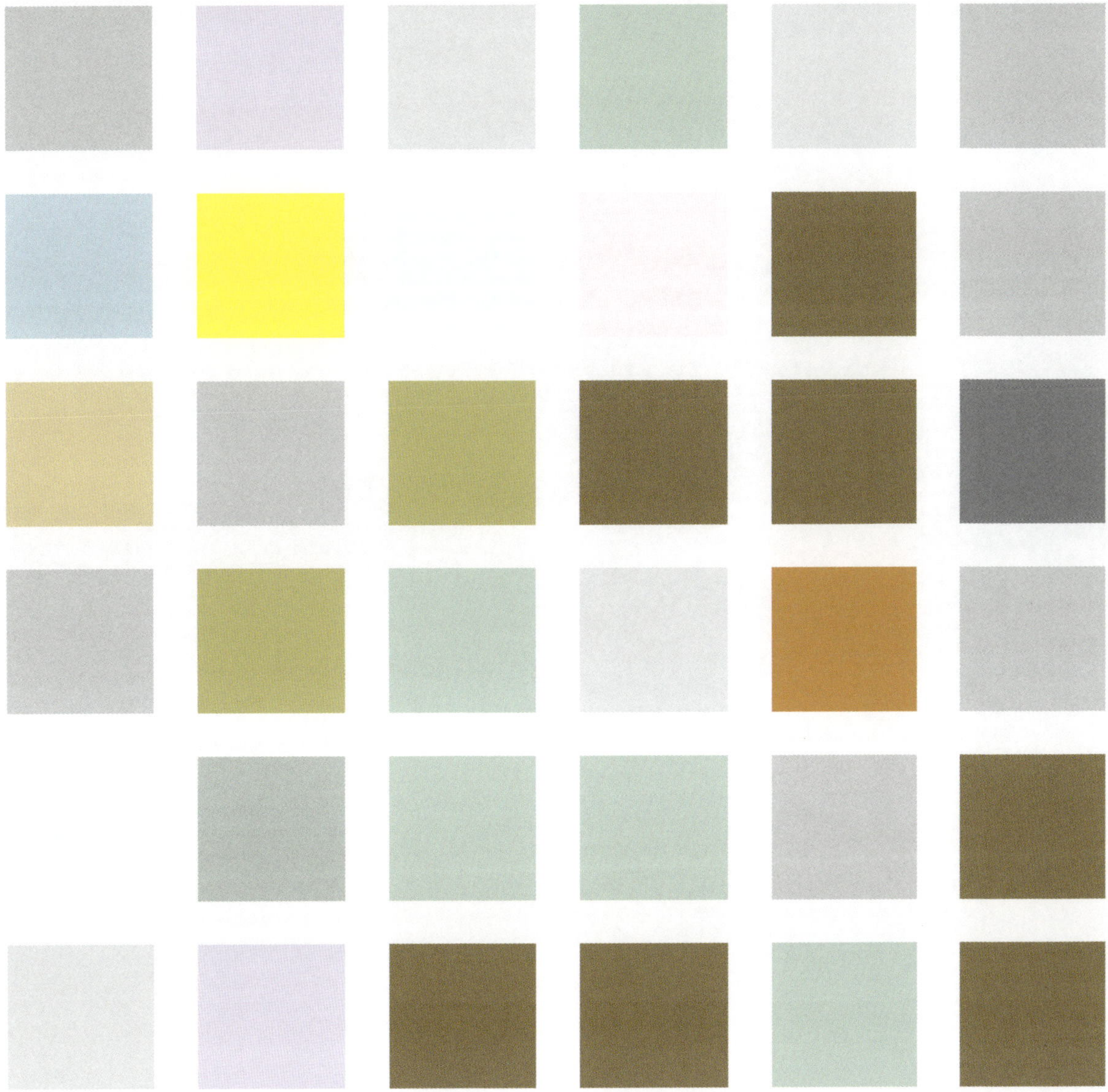

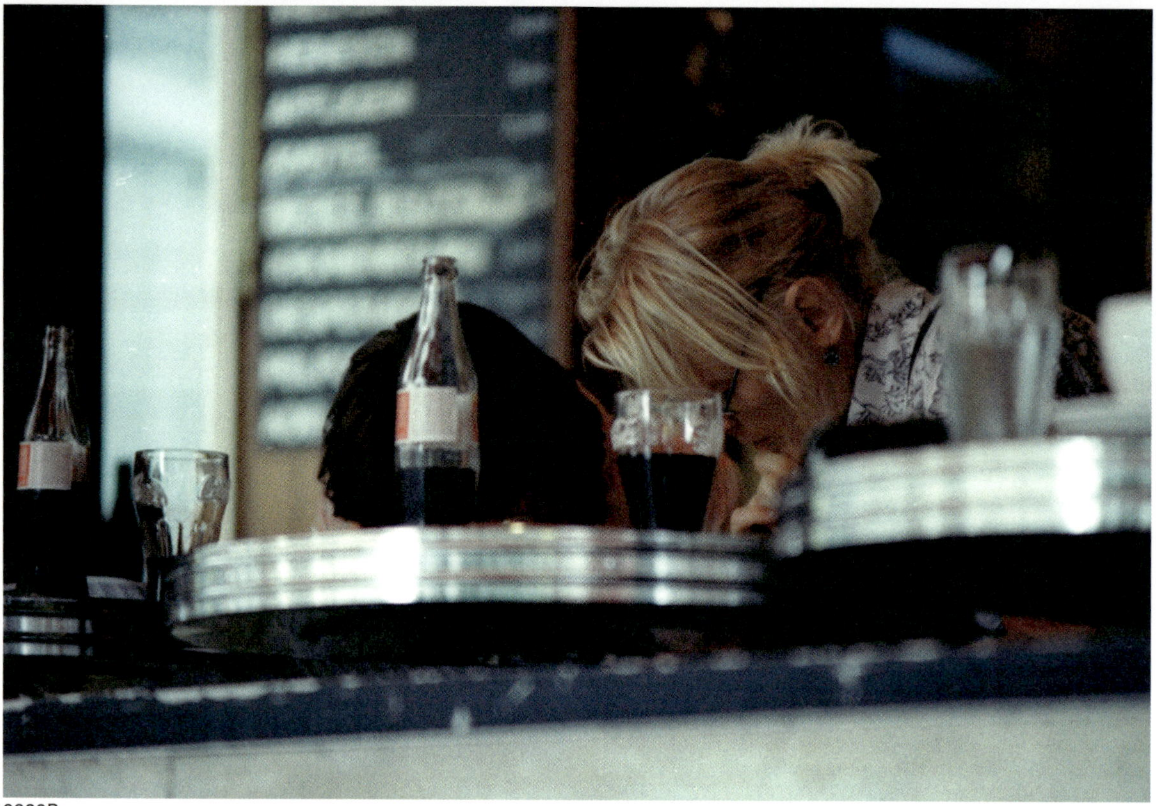

0220P

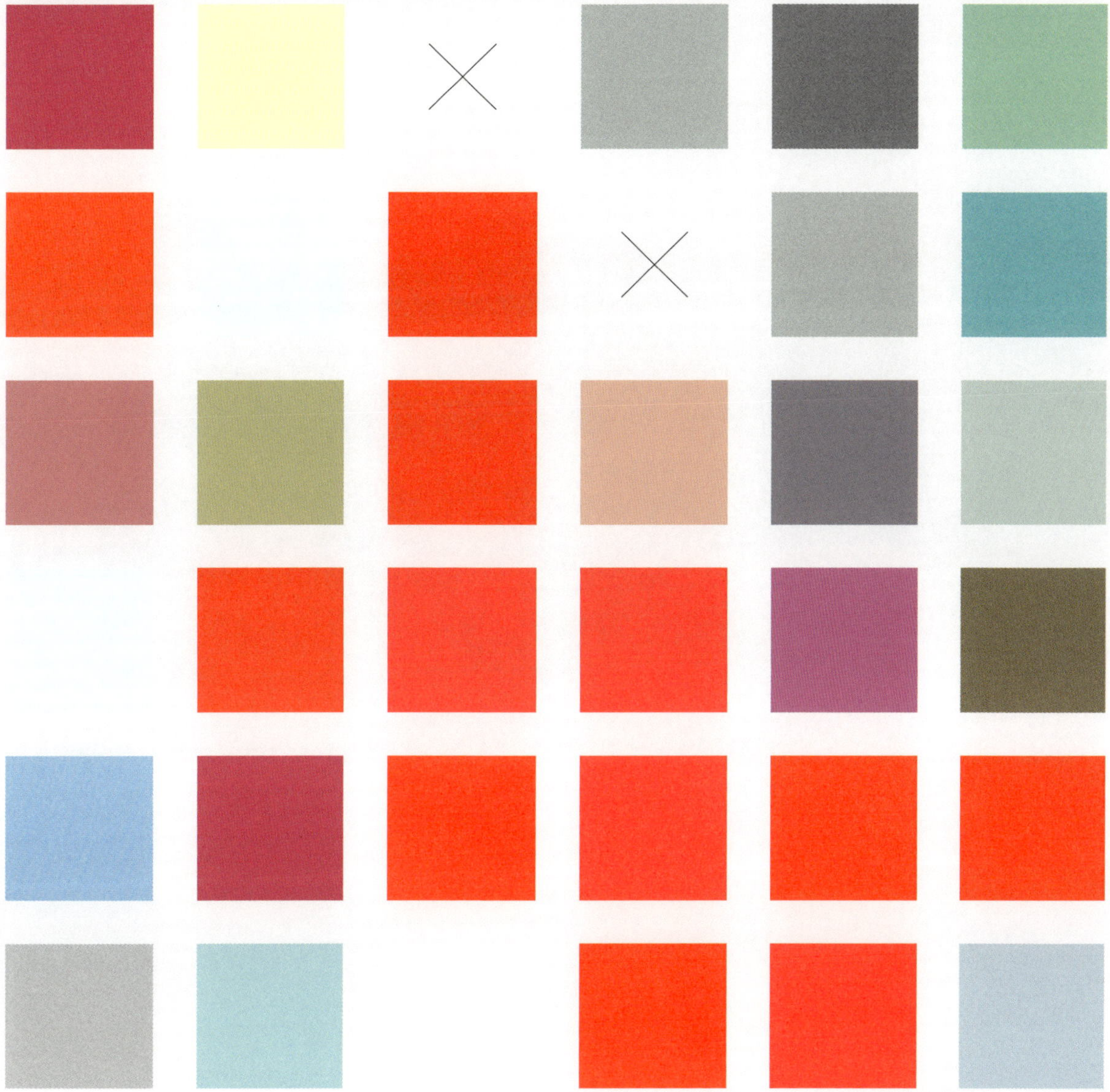

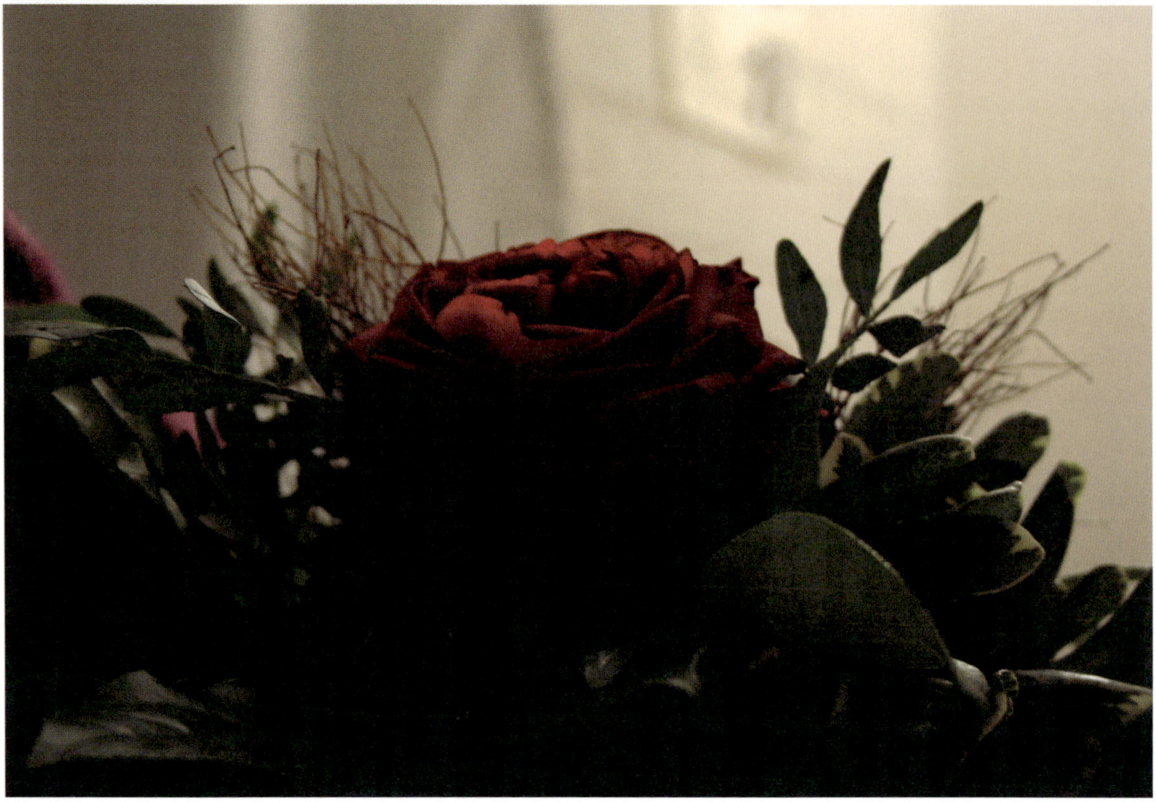

0248P

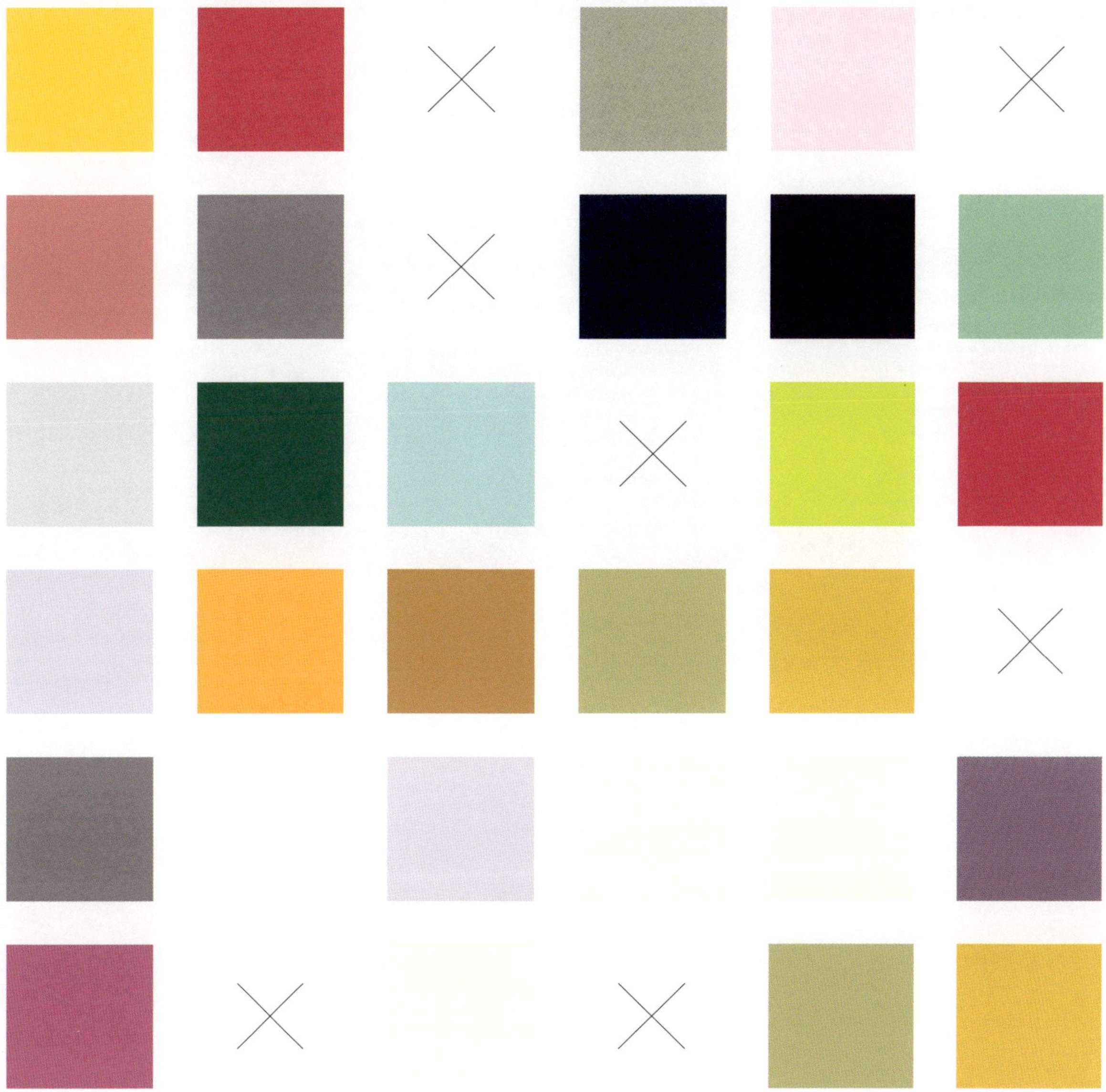

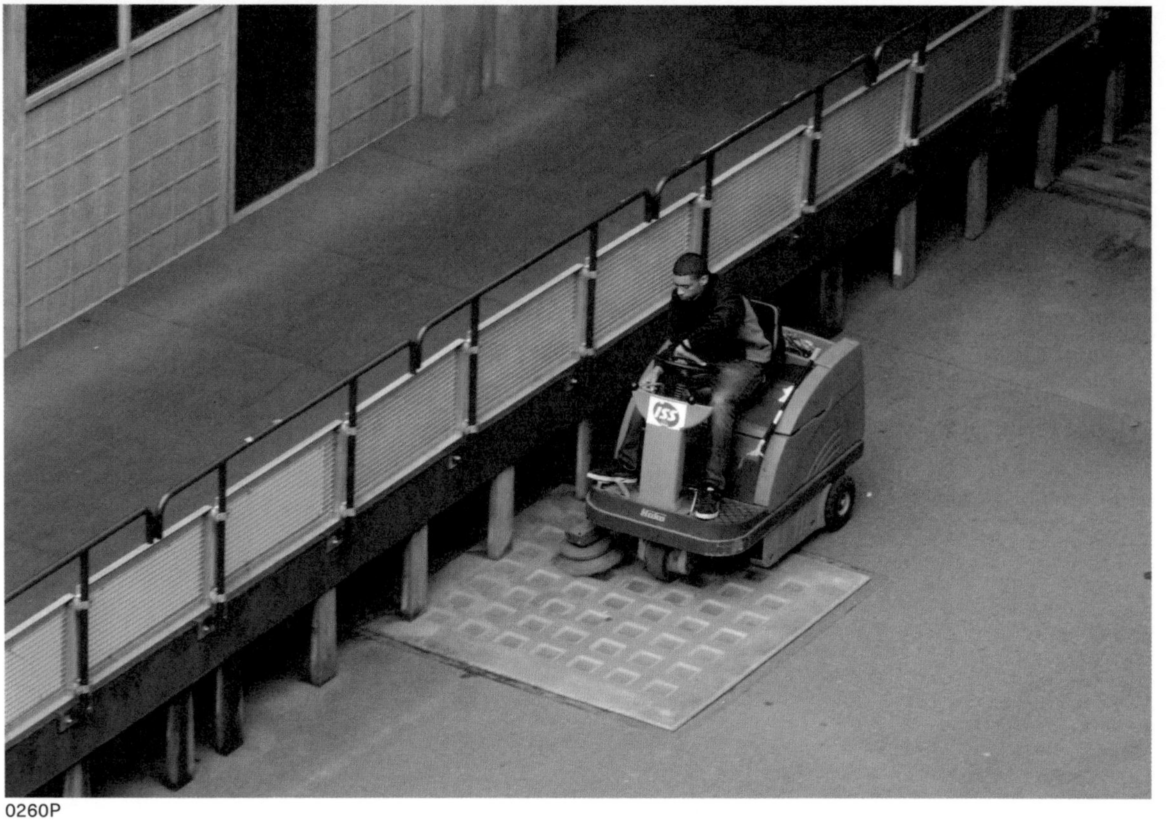

0260P

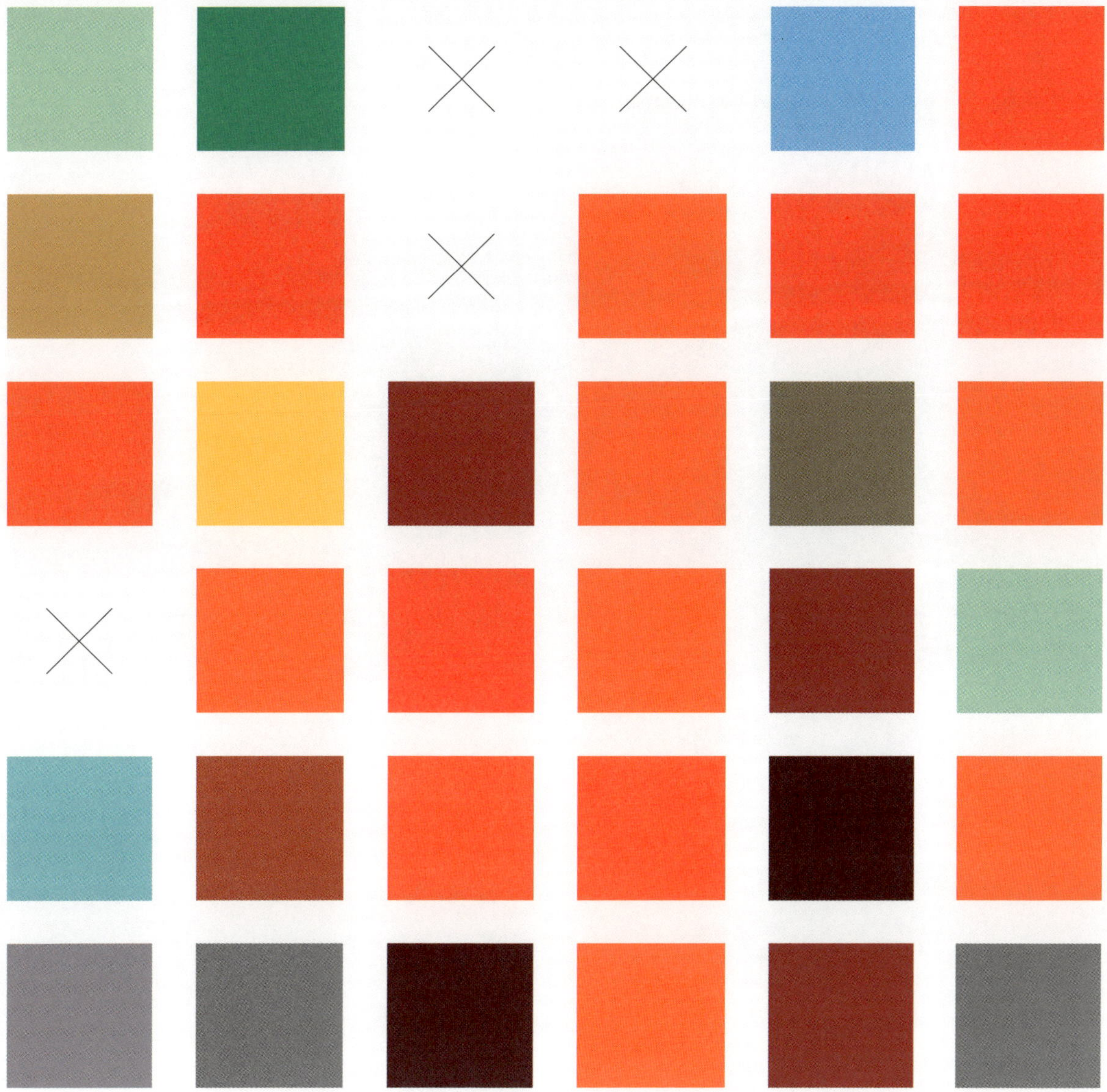

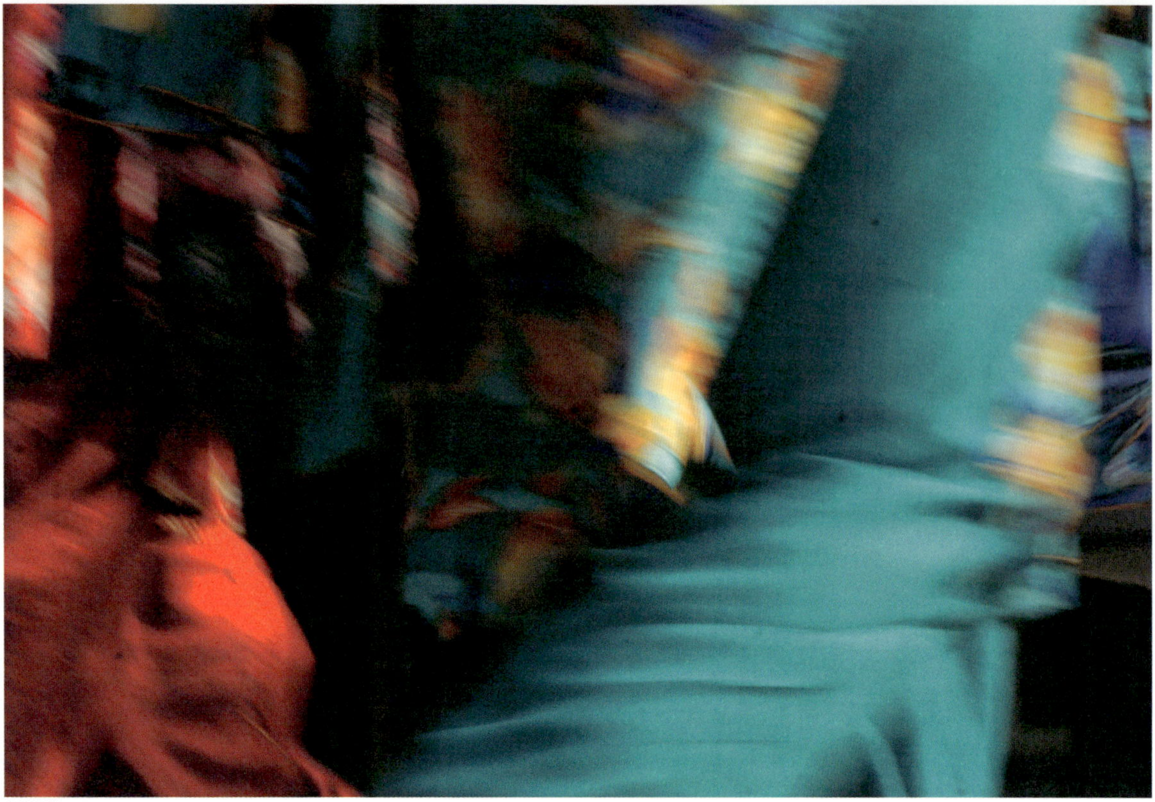

0290P

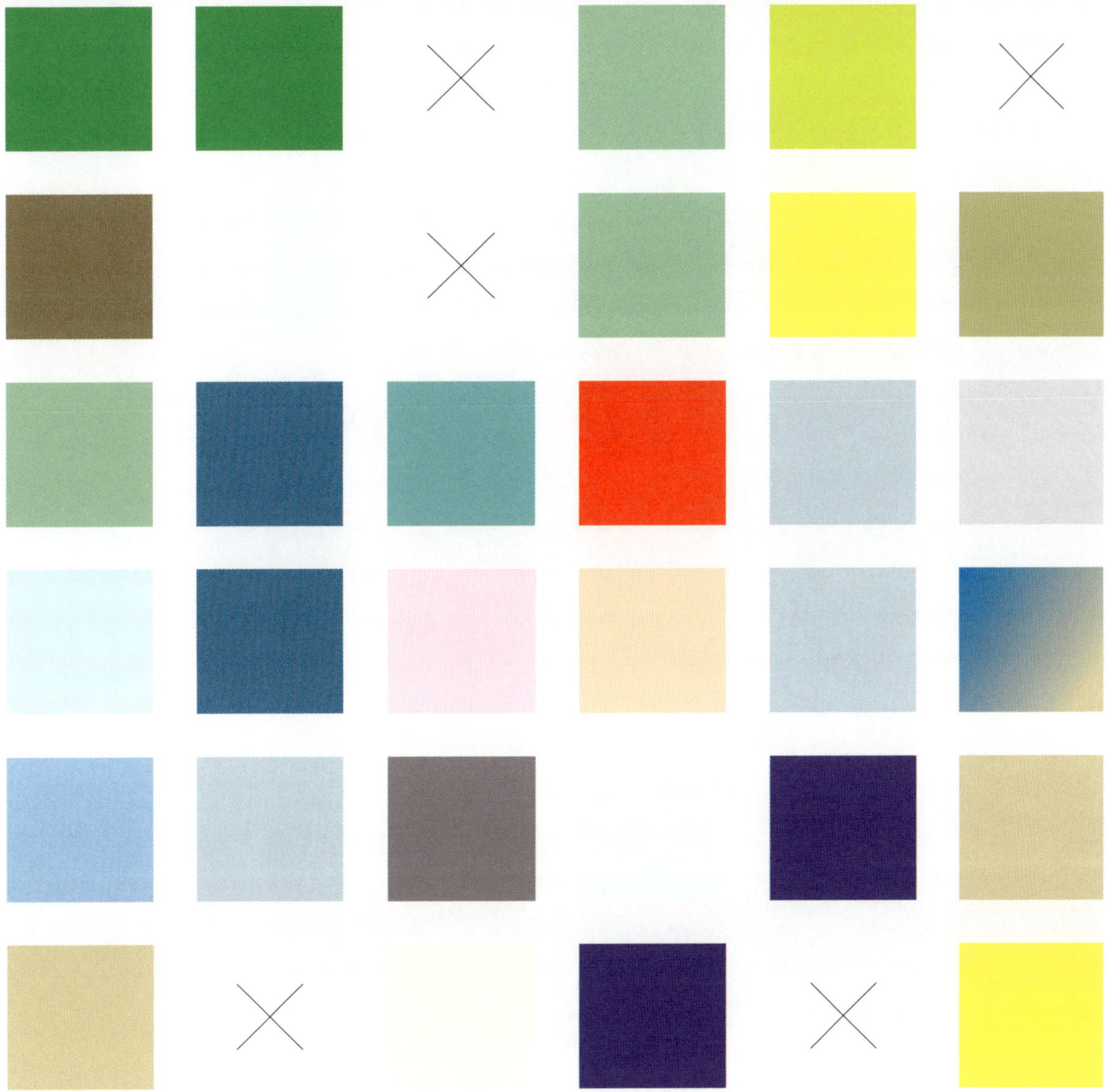

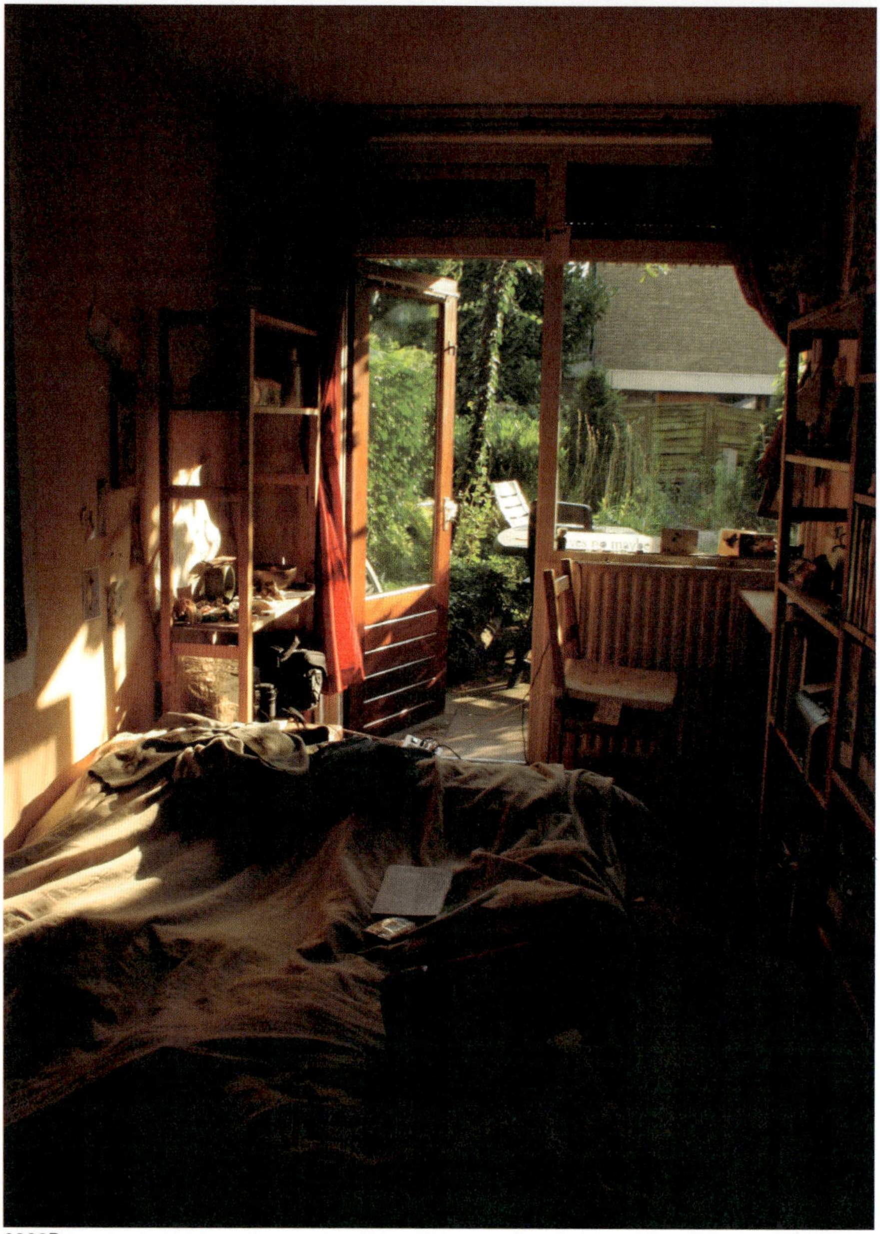

0366P

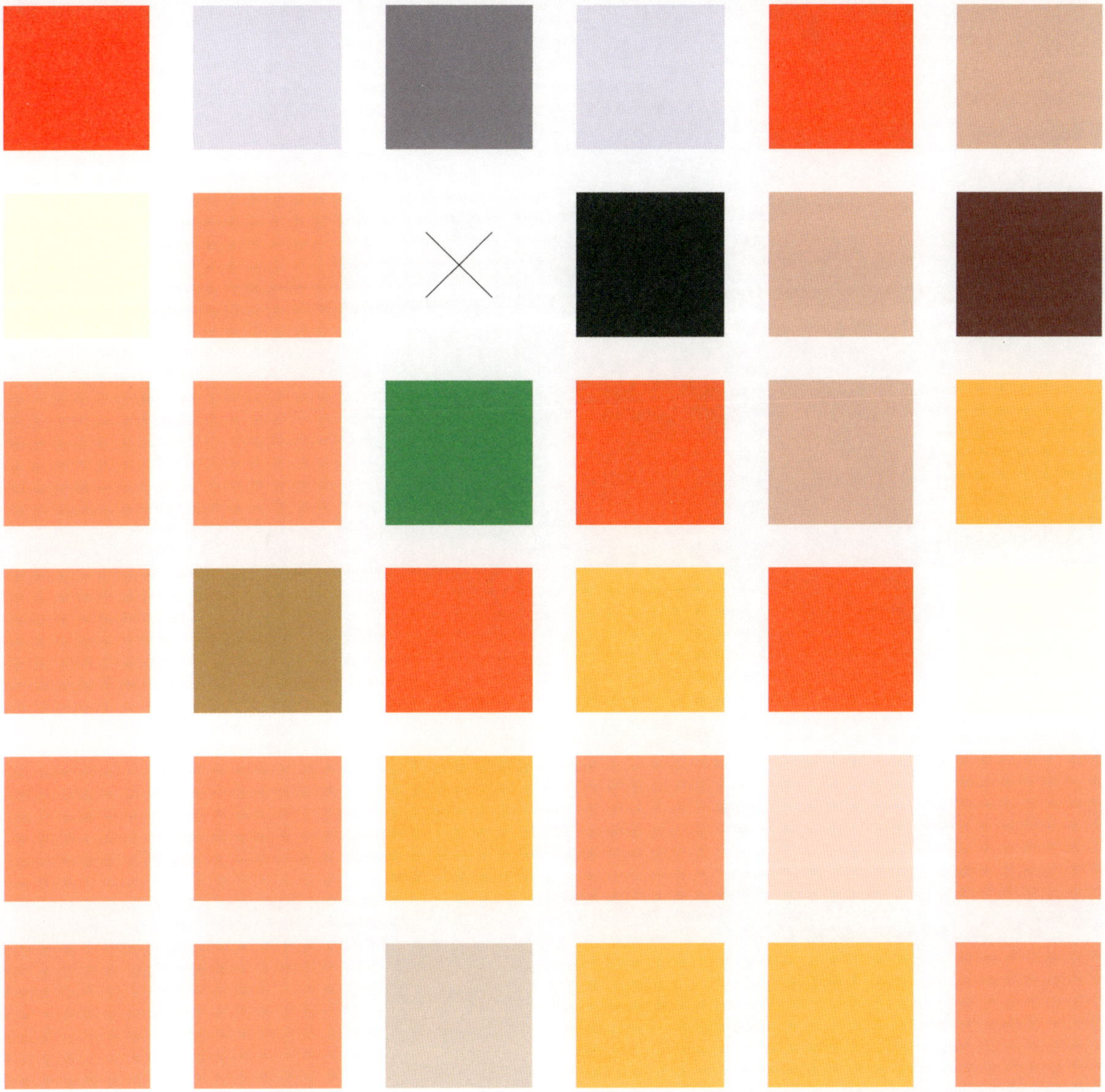

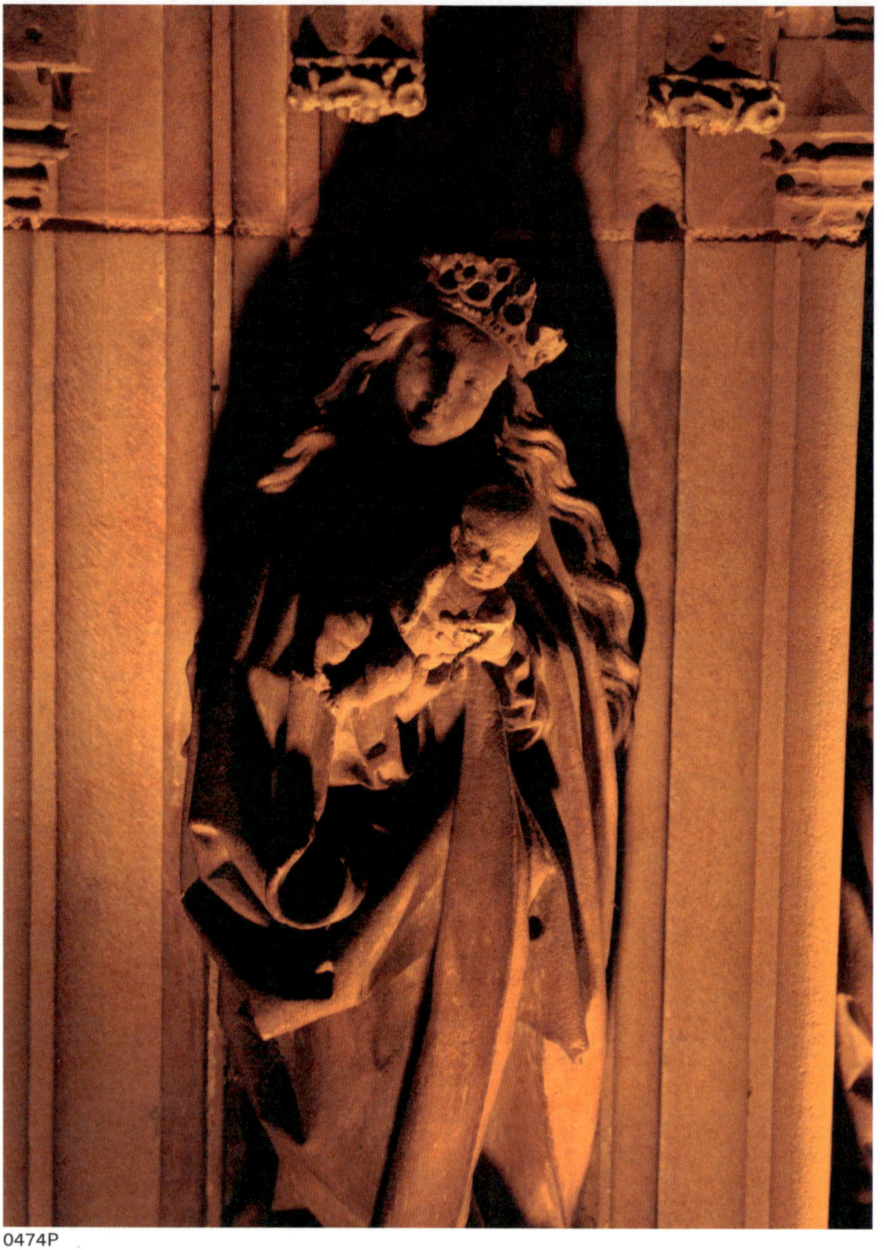

0474P

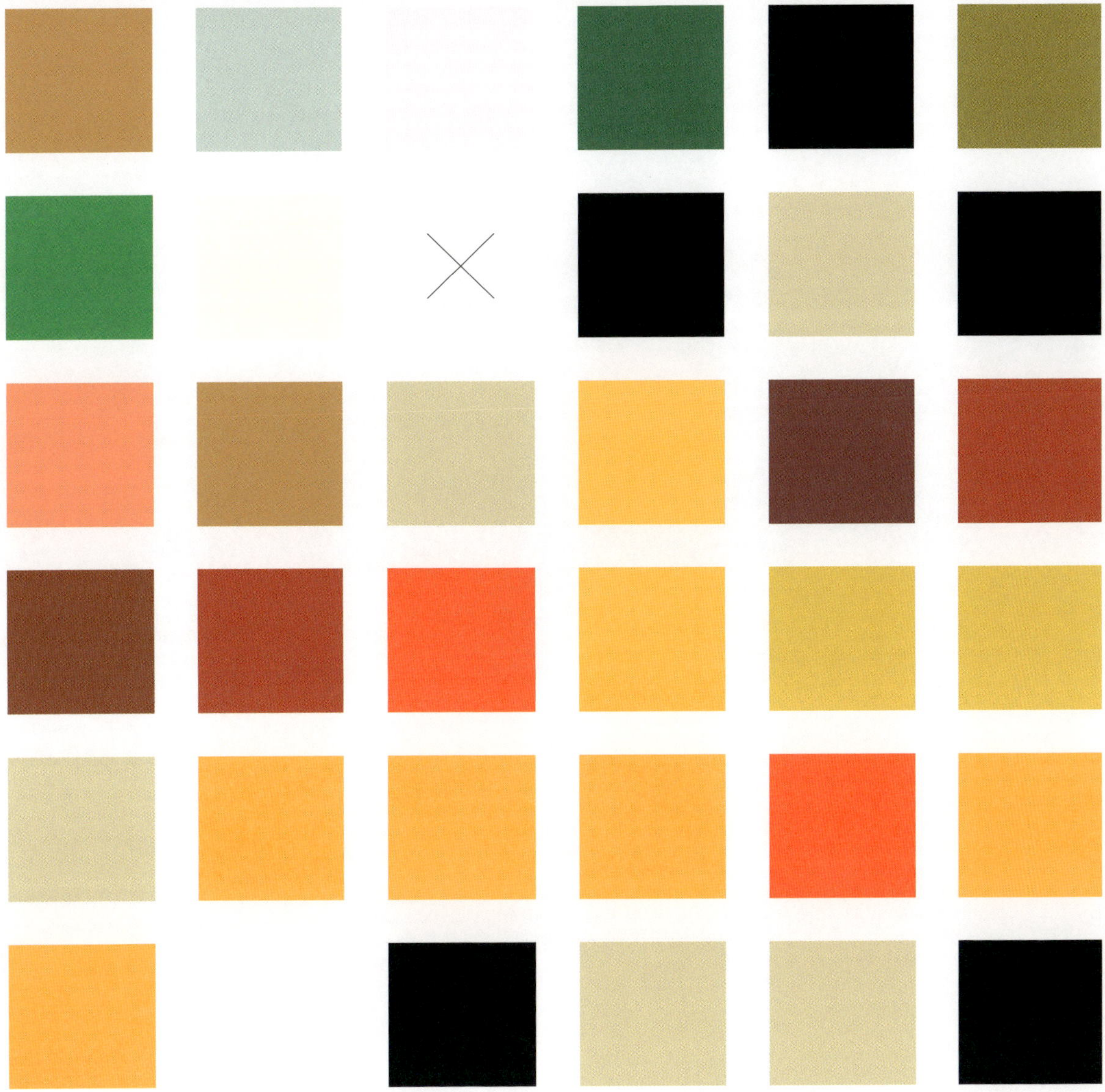

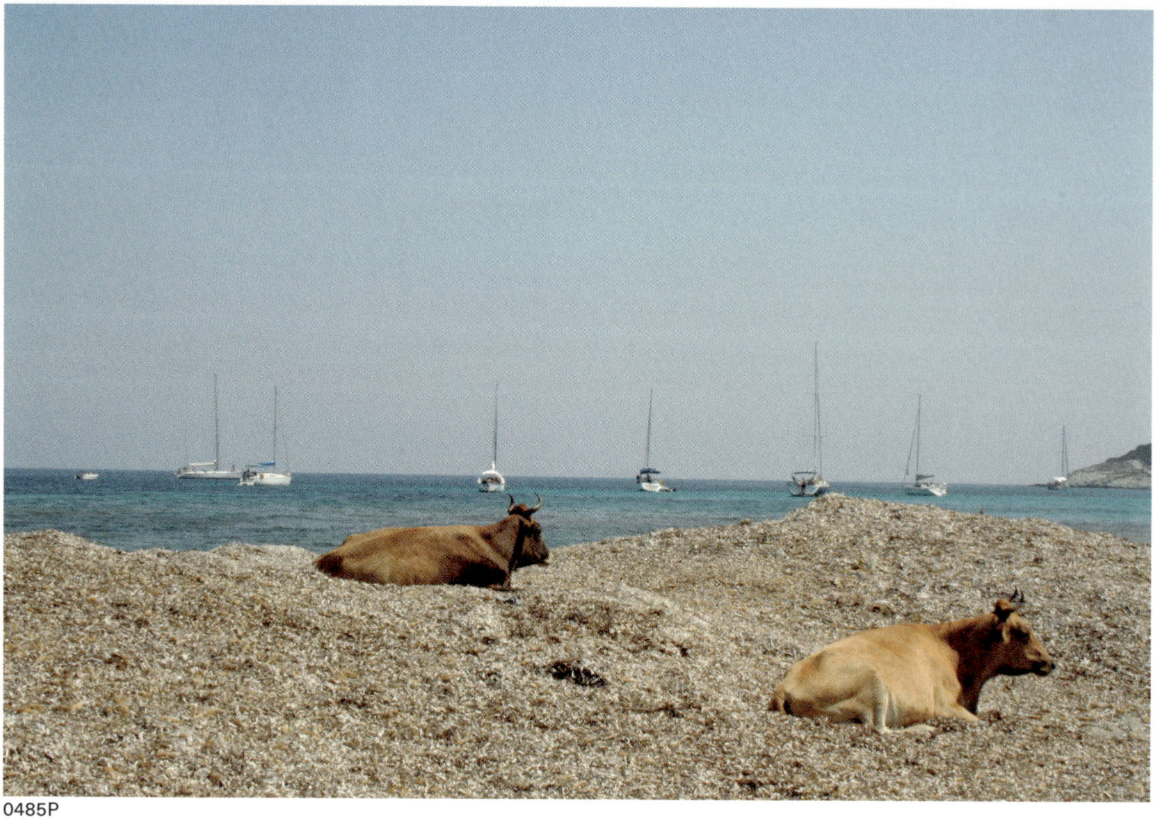

0485P

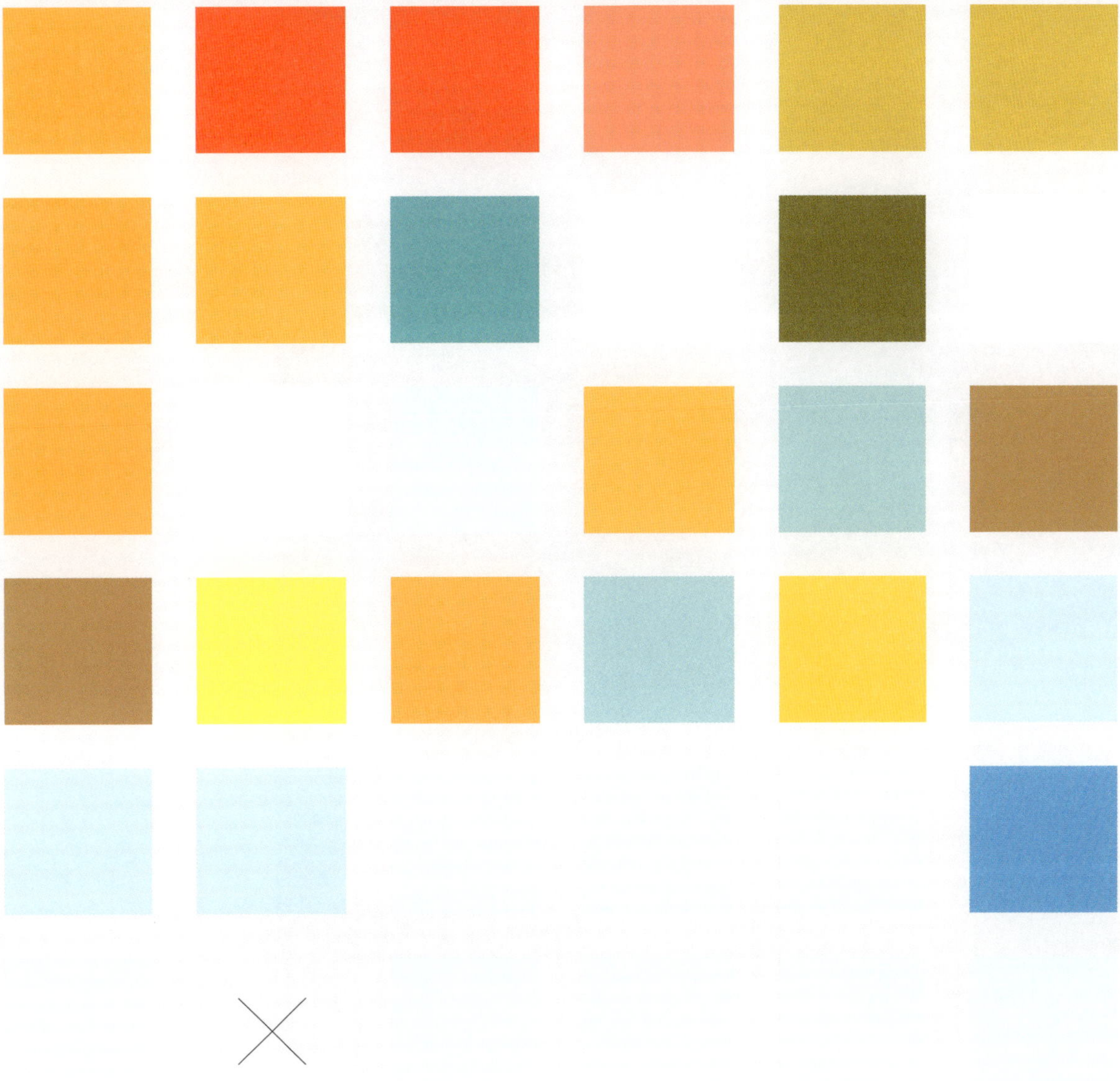

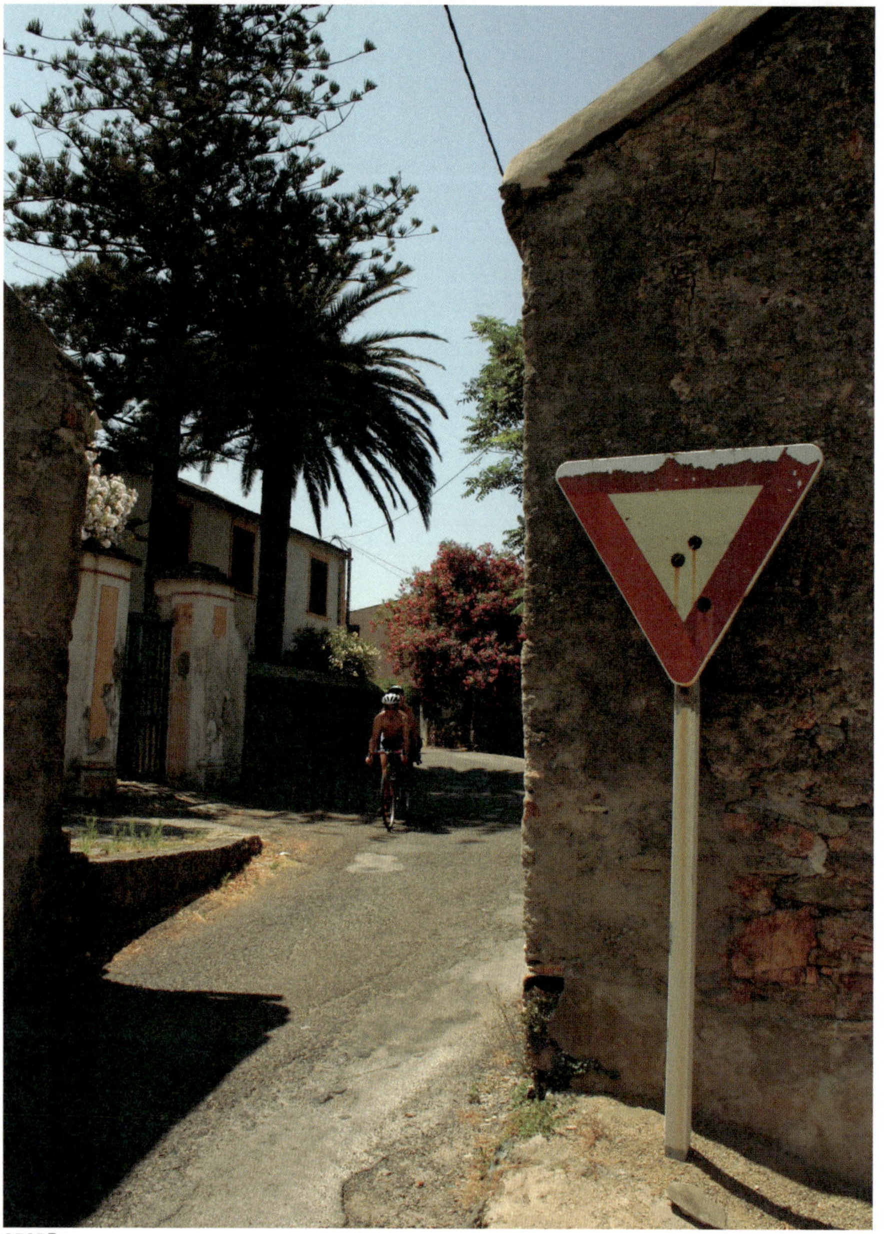

0505P

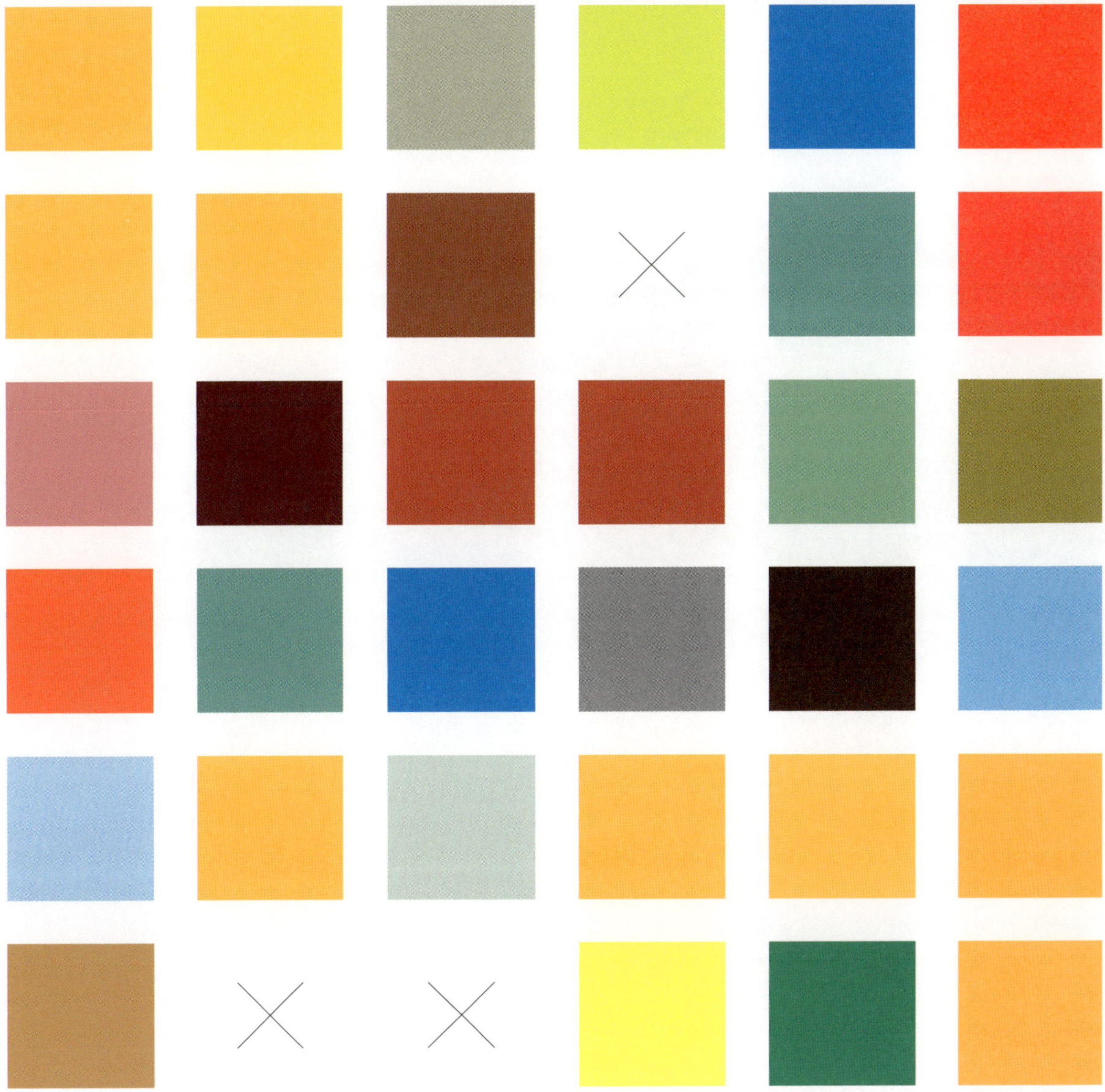

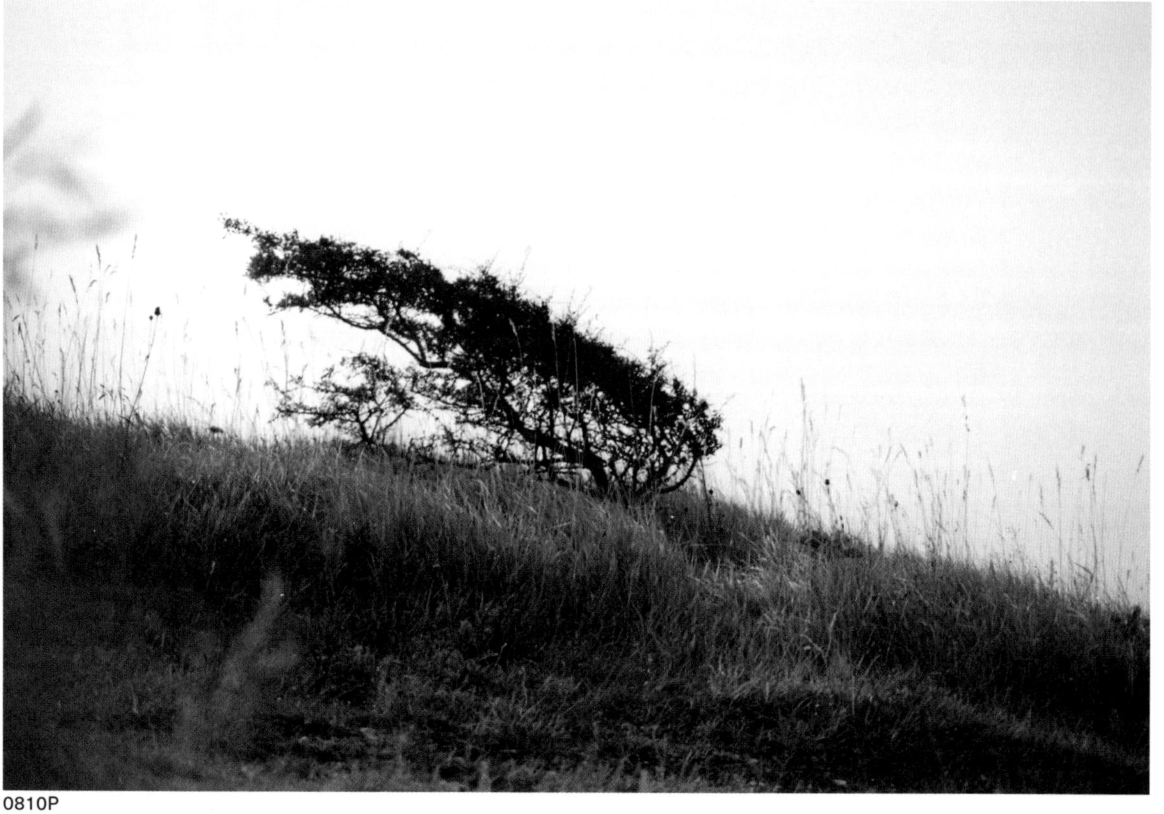
0810P

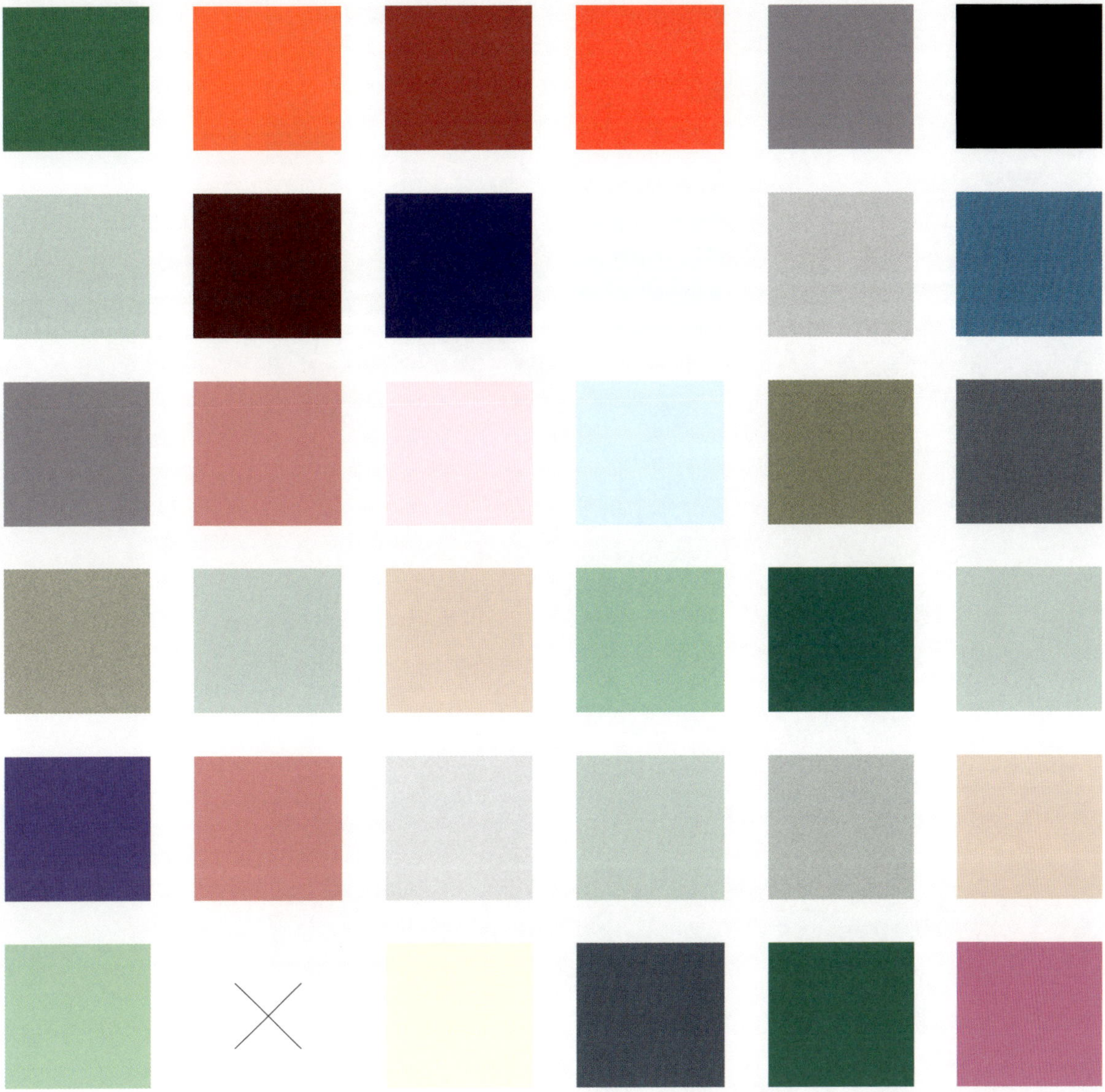

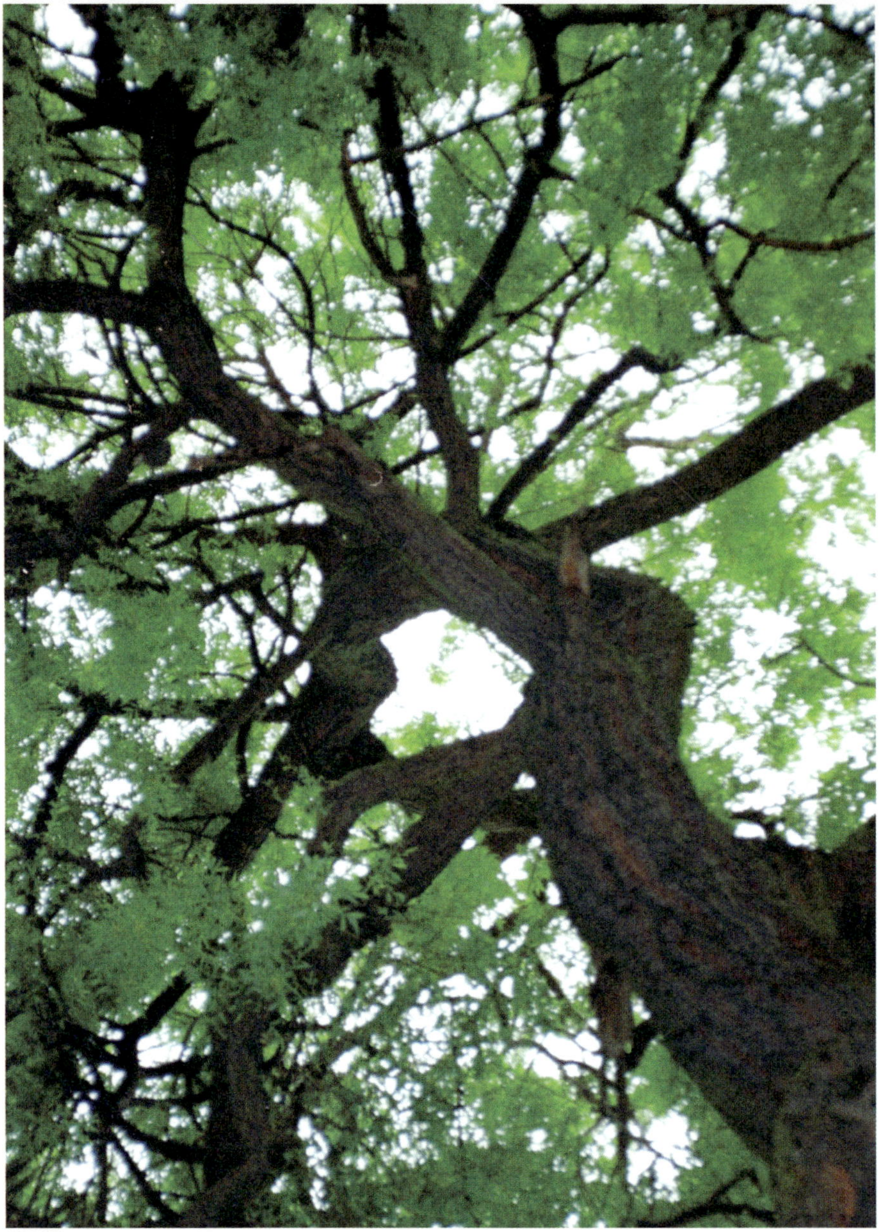

0820P

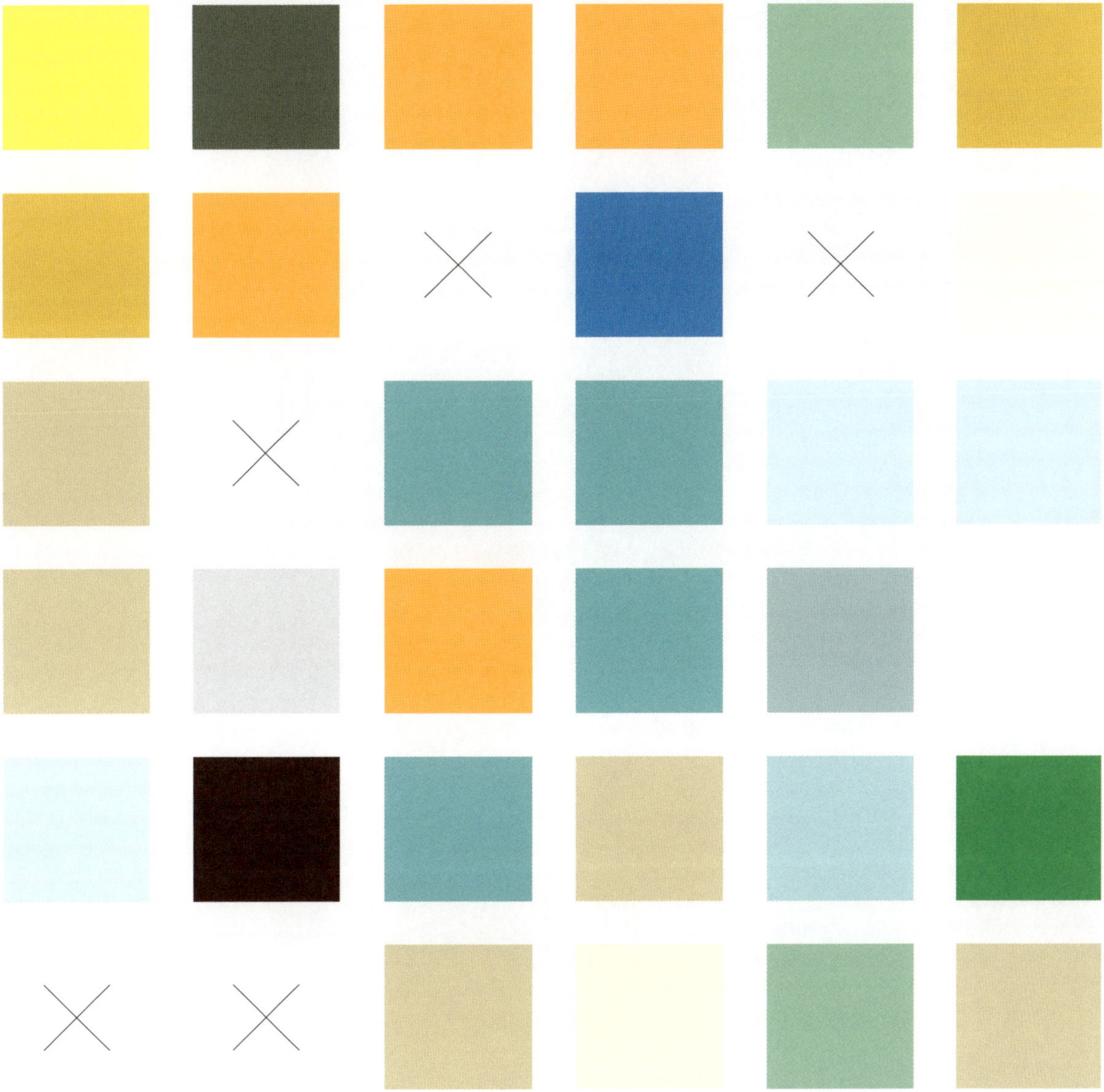

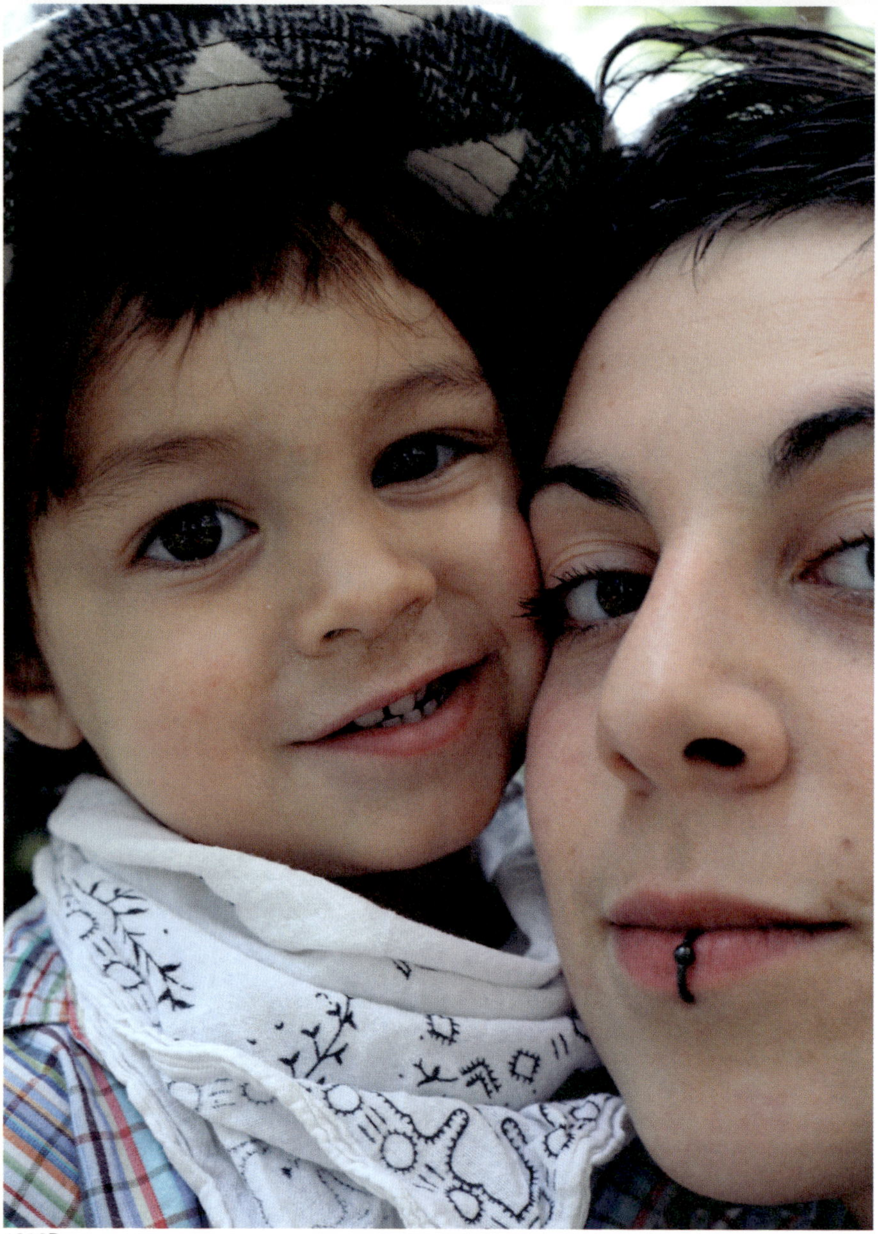

1216P

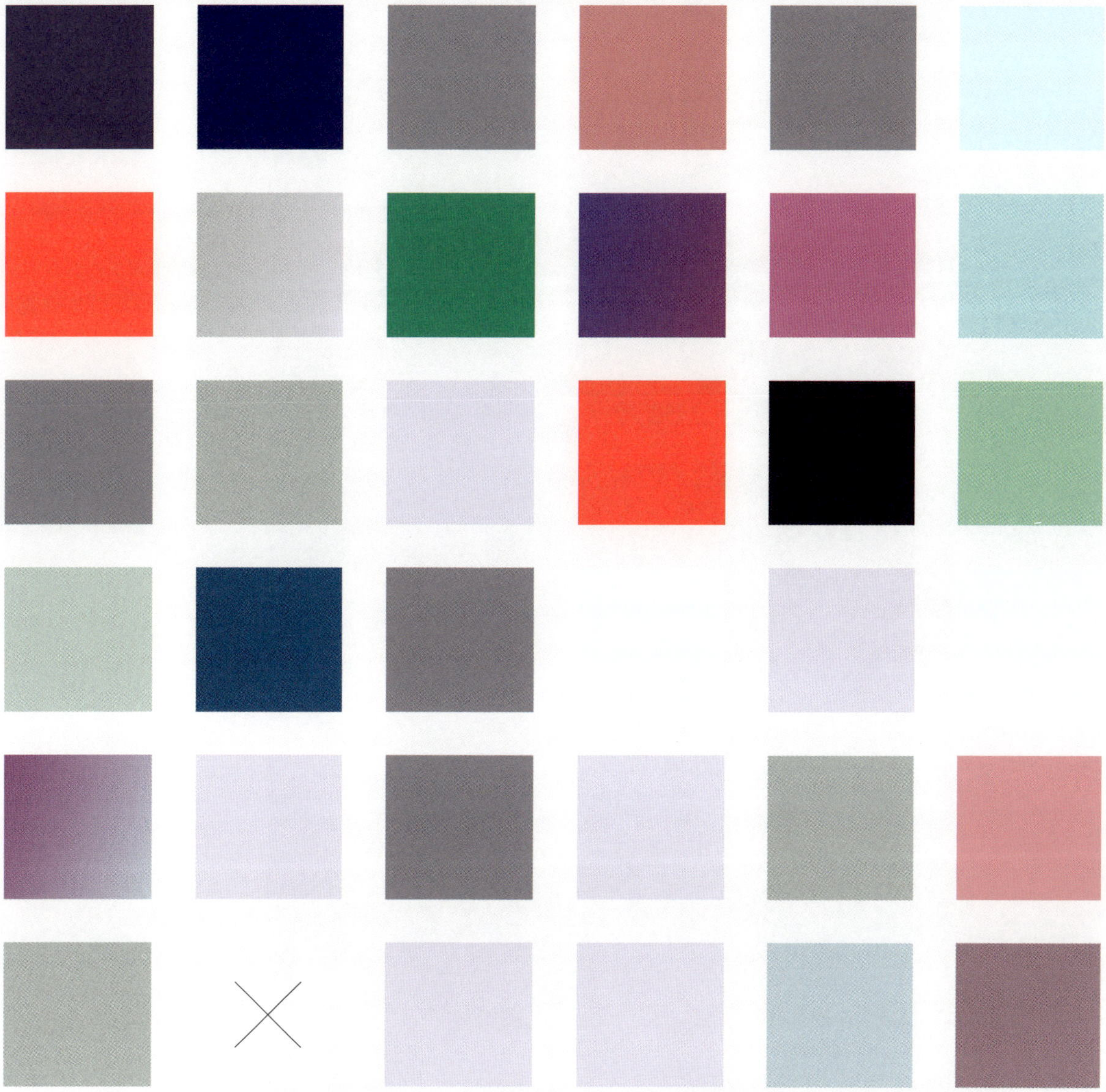

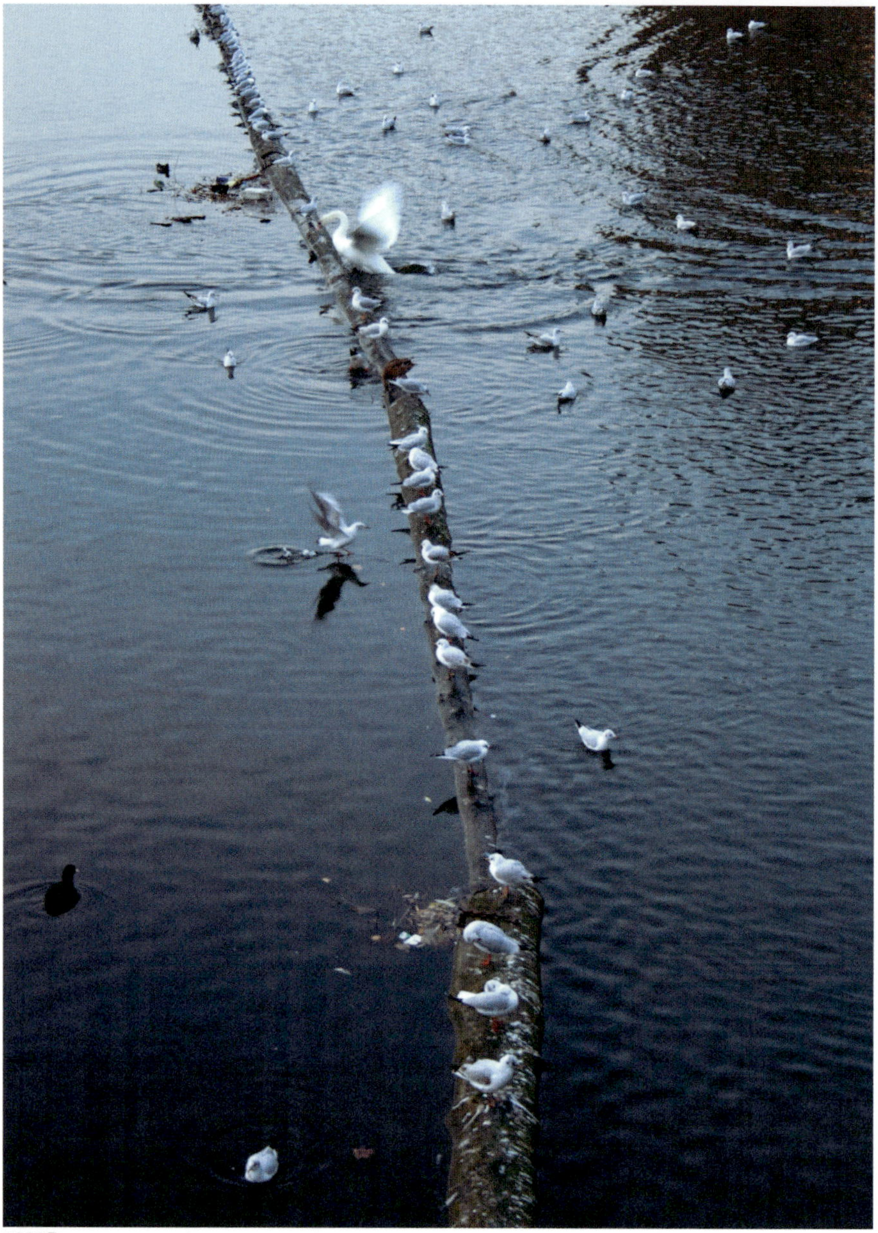
1228P

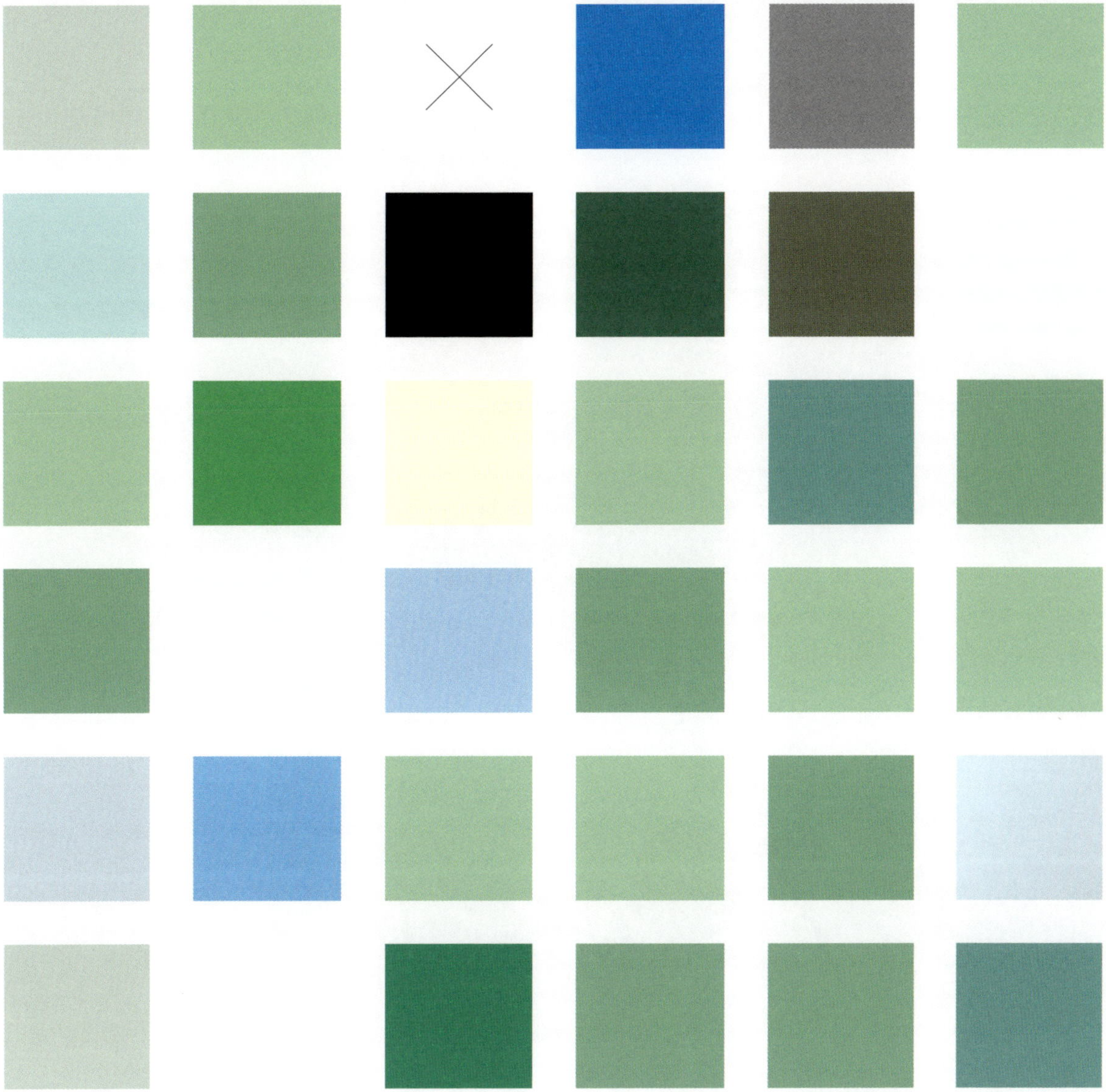

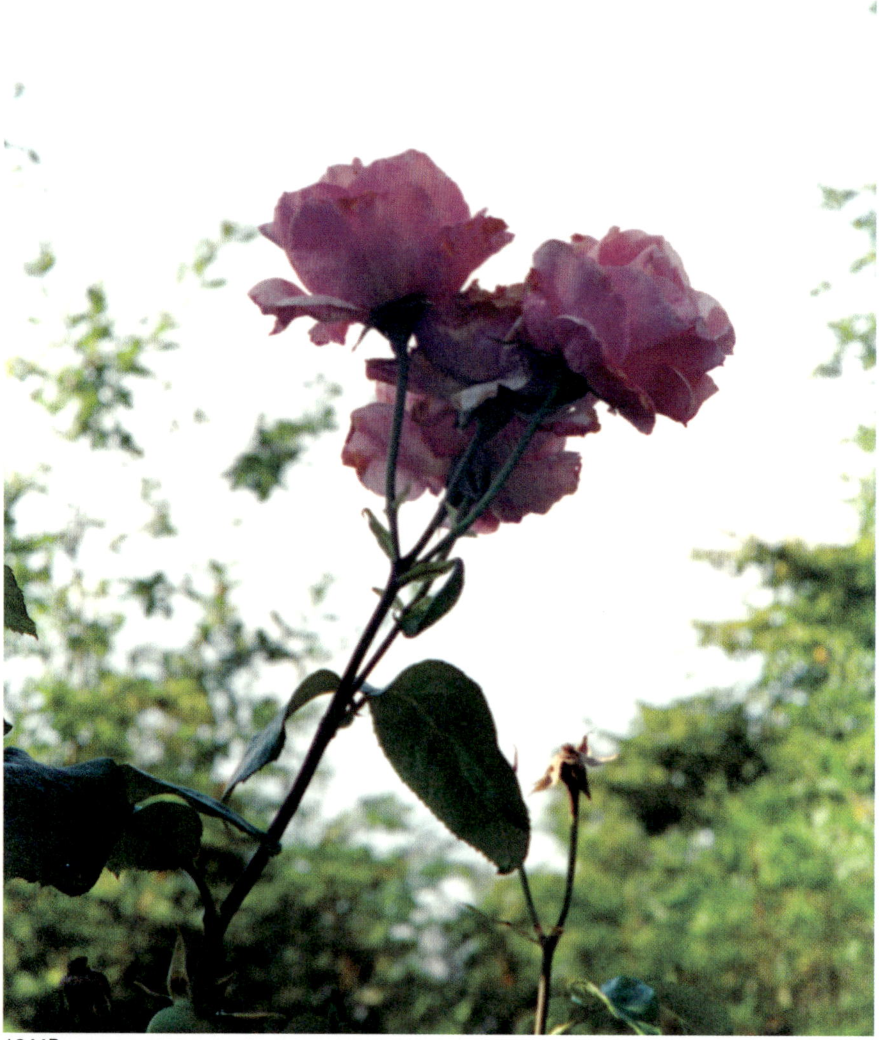
1244P

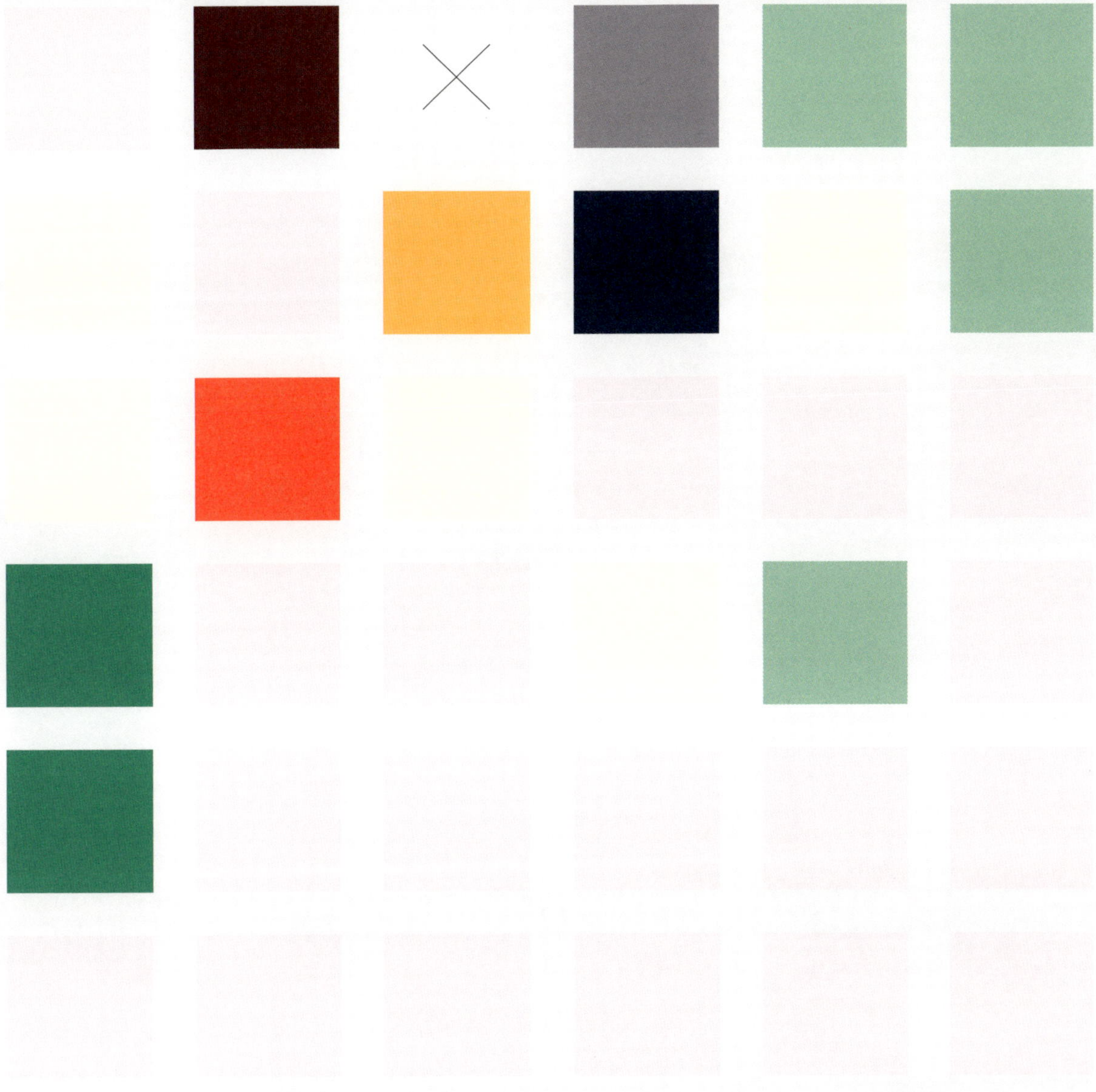

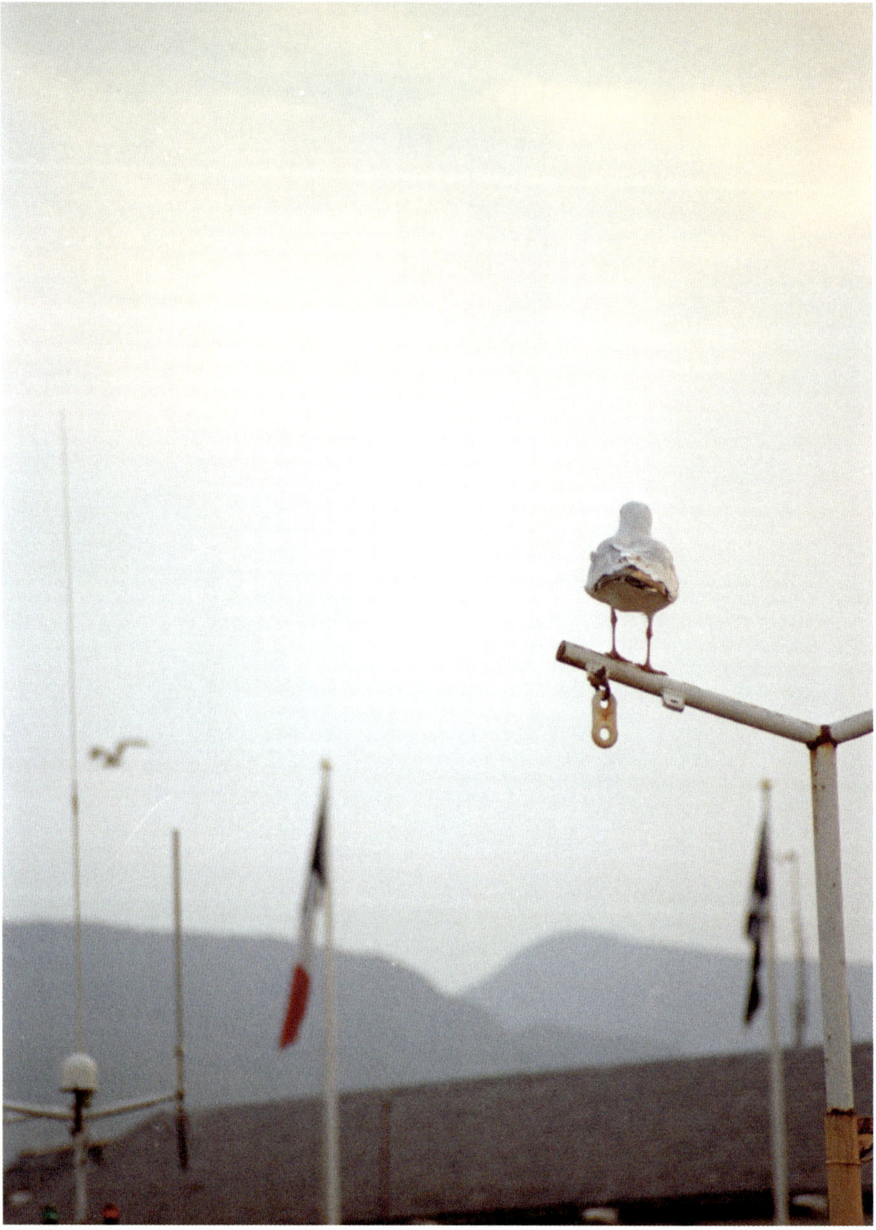
1264P

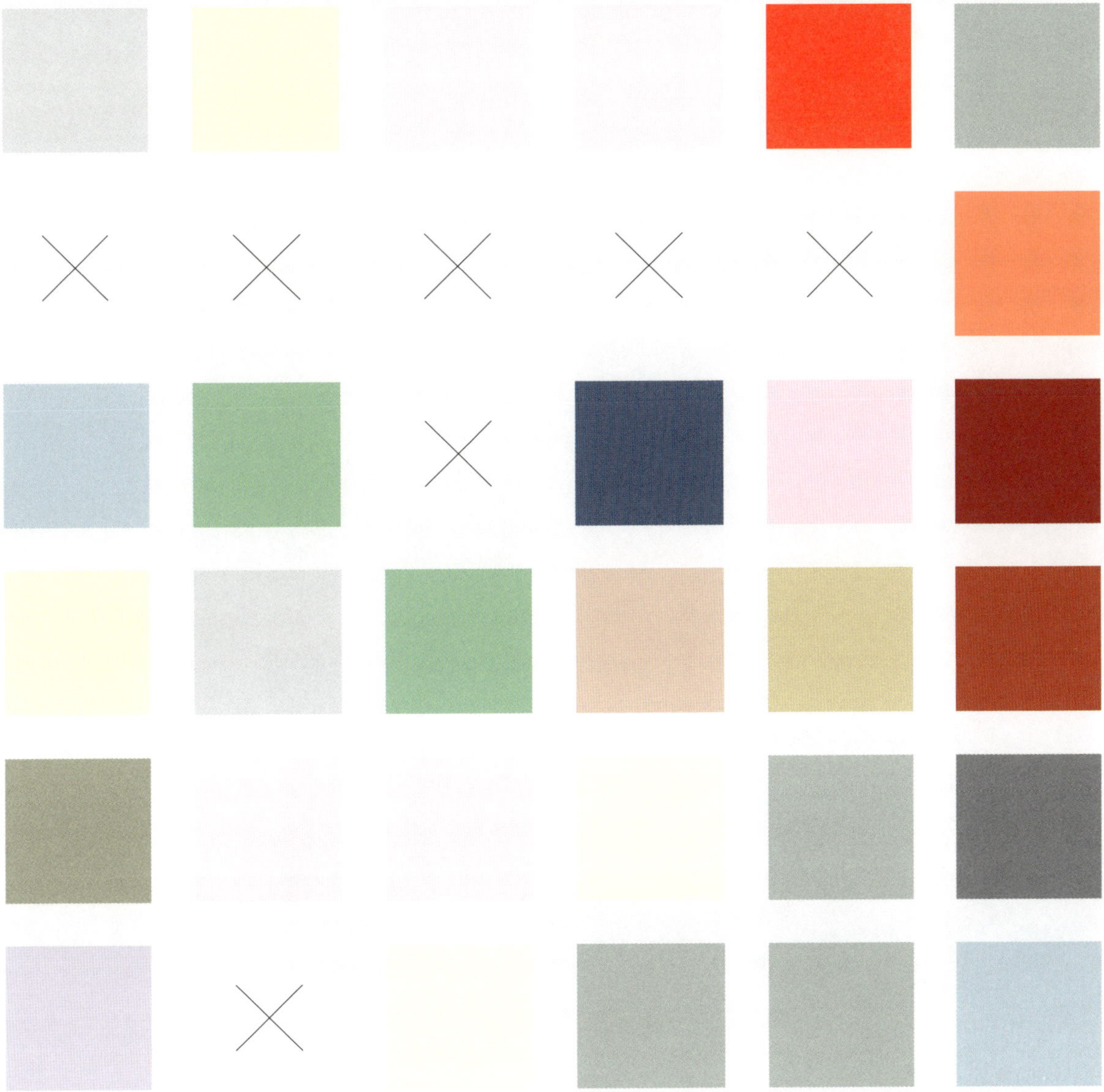

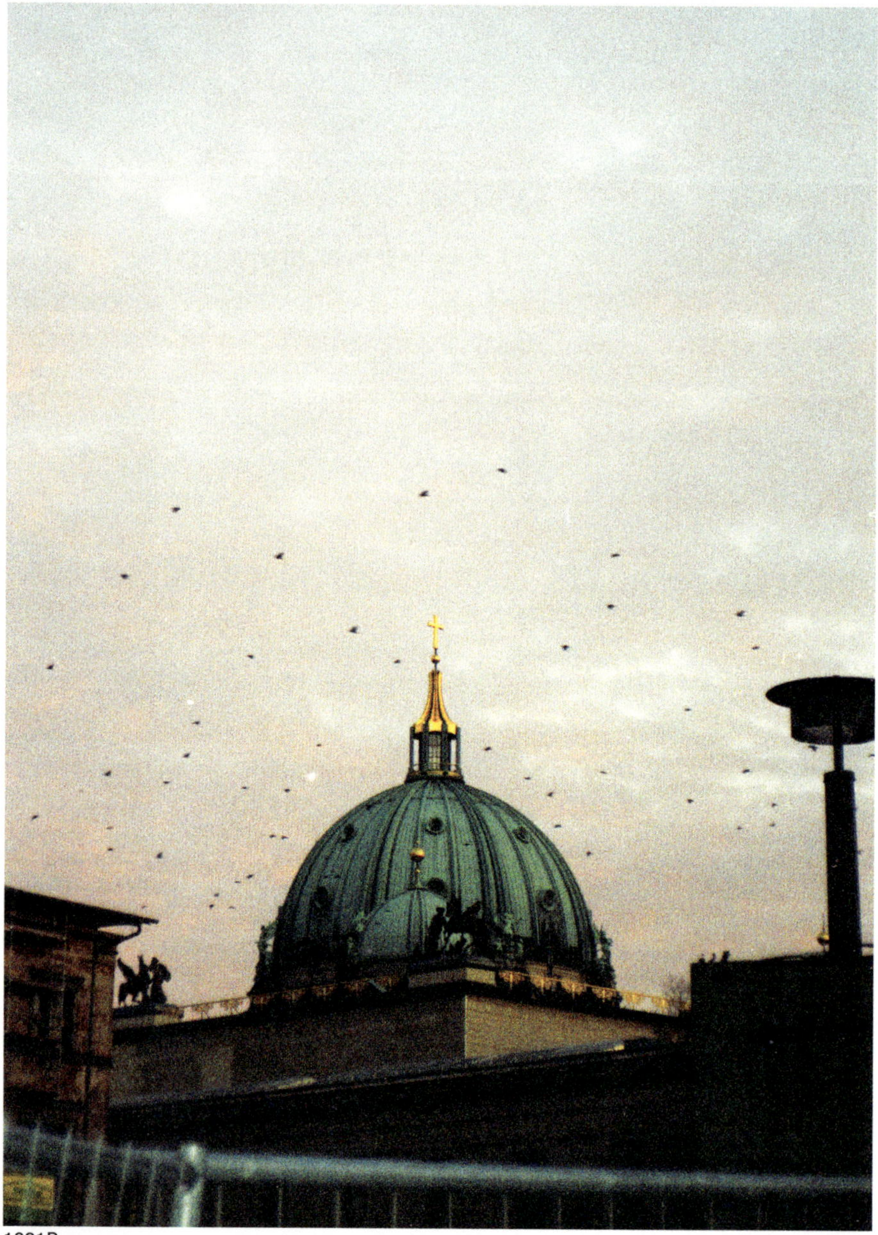

1281P

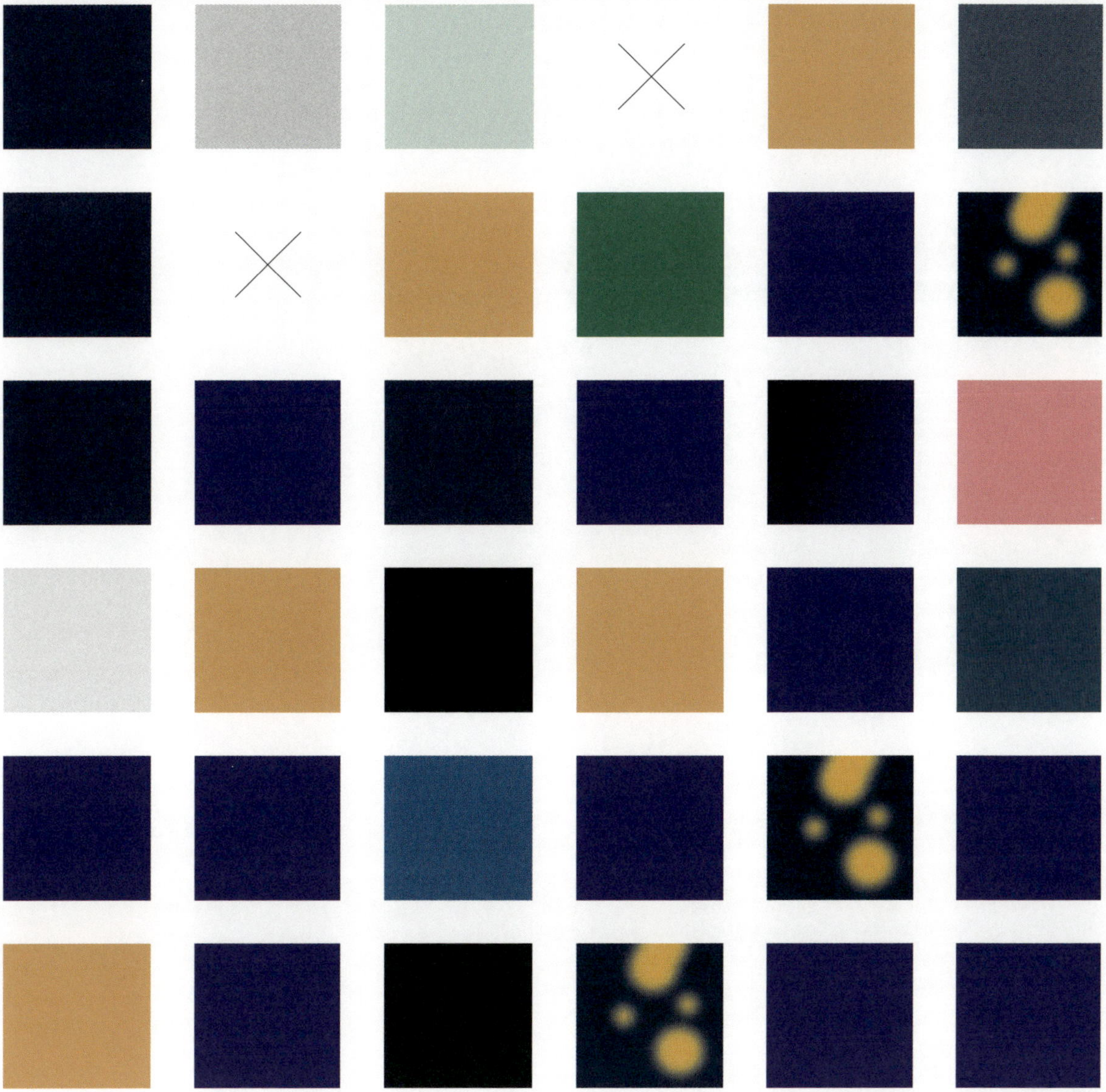

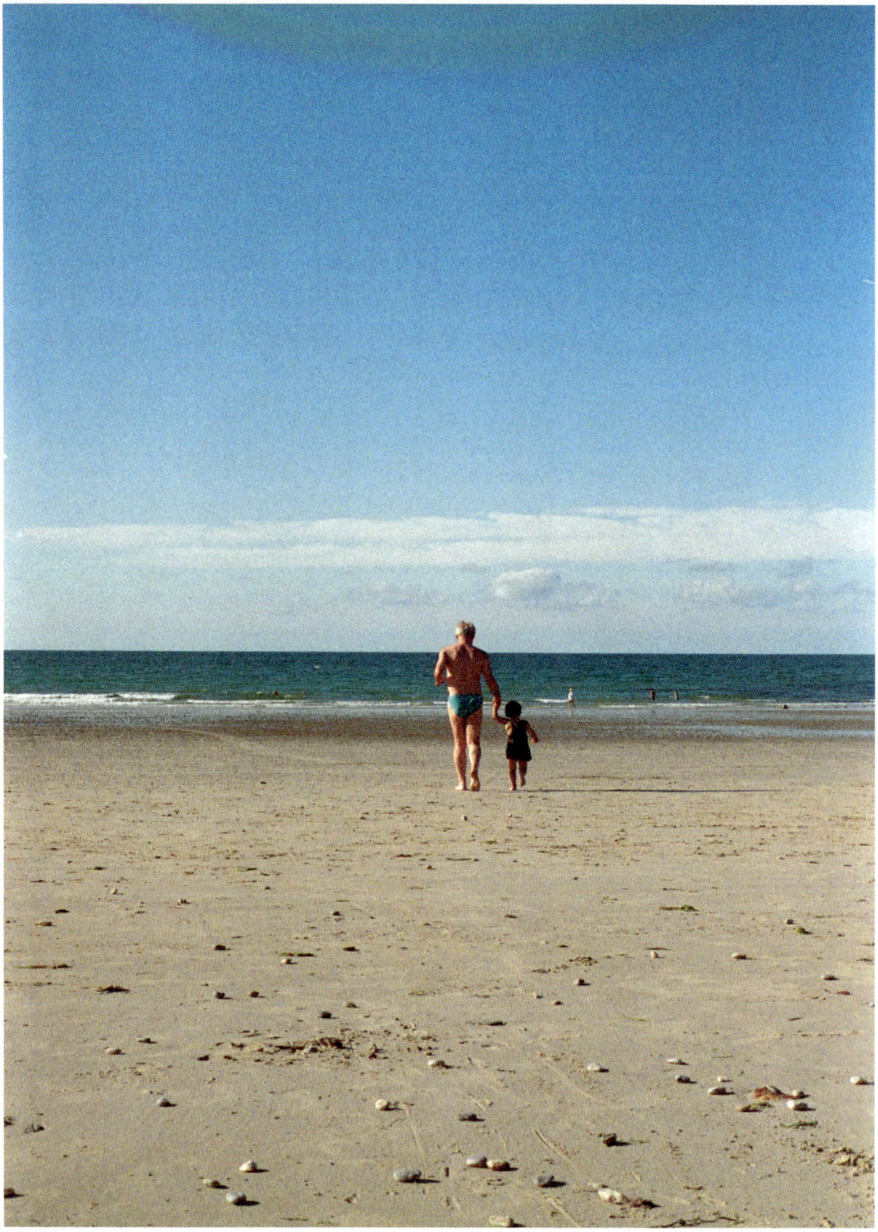
1310P

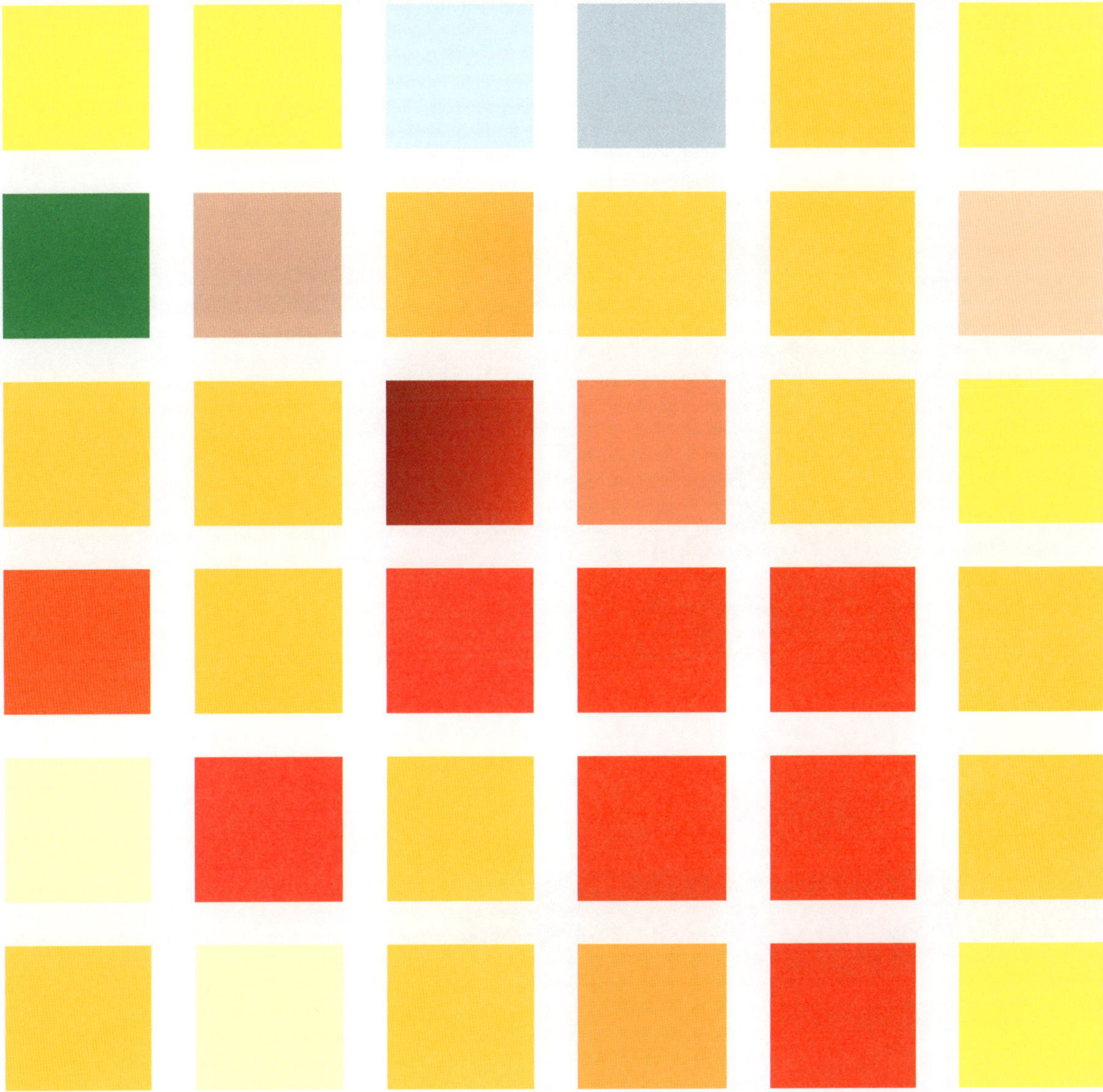

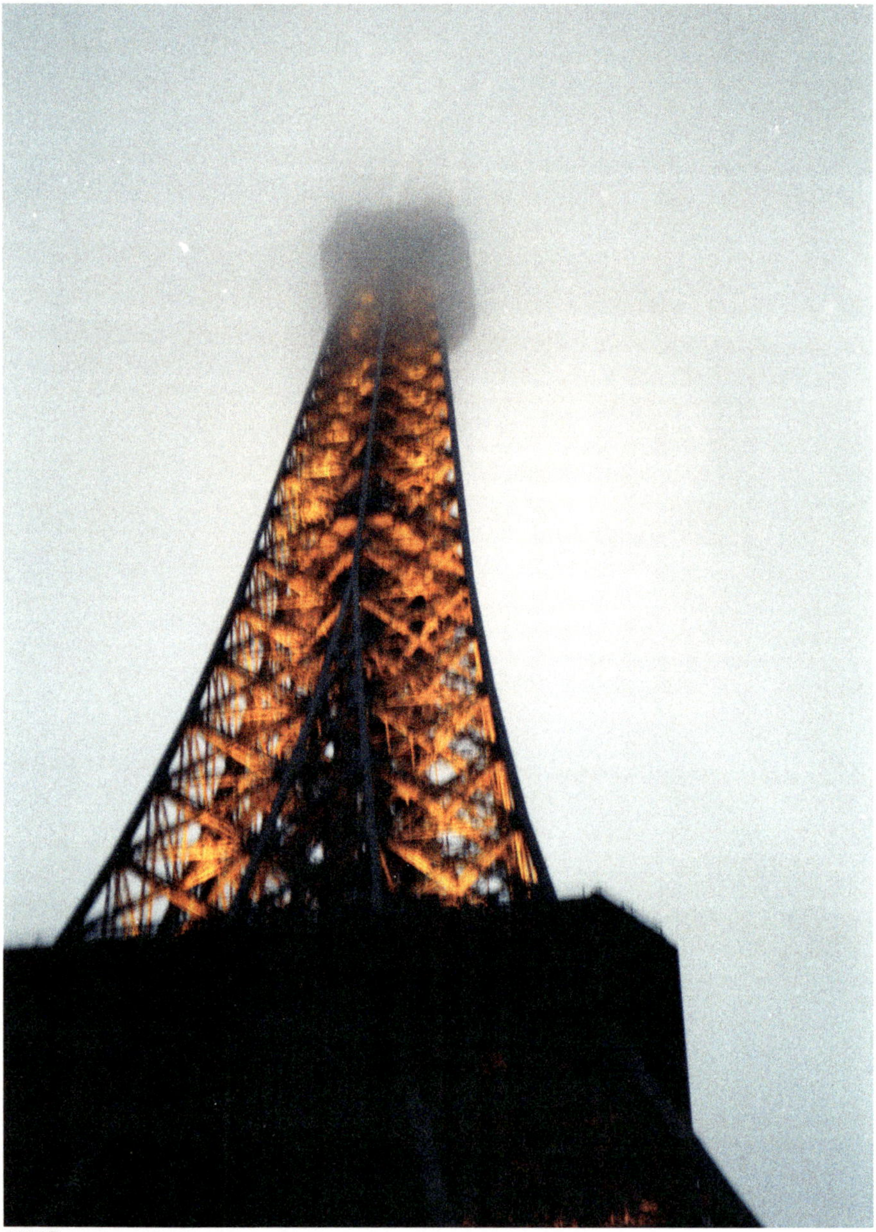

1311P

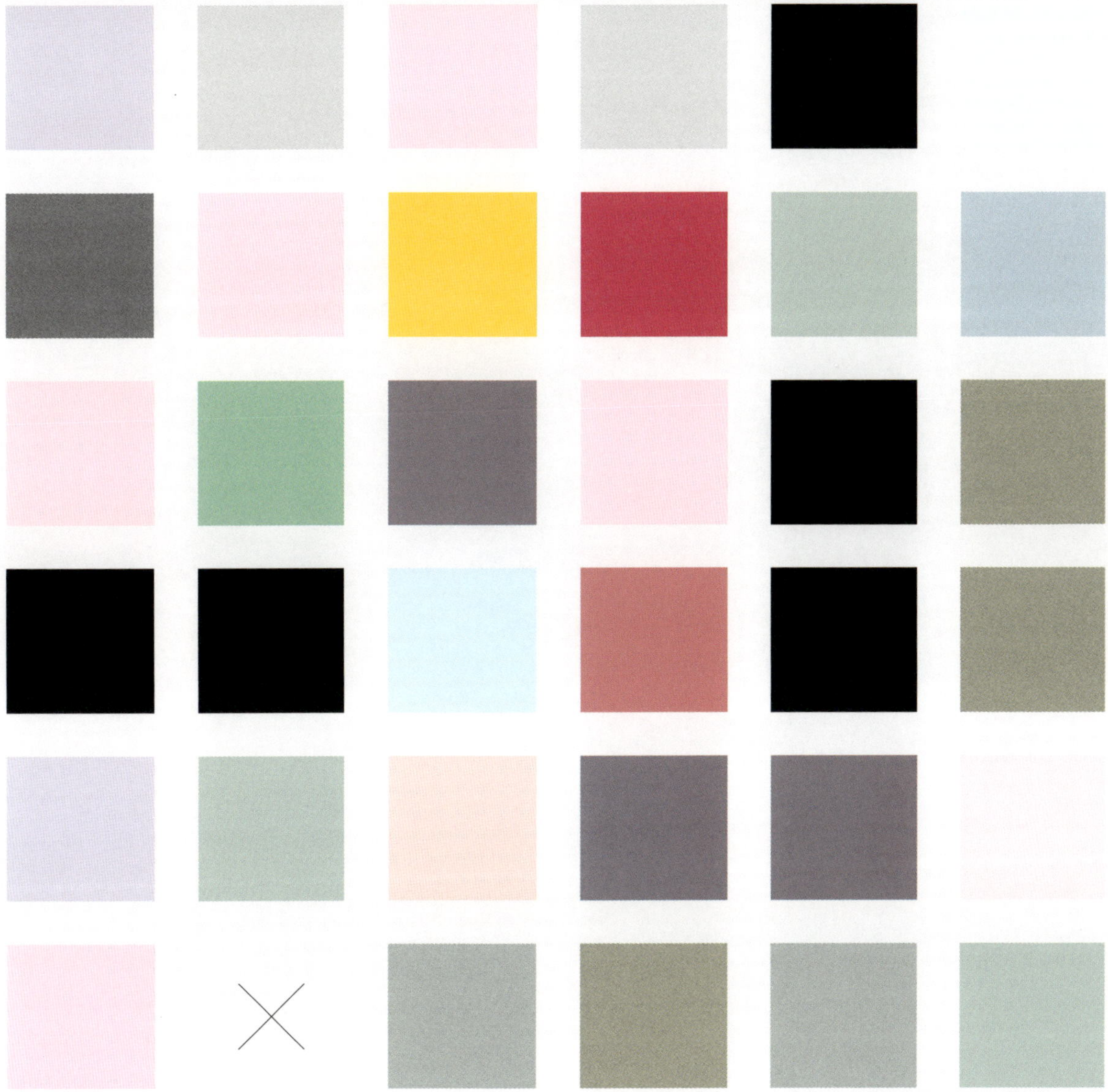

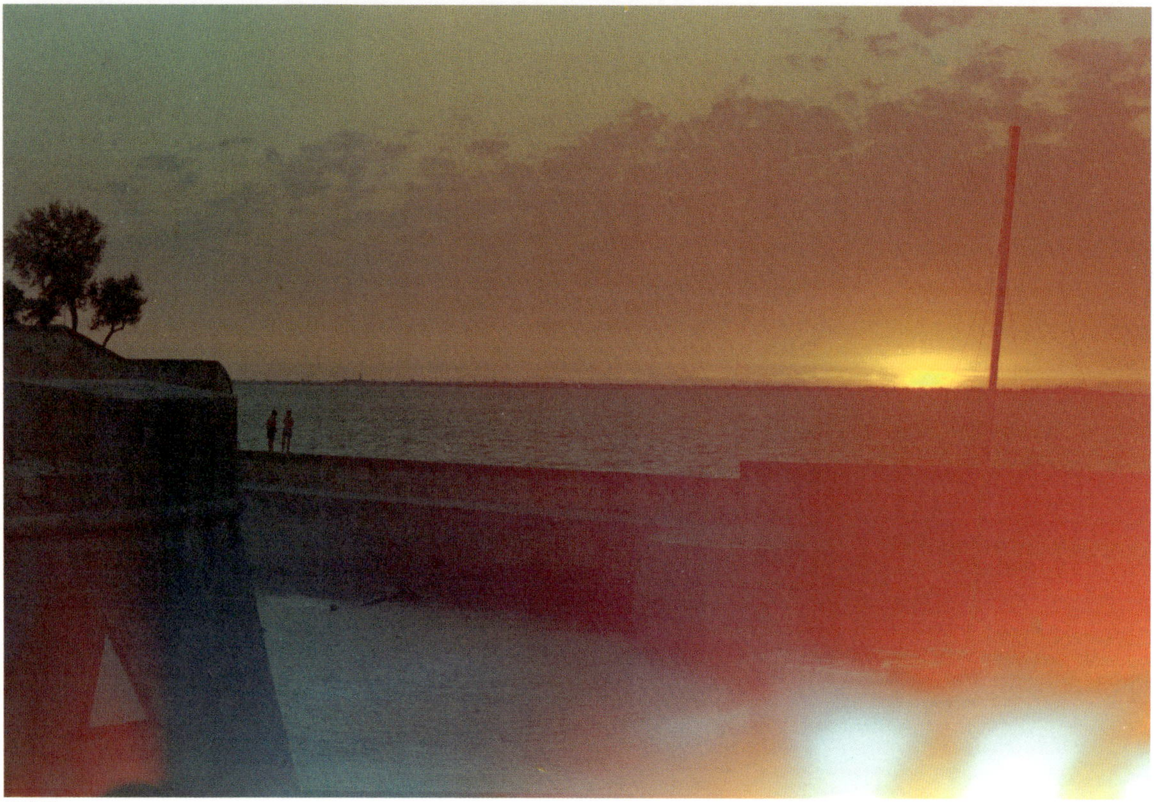

1343P

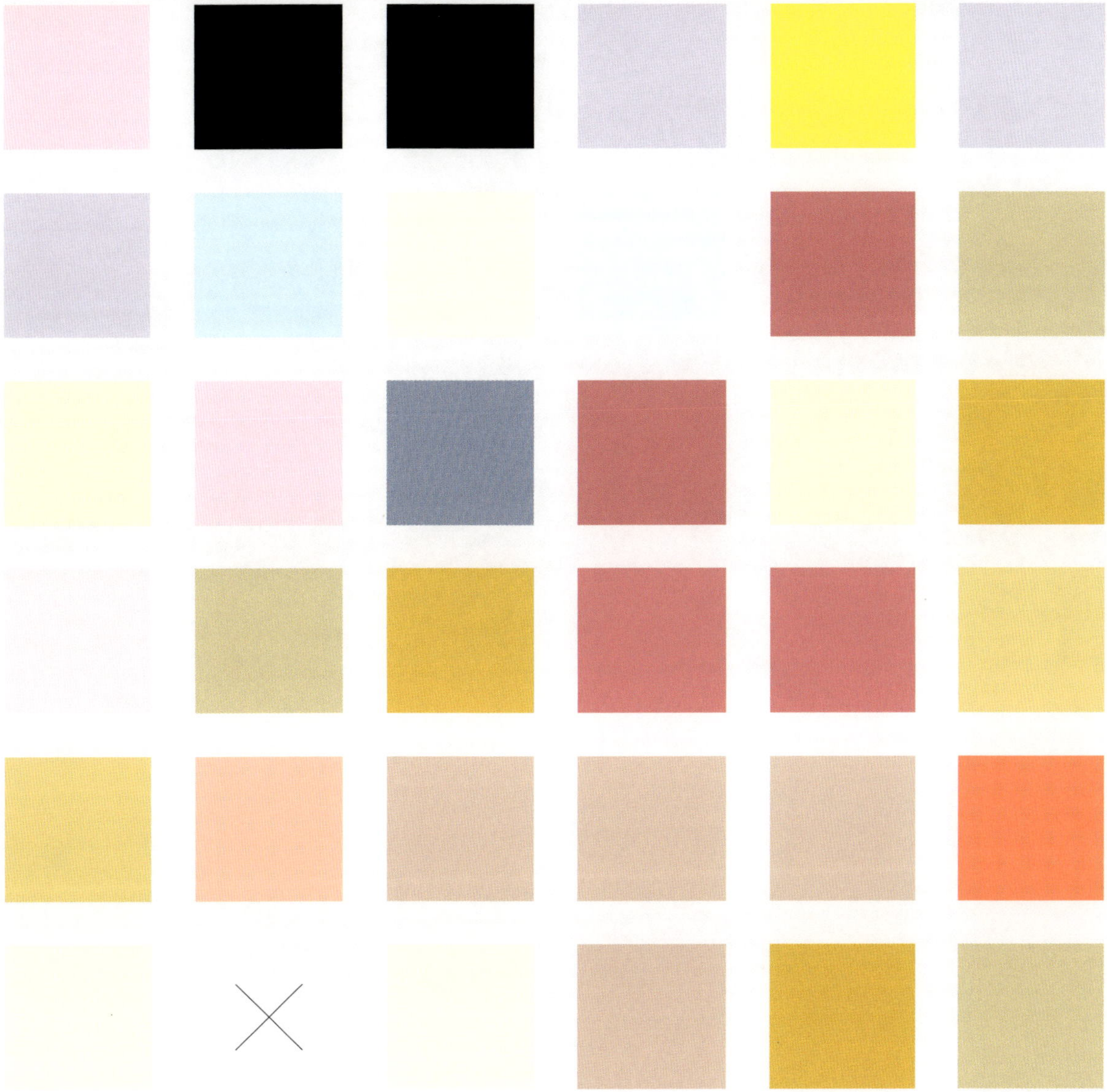

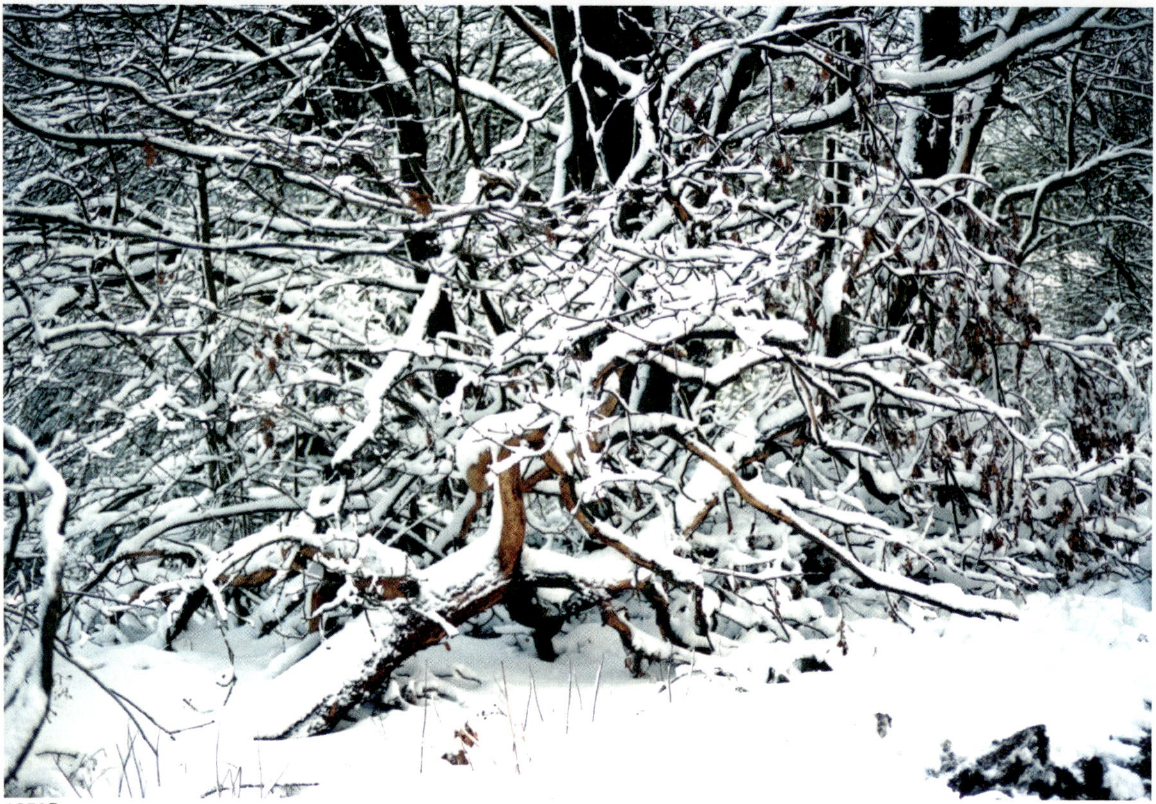
1359P

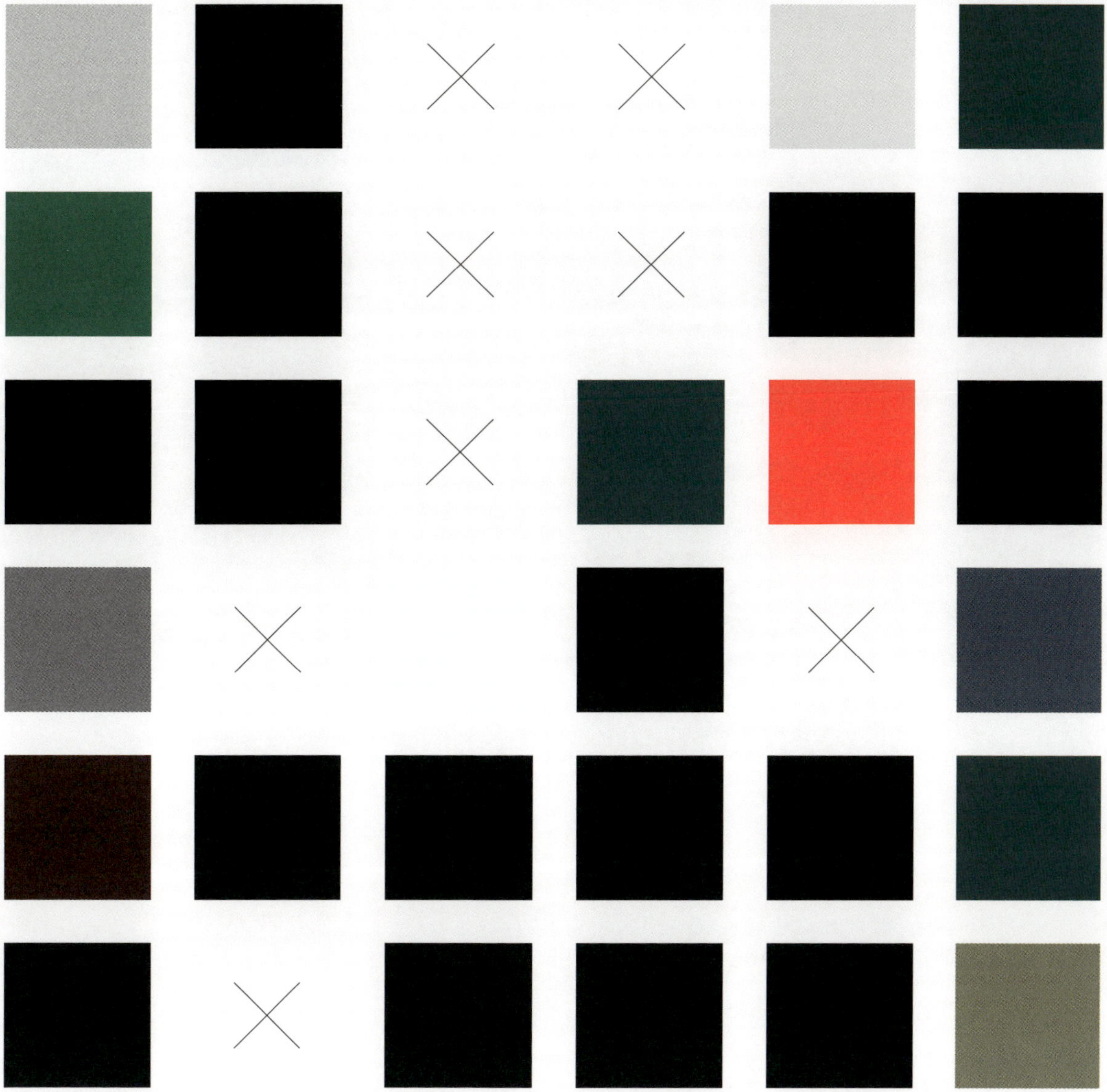

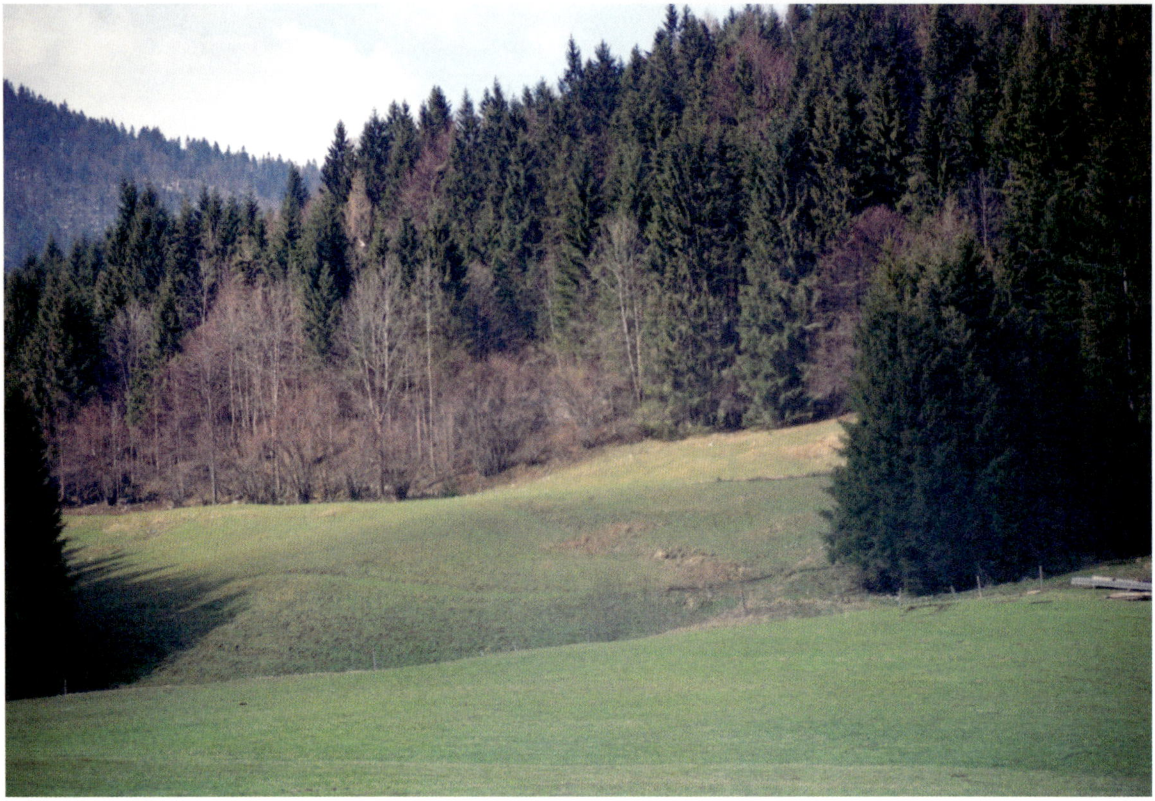
1381P

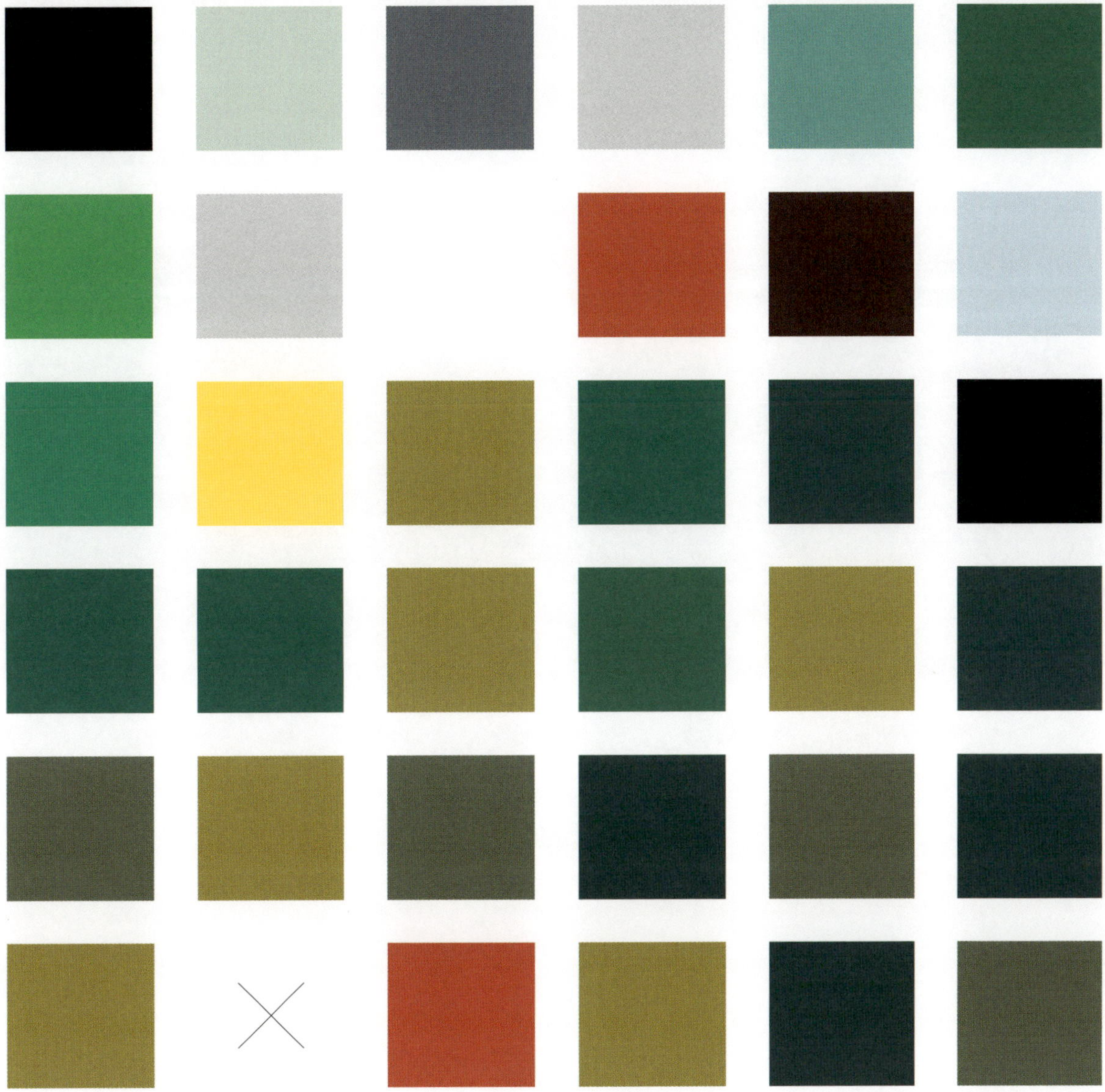

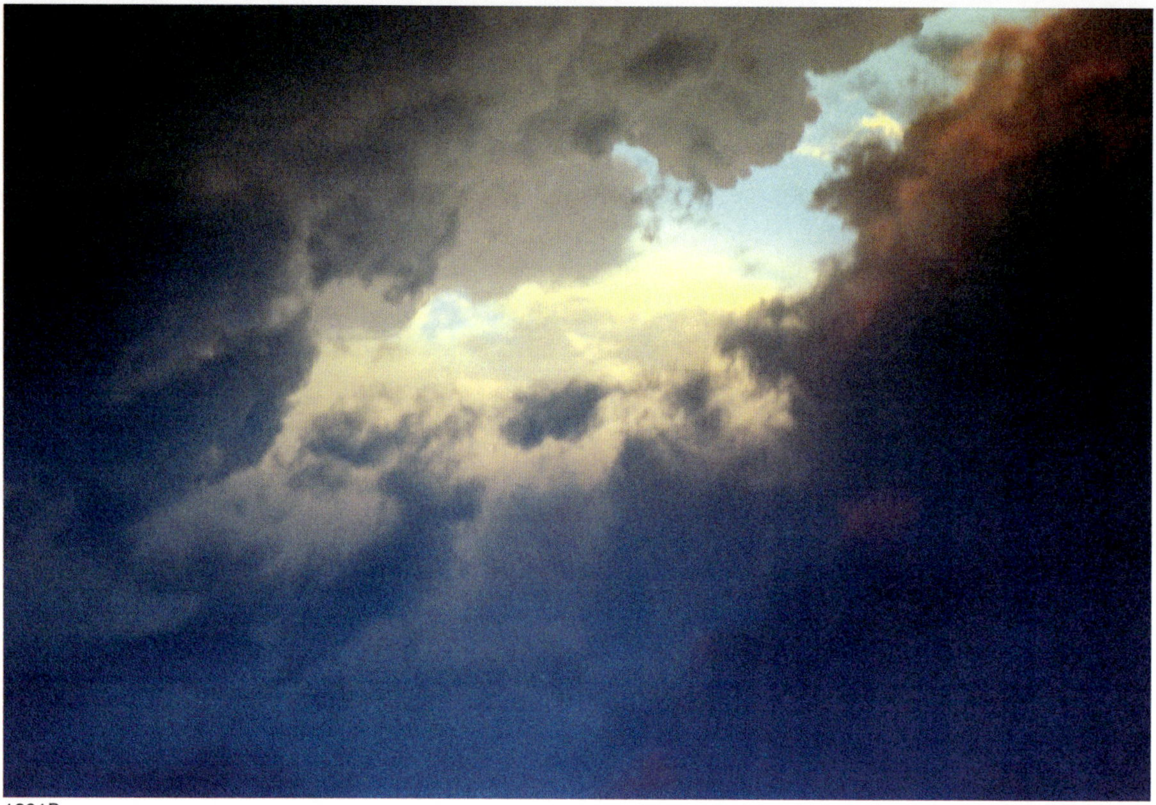

1391P

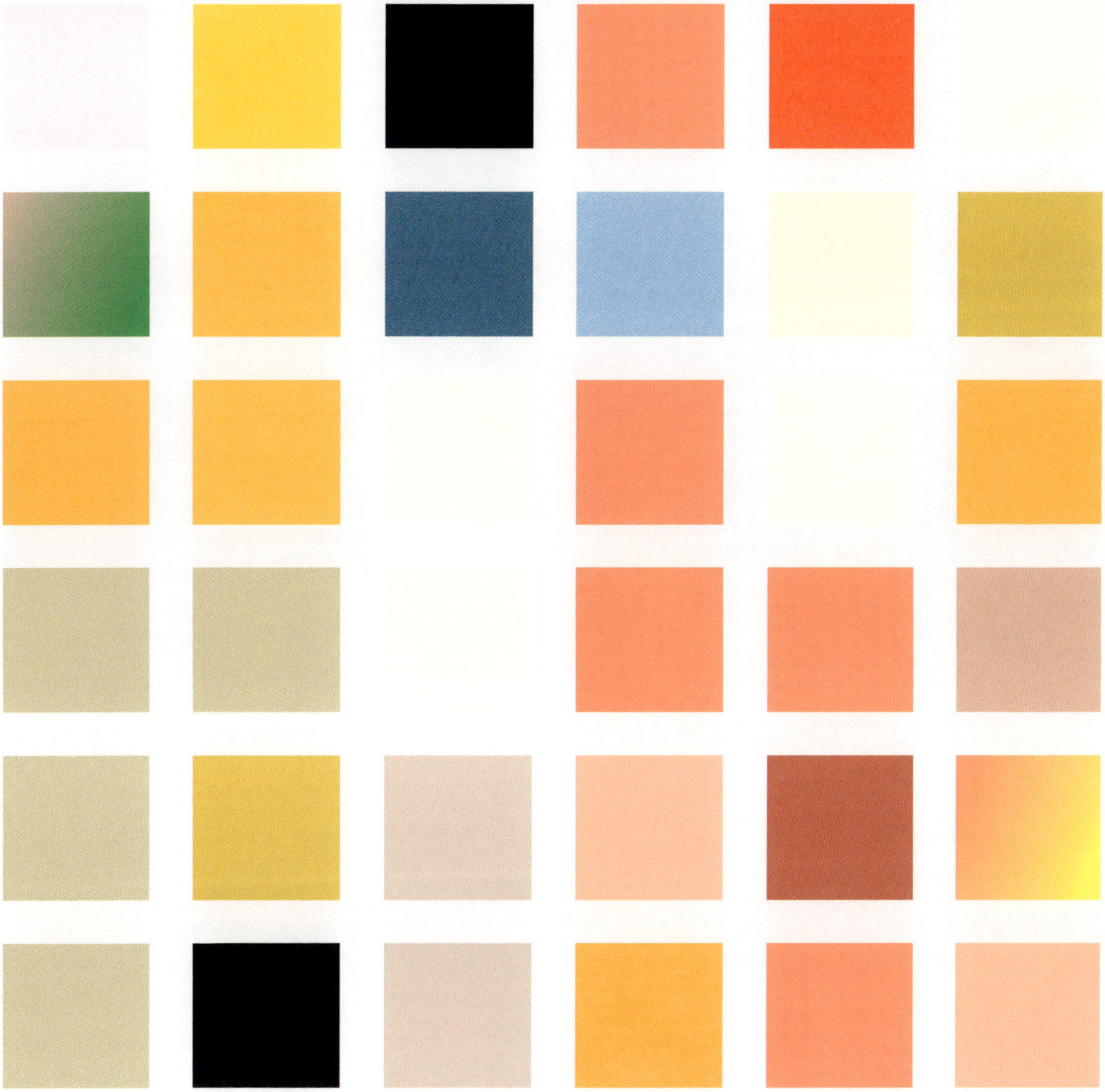

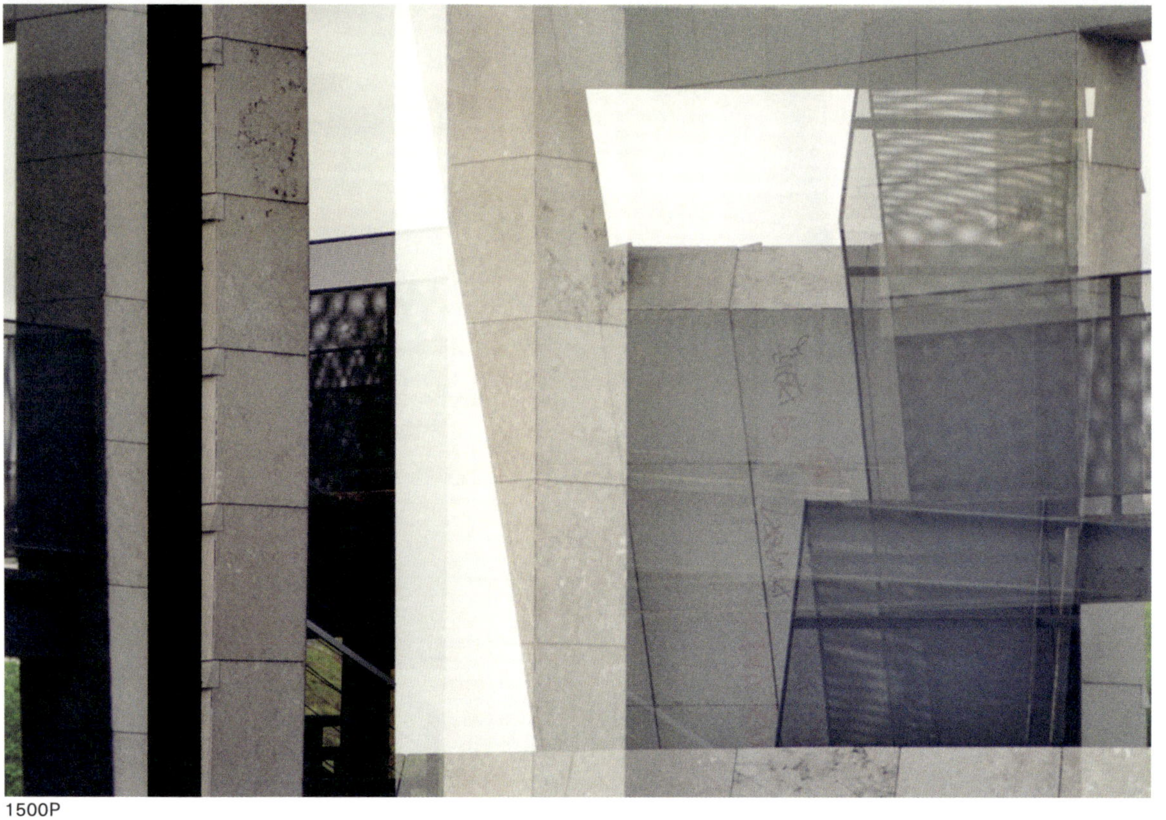

1500P

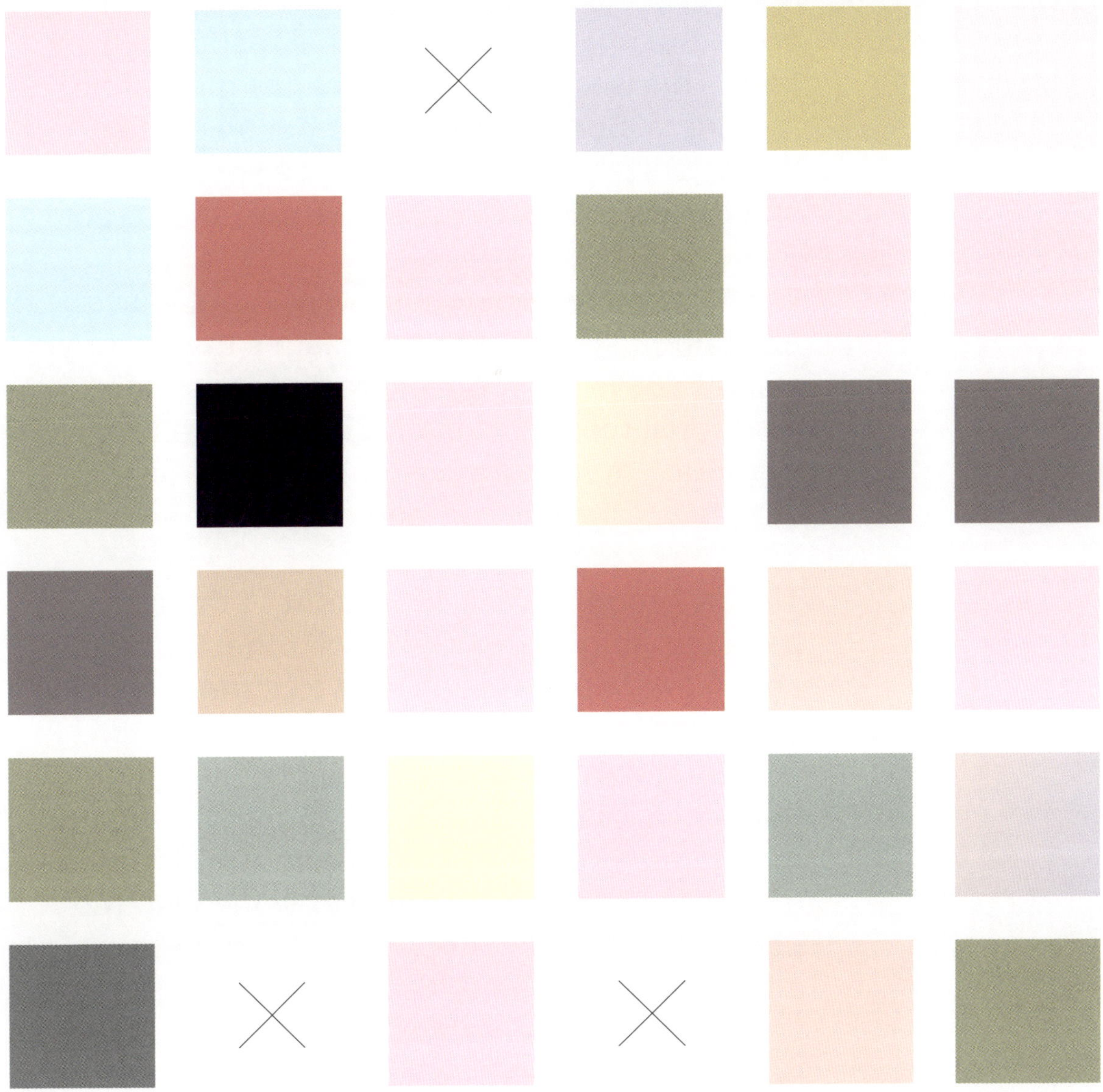

There are changes in emotional response over time. This is valid for some images more than for others. Some changed almost on a daily basis, while others seemed to be very static and coherent in the responses they triggered.

When trying to understand what I see, I have to distinguish between the images that clearly seem to have a strong and rather consistent emotional response, such as the portrait of the young man in the bathtub, and the ones that do not have any strong emotional response.

1

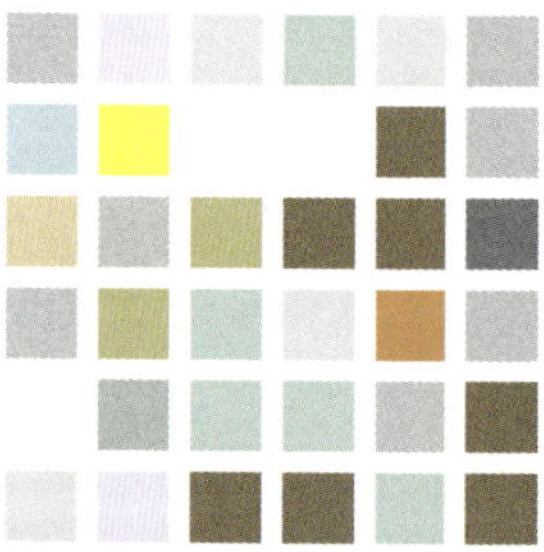

0036P

2

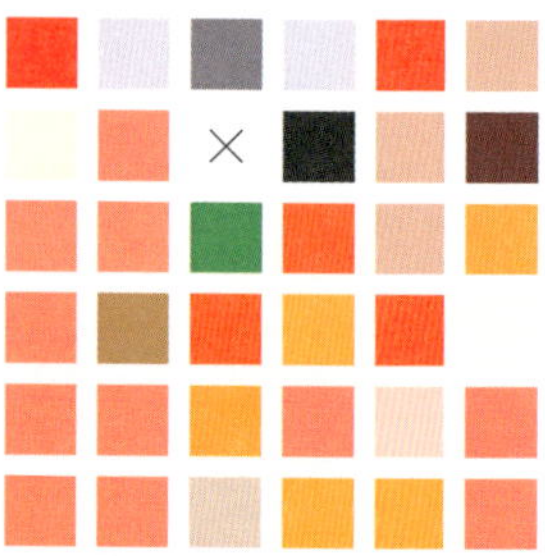

0366P

The bathroom shot is consistent,[1] I think, because we did not know each other very well and did not spend much time together: I quickly realised that he was not to be trusted, which is reflected in the grey shades of uncertainty that I feel. The time we spent together and thus the memory of him are limited, that is what I think makes the emotion so precise. Same with the photograph of my room:[2] I only lived there for a few months. It was a very exhausting time and I remember how soothing that place was to me, how much I enjoyed its light and peace. Both seem to produce a consistent colour range as a response because the memory evoked is precise and emotionally limited. I do not experience a multitude of feelings when I remember that person or that room.

Other images do not seem to trigger any consistent response, such as the photograph of my father, my son and my uncle together in Paris.[3] This might be because that scene unites many

3

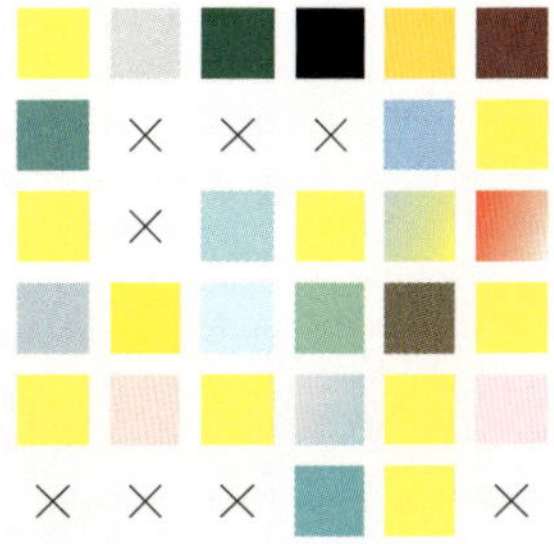

0004P

motifs and therefore triggers many different emotions: it makes me think of Paris (mostly blue-greyish and soft pink shades), but it also shows three individuals that have different colours to me. Furthermore, it symbolises family life, which makes me feel like a child, embedded in that structure—the colours of my childhood come up: warm orangey-yellows, soft tints that expand in space. All of these colours recur in the emotional response. It seems as if depending on my focus on a given day, I had different emotional colour responses: one day, for instance, I saw Paris in the shot, another day my father.

Sometimes the colour associations would grow more consistent during the test period. One example is the young man who is cleaning the floors in a building in Rotterdam:[4] a complete stranger, but every time I saw his image he became a bit more of a personality. A scene that looked more grey in the beginning because it was an unfamiliar place turned slowly into a pulsating red, changing from a warm orange to an almost brownish dark red—the colour the young man slowly took on inside my mind.

4

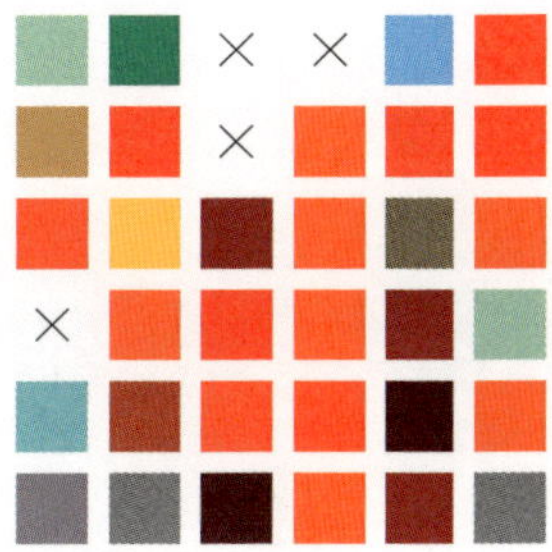

0260P

Many synaesthetes that associate colours with people describe that there will only be colour associations with familiar people or people that arouse strong emotions, such as for example instant sympathy. In this case, we might see a shift from a silvery shade of distance to a more or less constant response of an individual colour attributed to someone because that person on the image became more familiar.

5

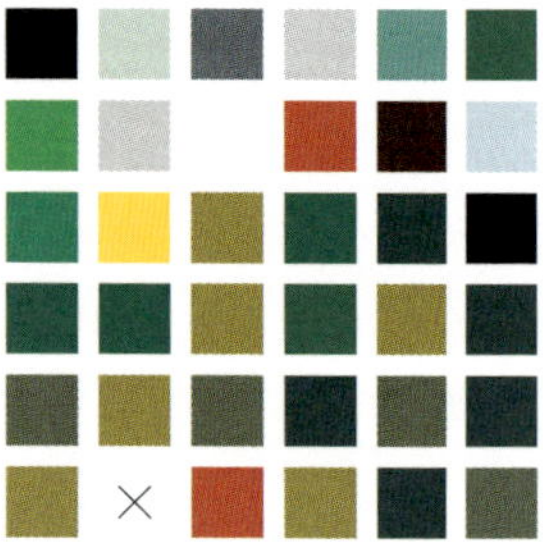

1381P

Another example of that phenomenon is the photo taken outside the window of my parents' car in Bavaria:[5] it brings back memories of my holidays there as a child, and especially of my father who would take us there regularly when my mom needed a bit of time off. The more often I looked at the flushing trees, the more the memory came back, and with it the smell of the car and the smell of my father's leather jacket that he would hang in the back next to me (which always made me feel sick). It was an old Mercedes 300 Diesel; I remember the little dots in the ceiling that would start to dance when I would relax my eyes, which would worsen my nausea. The dull green that became more and more consistent during the test period is the colour of those travels on the road: the green car, the green jacket, the subtle unease, the sickness, the landscape.

Considering the lack of consistency, I think several other aspects should also be mentioned, such as the time of day, the level of concentration (making the colours appear stronger or weaker), the inconsistency in display of the reference colours (different screens and prints), and the emotional state I was in when doing the test.

I checked the emotional response on different times of the day, more or less focused with differing colour ranges to compare them to, and some "heavy" emotions playing in the background due to my challenging private life. The latter seemed to have influenced the testing considerably. Around day 9 and around day

32, there is a considerable decrease in colour associations. Both times, I was going through a situation that I perceived as threatening for my child and me, thus creating major stress. Nine photographs fail to evoke a coloured reaction on day 9, and as many as 14 on day 32. Some images even show a decrease over several days around these events.[6]

6

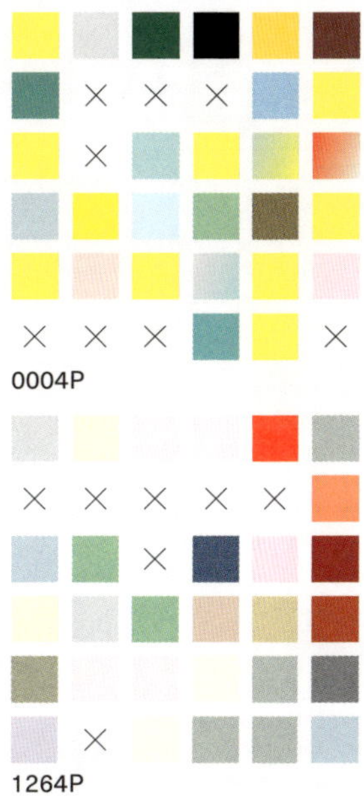

From experience, I know that my lifestyle has a huge impact on the colours. When I am calm, confident, happy, when I eat healthy and do sports, yoga and meditation regularly, the colours are omnipresent and clear. Being overworked, stressed, tired or worried makes them fade; all of these together can make them disappear completely. Or rather: I feel their presence, but as if their layer and mine are separated; I cannot access them. When I was depressed for many years as a young adult, the colours were gone altogether. I only noticed when they started coming back that I had been missing them for a long time.

This is also valid for hypersensitivity, which is increased when I am tired or stressed. My brain gets easily irritated by signals other people do not even seem to notice: strong smells, strong lights, background noises.

The inconsistency expands not only into quality (the change of a specific colour association), but also in quantity (the speed with which the colour perception was perceived, the intensity of the colour perception, the extent of it as perceived

7

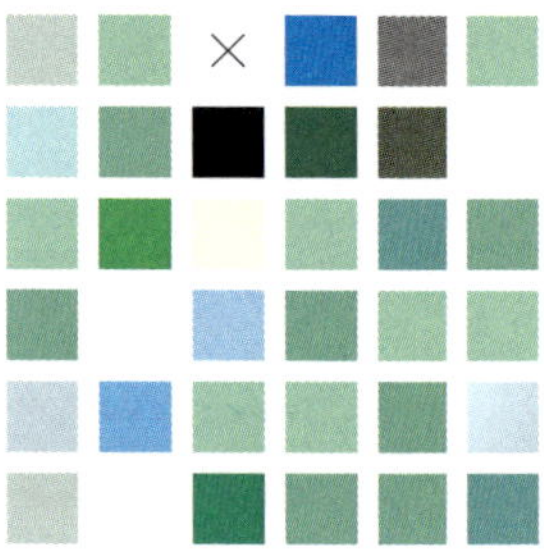

1228P

8

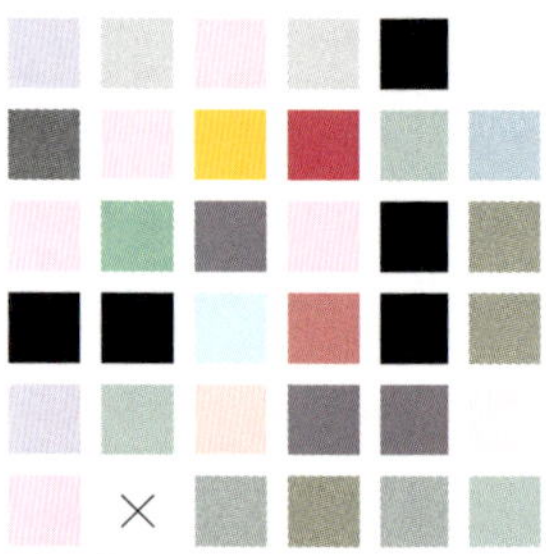

1311P

9

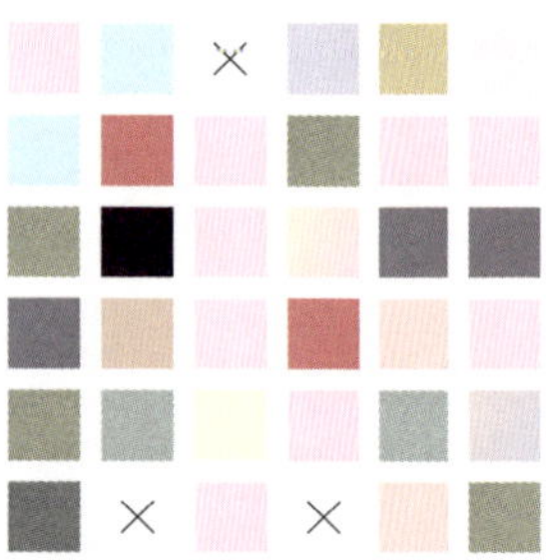

1500P

in space). This can, if I take earlier experiences into consideration, be explained by the strength of the emotion. An insignificant feeling will be barely perceived in colour, be quite weak, and I will feel it only inside of my mind. An overwhelming emotion, however, will appear instantly and I will perceive it as a strong and pulsating colour, expanding outside the boundaries of my body (I feel it occupies the space around me as well). In this chapter, I only researched quality consistency, not quantity. This might be looked at in another study to draw more conclusions on the origins of the colour associations.

The inconsistency can also be explained when we consider the concept of ideasthesia, which states that the trigger does not need to be the trigger itself (e.g. the letter 'E' or a certain person) but the concept of it. This recurs in the photographs that symbolise a certain place such as Prague or Paris—the colour I feel could well be that of the place and not that of the subject, such as the birds on the Moldova having the green/grey of Prague,[7] and the Eiffel Tower[8] as well as the Parc André Citroën[9] having the hidden grey and the pastel rose of Paris. If I take that into consideration, it might be that on some days the concept of the place is provoking the emotional colour, not what I actually see, and on other ones the emotion might be based on the literal content.

In contrast to what most research on synaesthesia considers a 'golden rule' in order to define it, synaesthesia can indeed appear as a non-consistent phenomenon. When I started reading up on it, I thought I could not have synaesthesia, since my emotion colours often are not consistent. More recent research, however, points out that the defining criteria for synaesthesia might have to be readjusted, since not all forms appear to be consistent, although they do meet all other defining criteria for synaesthesia. This could also lead to a much higher prevalence of synaesthesia than previously observed.

I encountered several problems during the research. First of all, the colour range: how difficult it is to capture my emotions on paper.

For this chapter, I worked with two different colour ranges: one printed out on paper, the other one in a pdf file that I would check on my laptop or on my phone. I quickly saw that the colours were not corresponding, and were inconsistent as well, depending on the angle at which I looked at them, the screen brightness and so on. I would, for example, be unable to find that minty tone from the day before, and the colour range often appeared very limited to me: in approximately half of the cases, I picked a colour from the colour range that came closest to the one I felt, but it would not quite be the right shade of my emotion. This was frustrating at times. Other synaesthetes might recognise the problem when asked to pick a colour from a limited colour range and being unable to find the exact shade. It might feel like cheating or not telling the whole truth, because the vocabulary is missing. It literally felt like I was betraying my reality by giving it an inaccurate description.

Furthermore, the problem of two-dimensionality made it hard at times to find an equivalent in the colour range: my emotions are not little squares in a neat row. Instead, they are really big inside of me, they float, they are more or less present in space as three-dimensional shapes.

So the results depended not only on my own emotional responses, but also on many external factors that I could hardly influence. This test might have to be repeated in a static environment, checking the photographs at the same time every day, comparing the emotions with a range of colours printed on one sheet of paper lit by a consistent source of light, with no distractions and, most importantly, with a consistent inner emotional landscape to guarantee the most neutral approach to the images. This is of course utopian, but it might be interesting to see if the results would be different if I were to approximate those ideal research conditions more closely.

In conclusion, I tend to maintain a careful approach: I did indeed expect the colours to be more consistent, and at the same time this very inconsistency made it clear to me how deeply my perception is influenced by my inner emotional landscape. We all like to think that we know who we are and that we base our decisions on what we perceive as a static outer reality. This research reminds me how subjective my reality is and how much it is shaped by how I feel in a situation. Almost as if I were to project my inner life onto the outside and still experience it as something detached from me, as some ancient Greek philosophers thought that vision happens when the light from our eyes touches the things around us.

It is clear that how a picture looks to me on a certain day cannot be separated from my condition in that very moment, and that what I perceive is always as much a product of my inner landscape as of my outer reality. I want to investigate this inner landscape further: where do the emotions actually come from? Is there a way to find out what precisely evokes the synaesthetic colour?

Chapter 5
Origins

To find out what happens in my mind when I look at a photograph, I wrote down detailed descriptions of my personal reactions to certain images that I used for this research. I hope to make my thoughts and associations more accessible by describing both the photographic impression and the emotional colour.

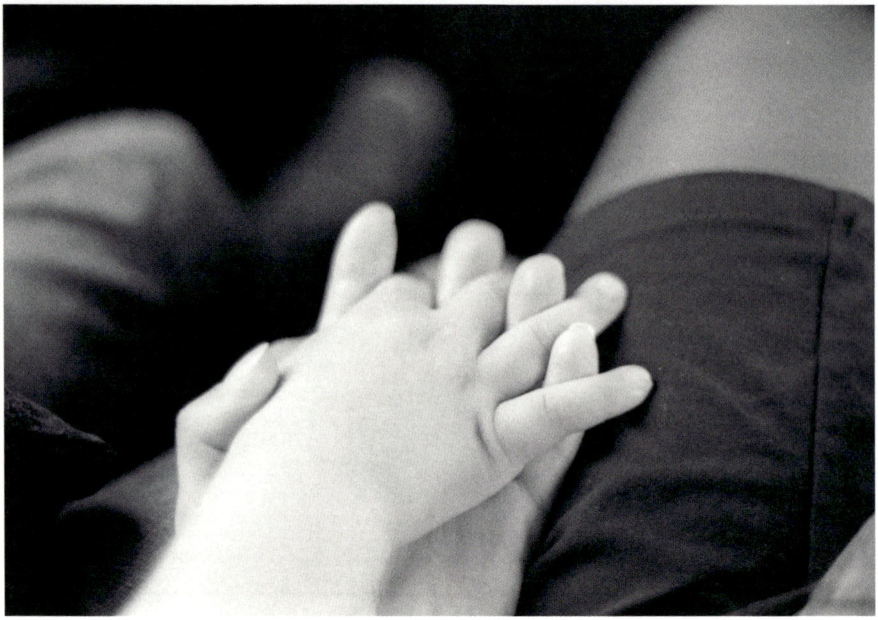

0058P

Germany (on the road), 2015

This picture is dear to me. It is the hand of my sleeping son in mine, on a trip home from a family reunion in Germany. It was just a few days after Sam and I became a couple, and I felt I could conquer the world. The endless struggle to protect my child from an abusive father felt so different in those weeks, as if my overwhelming happiness could carry me through anything. I feel a warm white, a comforting colour that tells of the happiness of that moment. It is a subtle shade, and at the same time a powerful one: a delicate colour with a strong meaning. When I am all happy, and that rarely occurs in its full extent, this is the colour that fills me up completely. It makes me feel stronger and confident through its warm and caring pulsation. Its presence is protection.

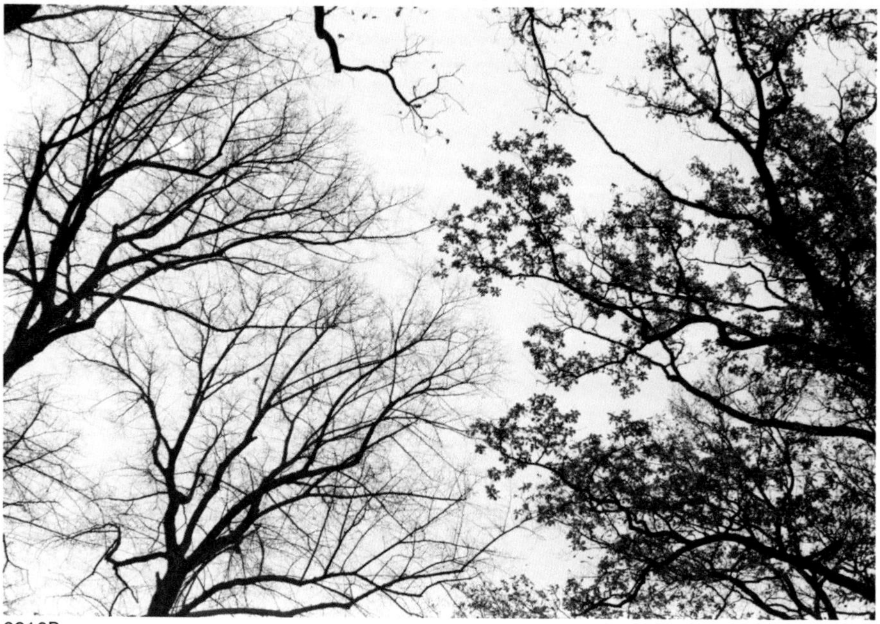

0816P

Breda, 2014

The trees around the St Joost Art Academy, end of 2014. A delicate grey, fading into a light lavender tone. I remember that time of my life as being easy and hard at the same time: in school I had started to develop an autonomy that did me much good, at the same time setting my standards higher and not knowing how to navigate that fragile artist identity. I remember those days as being an in-between-time, Zwischenzeit.

I knew that something ended — my uncertainty, my endless desperation at how little I felt I succeeded as a photographer — and that the confidence that emerged belonged to a new episode of my life in which I unexpectedly got the chance to do what I love and might just be all right.

I experience those periods of transition as a hollow shape, as if my presence needs to adjust to another structure in space that I have not entered yet. I feel I am there, but have not been handed down the codes of conduct for that space. I find myself tiptoeing around until I have found my position within these new dynamics.
The grey tone is not a cold one of distance and non-acceptance; it is a warm one that speaks of connection and has a notion of home. The places I feel I am allowed to enter because I fought for it always seem to me to be more accepting and also more precise, in a way, as if I would own them as inherently mine. I feel I fit better into them because there is a purpose in these situations, one that resonates with me. These situations remind me of how much I can influence the course of things. That certainty gives me comfort. I remember shooting those trees with a light-heartedness that surprised me, because it was new to me to feel like this at the academy.

I remember feeling I belonged there for the first time in over two years. This confidence carried me through the last months. I still feel thankful for it.

0564P

Bonn, 2011

My sleeping room in my first apartment in Bonn. My first place of my very own, where I arrived after a long period of uncertainty. I had been through all kinds of emotional and physical abuse and felt safe for the first time in a long while. It was a shelter for my child and me. I do not remember taking this photograph, but I know what it shows: shapes and tints that speak to me.

Why is the emotional reaction to this picture strong—a thick off-white?

Maybe because this photograph shows what kept me going through these years of early motherhood, during which art felt more away than ever. My time was swallowed up by taking care of a little boy and fighting for his safety. Feeling all odds against me, I felt my creativity was basically dead. All my energy went into living in survival mode, I doubted I would ever go back to photography.

Blanc cassé: off-white is one of my most cherished colours. It is the colour of trust and confidence. I think I feel it in those insignificant shapes because of what they communicate to me: how I never stopped photographing, how I kept a connection with it, no matter how subtle, through difficult times. It felt reassuring that even when I was feeling really depressed and helpless, I would be able to make something delicately personal. Something that had no meaning for anybody but me, something through which I was speaking to myself.

It is true that art needs strong emotions, and that strong emotions are how art should communicate to us. I still have major doubts about my path as an artist, but when I see images like this one, where a whole story pours out in front of me the moment I look at it, I can identify an authentic need of mine that has always been there: a strong tendency to hold on to moments, and to do so in my own language.

Throughout my photographs, I can see the need to frame reality exactly the way it presented itself to me, in that very moment. I never thought about showing any of these images to anybody else, they were always nothing more but my diary. That urgency, as photographer Mark Cohen would put it, is much more relevant to me than creating something that is universally meaningful. An artist should, I believe, first and foremost create because it feels like the right thing to do in that very moment. If through that authentic and honest urge to create we can touch other people, we might call ourselves artists.

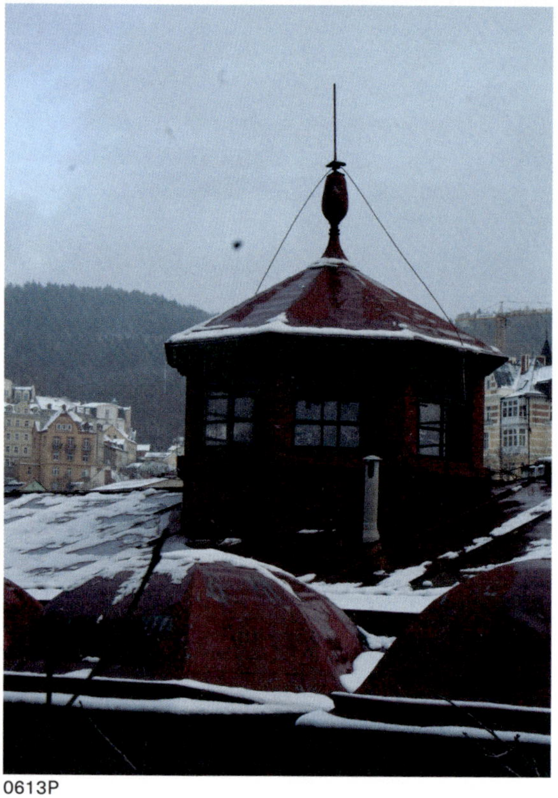

0613P

Karlovy Váry, 2009

This picture is a strange one. I was walking through Karlovy Váry, where I went with my father on a cold January day. I was five months pregnant and feeling very sick from the coach journey. 18 years after the fall of the Iron Curtain, the traces the Soviet regime left on that country were still visible, the most obvious being the condition of the buildings. After being part of a famous health resort around the turn of the 19th century, the big hotels and villas were in an abominable state, despite the money of the Russian nouveaux-riches that started flooding in. There was something so desperate and depressed about that place—I remember vividly how walking through the streets made me feel all void and sad. It was as if the energy of that place would not make sense. I could see all the glory and waste that took place here, but the actual scenery was far from appealing.

This building especially aroused my interest, but for a completely different reason. I had a recurring nightmare as a child in which a scary entity resided in a building very closely resembling this one. All I knew was that it was pure evil.

Children's nightmares are absolutely horrible; at the beginning of our lives, we have not begun dividing imagination from so-called reality, which means that both the fantasy and the horror are real. This impersonated evil which I usually called 'the man under the bed' would come back in many a night, and I knew I had to wake up before he got me.

In rare cases, I would find a building somewhere, either outside or even within our own apartment, that looked similar to the one in the picture. Something drew me close to it, but I knew that entering would mean going through something immeasurably evil. Even after so many years, I cannot find any better words to describe this absolute horror.

I took this picture with a morbid fascination, maybe a comforting sensation that it was up to me to hold on to it or not, to decide whether freezing it would make it live on, continue to exist within my photographs, within my life. I felt like capturing could banish the evil, because I would ultimately decide about its existence within my life. Is that not a hidden motivation leading us to take a picture: to give a scene space within our own narrative, and feeling the comforting power we exert through that gesture?

I think that the fresh grey colour that I feel is more related to that latter memory of mine than to how I felt when entering that situation. Fresh grey, to me, is the colour of despair, of evil, of complete disconnection. It literally makes me feel sick. It is a colour that does not appear often.

I might have already felt like that through the fact that the place depressed me, that I did not feel well because of the trip, but I think this photograph puts me into the mindset of the nightmare more than of the actual place where I took it.

Even now, writing these lines, I feel sick to my stomach. The evil man is still a vivid memory after all that time.

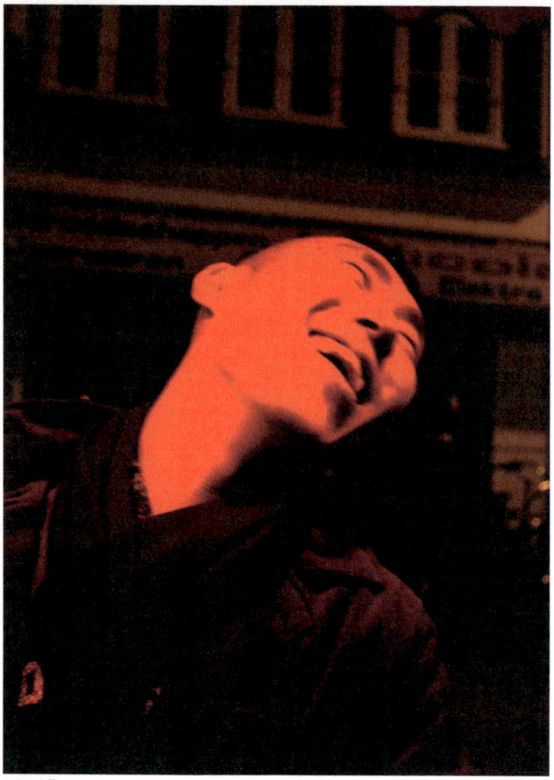

1210P

Amsterdam, 2008

Lee came over from Japan, where we had met one and a half years earlier. We lived together very briefly in Kyoto. He came to the Netherlands to perform with his band; I came over from Germany to see him and we spent a few days in Amsterdam strolling through the city.

He brought his enthusiasm and happiness, and I was glad to meet him at a difficult point of my life, just after giving up my place at the Royal Academy of Arts in The Hague because I chose to give birth to my child before getting back into the arts. Communication was still difficult, but we had maintained a strong bond. At a time in which my freedom, my lust for life and my own urge to explore all that was waiting for me felt far away, it felt empowering to meet him and his friends.

This picture means a lot. It was and still is a symbol of the gift of life to me, in many ways. It is a spontaneous portrait of somebody that cared for me, showed me how not to worry, to enjoy life day by day and be happy for all that is there. When I was in Kyoto, Lee got me a job and showed me around. I followed his band to concerts, and went to places I would otherwise never have seen. I practiced independence and learned that it suited me.

Meeting Lee was like coming home in Japan. It was finding a place where I belonged. I had a roof over my head and a job, and even if I would not stay long, it was a life of its own unfolding. This experience has carried me a long way since then: the confidence that you can travel to the other side of the globe looking for purpose, and within a few weeks see an entire microcosm expand that becomes your reality within the blink of an eye. Japan has been in my life for a long time, and in this situation, sitting on an early-autumn Amsterdam terrace, I was thankful to be reminded that this part of me, however distant it felt, was still there: the hunger to discover all that lies ahead, combined with the confidence that it would be waiting for me.

The colour this portrait evokes is a pale greenish tint, one that recurs in many pictures from that period. It is a shy colour. In its subtlety, it reminds me of the nature of this country. There is a serenity, a calm to it that I associate with the gardens and temples I spent much time in, just sitting and thinking. I perceive the colour as withdrawn in a way, and I wonder if this has to do with this journey being a conscious withdrawal from my busy life back then, when I needed to be alone and see what would happen if I were to be thrown back on myself.

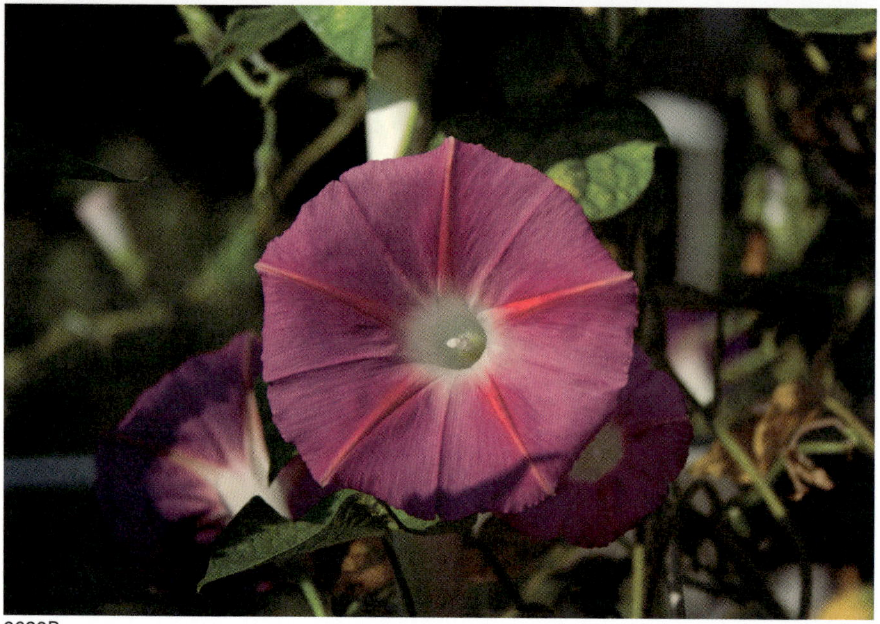

0630P

Bonn, 2008

This is a photograph of a flower that fascinates me. I have always been interested in plants that have been and still are used as medicinal plants. Morning Glory Calistegia (Ipomoea tricolor) contains LSA, which is the natural equivalent to artificially produced LSD. Morning glory grows all across Europe, and is one of those nice little flowers that you might know but never pay attention to. LSA is a psychoactive component which, albeit weaker, induces effects similar to LSD. Psychoactive substances have been used by humanity for thousands of years, mostly embedded in healing or sacred ceremonies to access the divine through the use of natural substances. Psychedelic experiences were not unknown to our ancestors and are not exclusive to indigenous cultures on other continents.

The colour I feel when I look at this beautiful flower is a powerful blue, a strong colour that contains confidence and knowledge. Psychoactive substances can induce very strong visual experiences and the perception of multiple dimensions. If someone is under the influence of a psychoactive component, they perceive what they see and experience as reality. Many who have had such an experience are left with the question of where reality lies, what defines it and how much we can trust our senses. A lot of synaesthetes recall psychoactive journeys as altering and enhancing their synaesthesia, just as many a person without synaesthesia might suddenly see sounds or perceive smells as geometric shapes. I find this fact fascinating, as it invites us to reconsider one of the most common criteria of synaesthesia (you either have it or you do not) and teaches us that an altered state of consciousness can allow access to new sensory pathways. The blue shade I feel might come from my medical interest in psychoactive plants. To me, these substances are carriers of knowledge which should be studied even if we cannot approach and understand their workings fully through the eyes of Western science. They give me an impression of major power, care and truth.

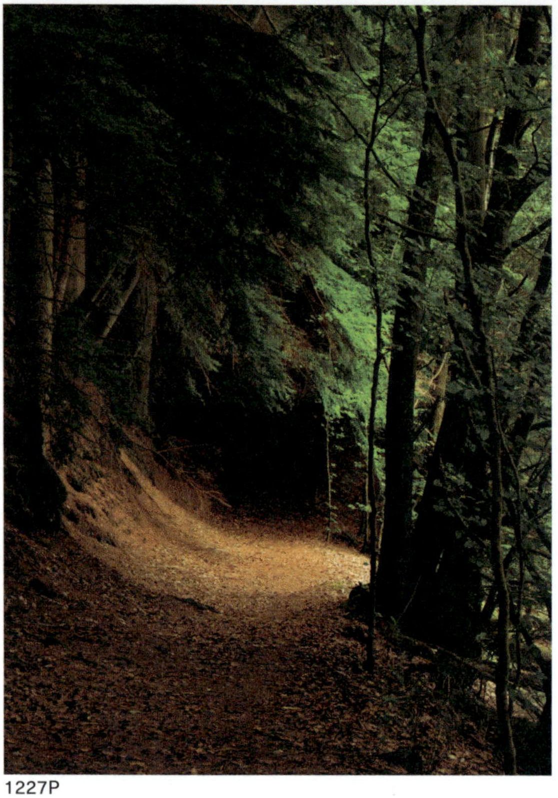

1227P

Himmerod, 2008

When I realised that I was pregnant in the summer of 2008, weeks before I was scheduled to start studying at the Royal Academy of Arts in The Hague, my world fell apart. I did not want to become a mother, but it was not as easy as that to turn my back on the pregnancy. I did not feel any confidence to be able to make a decision so profound, with such long-lasting consequences.

I dropped everything and went off to a monastery near my hometown for a few days, seeking peace and quiet in an overwhelming situation.

I spent my days strolling through nature.

It is hard for me to write anything about that picture, because it symbolises that struggle; as the mother that I am now, I would prefer to opt for another narrative. One of pure joy, of firm beliefs, of simple acceptance and happy anticipation on the birth of a little someone.

When I see this photograph today, it reminds me how fragile I felt back then and how, in contrast, this little life that I carried seemed so strong and present. I remember how it surprised me that there was a real someone where others see only a bunch of cells, how this presence seemed to have a dynamic of its own. There were all these moments in which my mind told me one thing, while my emotions told a completely different story deep down inside of me.

It was a good lesson about ego, about how much fear and stress can influence our mind and how we get to know what we really want when we close our eyes and listen.

Cold white is a colour that only appears as a pure shade when I feel disconnected to everything, when I am literally lost. I think the white comes up when I see this scene because this is how I felt back then. Other shades of white can be all gentle and soothing, but this is a cold and distant white that feels very powerful, that dominates everything. It was one of the hardest episodes of my life up to now, when I felt unable to cope with that responsibility and frozen by the idea that my life would never be the same, no matter which way I chose.

I was overwhelmed by the situation and could not think straight.

That helplessness is evoked through the cold white that I feel when I see this image today.

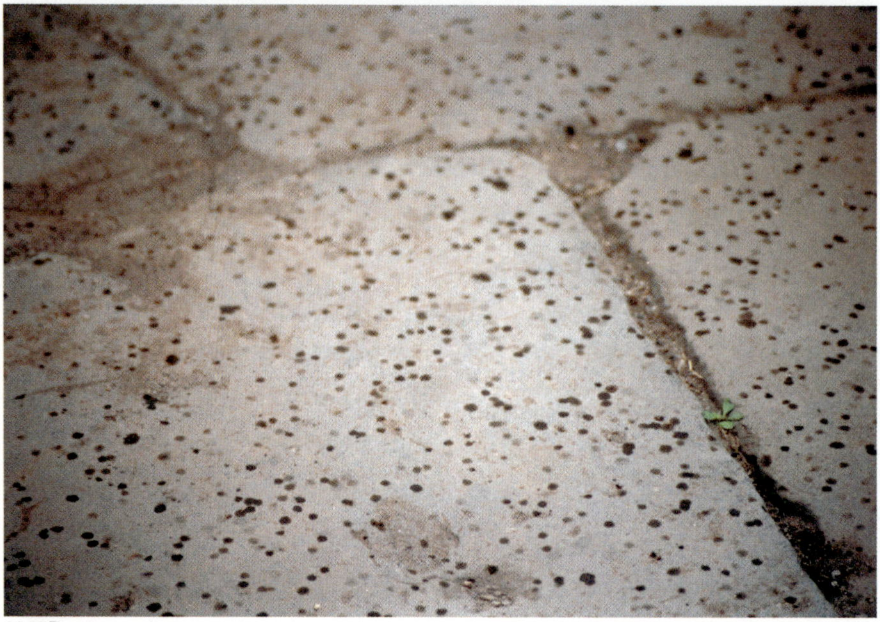

0848P

Rome, 2004

I love this picture. It has a rhythm and lightness that I feel I can rarely capture with my camera. I remember why I took that photograph: something in it struck me as perfectly representing a specific state of mind, one of calmness and serenity. A state of consciousness that implicates all kinds of diverging emotions and, without being indifferent towards them, embraces that very diversity as the essence of life.

I remember the stones were still warm, and that I liked how the raindrops did not interrupt the scene but added a layer onto it: the notion of transition, the knowledge that nothing is static but in a state of constant modification. It is a humble image, in some sense. It reminds me that all that makes us human moves between opposing poles and that we move through life more easily when we do not forget that duality creates meaning.

The colour I feel is a warm, deep shade between ochre and terracotta. It has a duality to it. It assembles diverging emotions, causes them not to diminish in intensity but to combine peacefully: love and pain mixed in a healthy dosage. It is a wise colour, one that I rarely feel when I am stressed out or worrying too much. I wish I could feel like that more often: so much respect and acceptance.

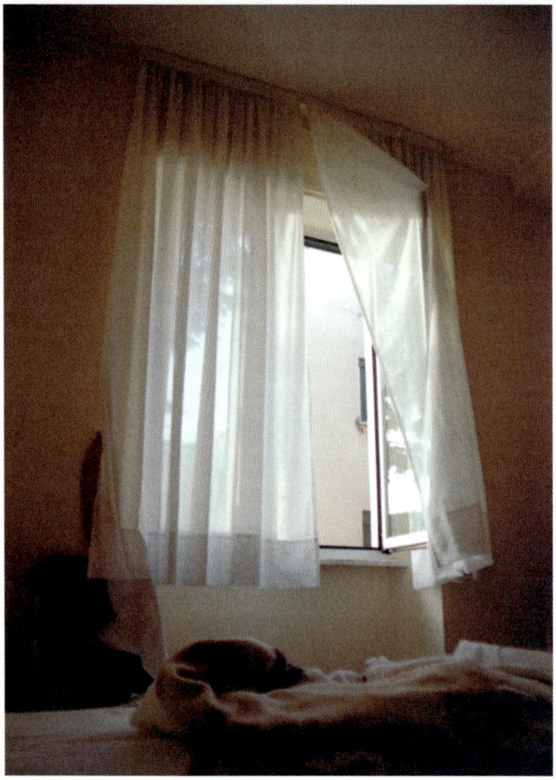

1256P

Rome, 2004

I do not have many memories of this trip, but the scene gives me that airy, soft, warm feeling of the south, la chaleur du sud. An early morning, a beginning.

I shot this picture a few months before I ended secondary school. It was a time when everything seemed possible. I wanted to travel a lot; I had many plans. It was an exciting time. I knew it was a unique period in my life, a transition that would give me a freedom that I would never have again. When I look at this photograph, I see all those elements: the city of Rome with its strong smells and impressive Roman architecture, the anticipation of freedom, the eagerness to discover, the travelling mood, the warmth of the south.

That particular photograph has the colour of the south: the warm, earthen, powdery tone that I get when I think of walking through pine forests barefoot, when we collected plants as children, the smell of my skin after being out a whole day. The dust of the roads, the sounds of the insects, tired limbs after being in the sea all day, late nights and night skies sprinkled with stars: these images come up when I think about our trips there when I was a child. There was so much to discover. I remember the smell of different herbs, the taste of freshly churned butter, the smell of tomato plants. It was also a time when we were really free. I read a lot, lying in the sun. There was so much time to just be.

I guess we all idealise our childhood memories. When I think of the south today, it is intrinsically connected to those memories. Current experiences and childhood memories merge; they become one in my mind.

The south has become my Arcadia, my place of existence where life was real before it got distorted and became the life I have to live as a grown—up, which feels so unnatural to me. Le midi, the place of dreams, invites all kinds of projections.

This colour is expansion to me, seeing open spaces that invite me, nothing reminding me of my everyday life, but also expansion in my mind, not affected by any tensions.

It is as much the colour of the south and of travel and expansion as it is the colour of "Fernweh", the longing to be somewhere else. Its powdery structure might come from the old buildings I see when I think of that trip, or of the south in general: the dust that dances in the sun at noon, the patterns on my hands after playing on the ground for hours.

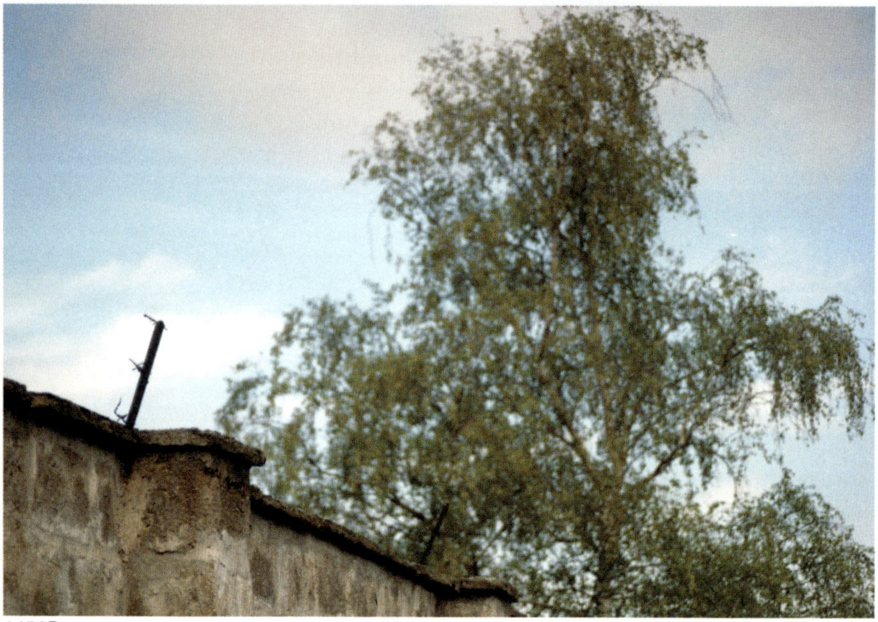
1456P

Sachsenhausen, 2004

This place is one of the most terrifying that I have ever been to. It is the former Sachsenhausen concentration camp near Berlin. I went there for a moment of remembrance when I was fifteen with a group of young Germans, Jewish Israelis and Palestinians. We visited the place and I remember feeling all void and weak while walking through the alleys and abandoned buildings. The place sucked all the energy out of me; when I had to read a poem out loud at the time, I almost fainted. I remember how my voice broke. This place silenced me.

It is impossible to capture the horror in images, especially not after so many years, but I took exactly two pictures that day: one of the gravel formations on the ground and this one of a tree. I think this image was taken in order to capture, probably unconsciously, what to me is the real horror of the Holocaust: that anything like this could happen while life was continuing unimpeded outside this hell. I often feel, as a photographer, that capturing the exact meaning of a situation is extremely difficult. Not in the sense of a narrative that would explain what is perceived in the image to someone else, but the meaning it has to oneself.

That day I was confronted with unimaginable evil done to humans by humans. I felt the only thing I could do to approach the essence of that situation was to capture the only beauty I could find there, because its contrast with the horrors would come closer to a description of evil than anything else I could have photographed that day.

The colour that I feel is absolute disconnection: cold, threatening grey, merging into the dirty white of pain and loss. A place like this takes everything away from me, all my positivity and energy, and leaves me with the bitter feeling that my love of life is an illusion, since it cannot stand against such horror and helplessness.

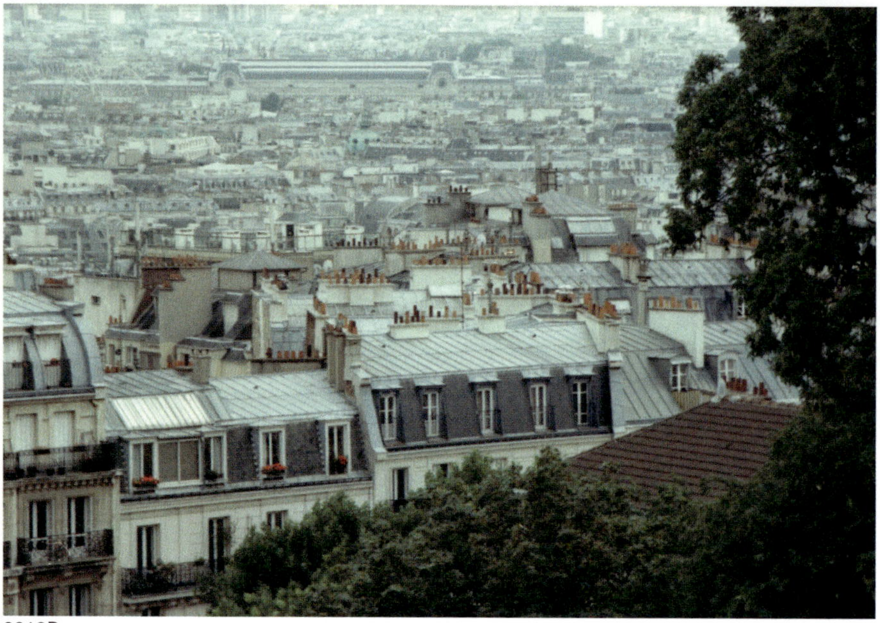

0019P

Paris, 2002

As a teenager, I used to stroll through the town on my own, eager to see all its diversity. My mother grew up in Paris. Despite never having lived there, I feel at home in this city, and still go there very often to visit my grandma. Growing up in Germany, I never felt particularly French. When I hit puberty, I felt a growing need to search for my roots, especially because I felt more and more lost with the years, not knowing any place I could call home with an authentic voice. There has been migration and displacement in my family, out of free will or forced because of war, and maybe this has reached me through my previous generations. I have always felt most at home when I was on the road, freed from the restrictions that define my person in my normal life. Thrown back on my very own self, I felt life presented itself with a more honest, more authentic voice. I always felt that if there was no familiar environment to respond to, I could perceive reality with a clearer mindset, a clearer view.

Paris is either a pale rose, sometimes mixed with a powdery, velvet white, or a pale blue.

Sometimes I wonder if my practice of using outdated film to shoot over many years has merged with my emotional colours that come up when I see certain places I have been to, because the memory itself merges with the photographs.

In the rosy shades lie comfort, absence of worries, and some sort of delicate confidence that, even if it might be invisible at this very moment, there is an underlying structure in life that sorts all these confusing strings of situations, decisions, encounters and emotions out into a pattern that constructs the shape of my life. Whenever I am in Paris, I have a great deal of time for myself. I go to bookstores, see magnificent films from the fifties in tiny art cinemas, walk through one exhibition after the other. It is always a time of input as well as reflection and maybe that allows me to be more at peace with myself, because my curiosity as well as my need for withdrawal both get the space they need. This quiet pace is likely creating the rosy shades of calm confidence that I feel when I am there.

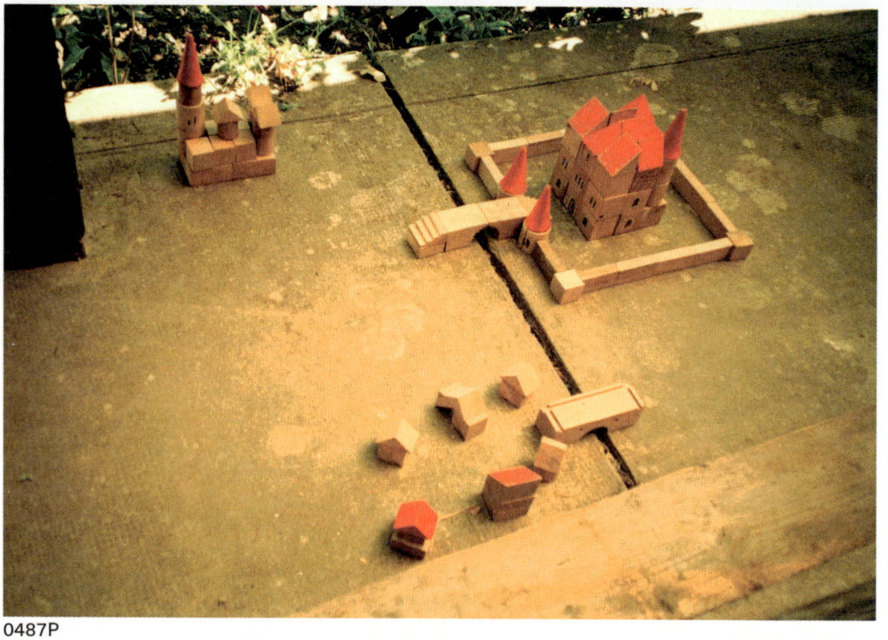

0487P

Feuguerolles, 1996

This is one of my earliest photographs. It is an early example of my need to frame my reality and my fear of losing out on bits of the world that were once important to me. It is my grandfather's house in Normandy, where we spent many summers when we were kids. The colour is a warm orange-yellow, the colour of my childhood. It is a light tone, a simple colour—one that knows no complications, no hidden intentions, no lies. It comes back when I think of happy moments I had as a child, and the summers in my grandfather's house belong to those fond memories. Seeing this photograph today gives me a painful little pinch in my heart. It moves me to see that I was already doing back then what many years later became my passion. I was already capturing my world with my camera because I could not accept that people, things and places disappear—an inability that still haunts me today. There is also an element of sadness to it when I realise that, even as a child, I was determined to hold on to things, which I interpret today as an inability to just enjoy what is there and to be fully mentally present. Somehow, to my 30-year-old self writing these lines, it seems inadequate that, at this age, one would hold on to what one has built as a game. On the other hand, I like the merging of the two layers: the fantasy world and the "real" outside. This process of taking my fiction into reality by securing it on film seems symbolic to me of a process that I remember vividly taking place at that age: the slow realisation that childhood would not last, that my world would soon change. Maybe this early concept of loss is the major driving force behind all the images that I took throughout the years.

In this chapter, I tried to give a detailed description as to how the stories behind the photographs influence my synaesthetic reactions. While trying to find the right words, I notice how much my perception is intertwined with the colours of my emotions, and how big a role these colours play not only in my perception, but also in the way I remember situations. My memories are connected to emotions that give them an individual tint. I wanted to investigate on where these colours exactly came from, and to guess which factors might possibly have influenced their appearance. I found out that many pictures that I cherished along the years have very distinct colours. And I understood how much I love certain photographs because they make me feel colours that just perfectly describe certain states of mind, such as the stone pavement[1] or the red shades of my room in my first apartment.[2] I wonder how much these memories have slowly gained a significance of their own through pairing with the synaesthetic colour responses: would I remember certain places as fondly were it not for the photographs that I took of them and that I have looked at so many times, deepening their meaning by repeating the colour reaction? Or is it just this phenomenon that we all know: that through images from our life, people and situations are vividly recalled, up to a point at which the real memory fades away and is superimposed by the projection of our current self onto a situation?

1

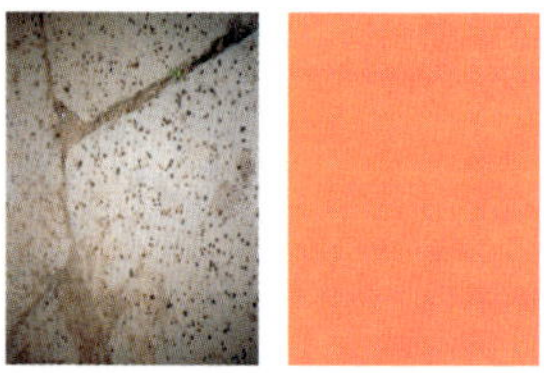

0848P

2

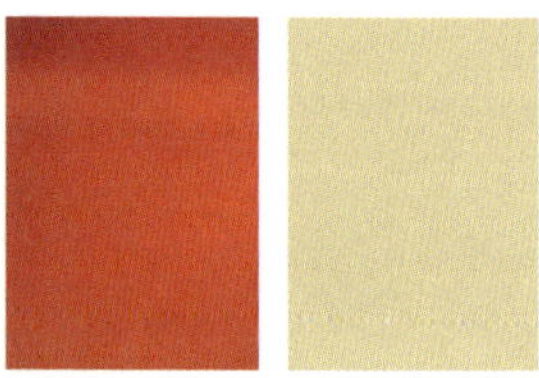

0564P

I know that I have kept many printed photographs of mine to put them on the wall, and many followed me through a number of apartments without me ever questioning their presence. I suddenly realise that many of them have stayed with me because of the colour they make me feel: some of them giving me soft white shades of comfort, others uniting loved ones in a colourful harmony. Writing these lines, I look at the photographs that hang on the wall of the atelier and I feel all these shades inside of me, subtle carriers of powerful emotions. There are comforting and soothing white shades, powerful reds and serene greens that speak of harmony and connection.

Some images do not even have a specific subject — I might not even remember where and when I took them — but they could still give me a very precise colour that I might want to have close to me, and therefore hang it on the wall next to my desk. Some photographs are only dear to me because of my synaesthesia: sometimes the colours that come up describe a state of mind or an emotion with such accuracy that I feel words fall short next to this immediate visual precision.

Another side effect of searching for words to adequately describe what I see and feel when I look at my images is that I noticed a growing sensibility for my own synaesthetic reactions. Over the last weeks and months, while working on the texts for *Layers of Reality*, I grew more and more aware of my associations. I would be much more attentive as to when the colours appear, and instantly question myself about it:

why dark red, why grass green? Why is this shade so much more distinct than another? Why are the colours so much stronger today than over the last week? Why do I have reactions to some photographs of a scene, but not to others?

In conclusion, I can state that many of my photographs give me precise colour associations, but that I was not aware of the complexity of their origins until writing this chapter. I am surprised to see that in most cases, I understand where the colours come from (or, at least, I can venture a plausible guess). I thought it would be much more vague, just as the consistency tested in the earlier chapter was not as precise as I expected. I found out that I have a fondness for some of my images more for the emotion they convey than for the actual subject, but then this is not a purely synaesthetic quality. My memories blend in with my feelings and become an undiscernible synthesis of fond remembrance and actual emotions linked to colour. Some pictures perfectly transport both a mix of ancient feelings and a colourful emotional reaction in the here and now. If these are in tune, I hold in my hands a precious tool. This is why I photograph.

Perception and Synaesthesia

Erika Schippmann
psychoanalyst

Psychoanalysis mainly covers the verbal and symbolic dynamics of how sensory experiences are processed. However, since Freud there have been authors pointing out that, in addition to the symbolic, there is an important preverbal area of mental experience that is based on processes of perception, adding to the processing of experiences in its own way. Those are the many preverbal experiences in a world of perception that leave traces inside of us: the sound of language (not its meaning); smell and taste that are connected to an entire world of experiencing; a sequence of motion that triggers certain emotions; forms and colours that imply a certain experience. All those perceptive processes can happen simultaneously, for example when a child is breastfed, which at the same time contributes to a certain inner experience—a warm, filled, soft, bright, inner well-being.

The traces of this world of perception follow us in many individual ways.

When someone perceives very early and with great sensitivity that the most important persons one has, the parents, are preoccupied with themselves and their own problems, there is a chance that that world of inner perception and the sensory experiences that contribute to it will turn out to have a certain importance. A child whose mother is unable or barely able to resonate with her children's needs could, in finding ways to help oneself, increasingly turn to the emotional objects that can be experienced sensorially to allow experiences that are lacking in contact with the child's mother.

Such children, similar to Ms Püschel, might develop a high sensory ability or possess it inherently. Perception allows satisfying, fusional experiences with sensory objects that are within reach: a cuddly blanket, a coloured ball, a rattle, and many more. The sensory experience coming from those objects is perceived by the child as reassuring or reviving. It feels protected from emotions of emptiness and loneliness, but also from feelings of losing itself or falling to pieces.

A dream that Ms Püschel told me she has had repeatedly expresses the child's unfulfilled desire for sensory interaction with the mother—perceptional experience within a dream can partly compensate for this: "I am calling my mother over and over again, I can't see her. Then there she is, turning around to me, looking at me, but then she turns her head back again and walks away as if there was nothing." This is about presence and proximity that are painfully missed. The dream succeeds in making the mother visible. The second the mother looks at the child, there is appeasement and recognition. Even if the touch the child longs for is lacking, the dream tries—through the reciprocal optical perception, seeing and being seen—to replace the missing proximity and physical interaction. Synaesthesia as a phenomenon and the question of its origin may be answered from a psychological view with this little example: Premonitory is an inner mental pressure, the longing for the calming, physical presence of the mother, that the child experiences. The mother's touch would lead to a relaxation of the muscles and blood vessels, improving circulation and conveying a sensation of warmth. This calming is what the sensory dream experience tries to actuate. The optical experience of the outer world is translated into an analogous acoustic experience, the hearing; both soothe. Many examples from our daily vocabulary illustrate that shifting between different outer sensations is a common process: a treat for the ears, sound timbre, eye candy. Here the acoustic or optical sensation is tasted or seen simultaneously, and by this the triggered inner impression is deepened and more differentiated. From the perspective of psychoanalysis, the process of perception—as well as the coordination of outer impressions, the synaesthesia—has to be understood in connection with the pressure of the mind to make sense of the search for comfort, satisfaction or liveliness. The synchronisation of different modes of perception, modes that are experienced as real during memorable satisfactory experiences such as breastfeeding, enhances and refines that experience. Consequently, sensory incidents can result that deeply and unconsciously correspond to the imagined maternal closeness. Those sensory incidents take on their individual meaning by being repeatedly employed for a certain inner experience, throughout continued development. Finally, it could be seen as the unique ability of art to build a bridge between an unconscious deep layer and the volatile world of exterior perception.

A Reflection on *Layers of Reality*

Olympia Colizoli, PhD
neuroscientist

Layers of Reality is an extraordinary glimpse into the unknowable: the subjective experience of another person. As individuals, we are trapped inside our unique points of view. Although traditional modes of thought teach us that there is one objective reality 'out there,' we must shift our paradigm to realise that each person experiences the objective world in *at least* slightly different ways. Differences in subjective experience are perhaps best illustrated in the phenomenon of synaesthesia[1], which cuts straight to the heart of *qualia*: the content of experience.

1
Rich, A. N. & Mattingley, J. B. Anomalous perception in synaesthesia: A cognitive neuroscience perspective. Nat Rev Neurosci 3, 43–52 (2002).

This wonderful piece of work sits at the crossroads of art and science. *Layers of Reality* is a well-designed and systematic inner exploration into the structure of one's own experience. Anna's work brings up fundamental questions in the fields of psychology and neuroscience, some of which have been a pastime of philosophy for ages. Is knowledge gained through the senses? Can two individuals ever share the exact same experience? What becomes immediately apparent is that Anna's reality is probably vastly different from yours. Anna's experiences are similar to other synaesthetes with the same sub-type of synaesthesia (e.g. emotion to colour[2]); however, synaesthesia is characterised by its idiosyncratic nature[3]: every two synaesthetes of the same sub-type will have *different* experiences evoked by the *same* stimuli (e.g. the feeling of 'love' evokes the experience of blue for person A and red for person B).

2
Ramachandran, V. S., Miller, L., Livingstone, M. S. & Brang, D. Colored halos around faces and emotion-evoked colors: a new form of synesthesia. Neurocase 18, 352–358 (2012).

3
Cytowic, R.E. Synesthesia: A union of the senses. The MIT Press (2002)

What makes a person a 'synaesthete'? While it is undoubtedly the case that individuals vary to extremes in terms of their subjective realities, not all deviations from 'normal qualia' are considered to be synaesthesia. From the perspective of a researcher, it is important to have strict diagnostic criteria to be able to compare across studies and make valid conclusions about a specific population. Importantly, the diagnostic criteria as inclusion for synaesthesia do not discount other unique types of experiences as valid in their own right. The use of criteria only highlights the importance of semantic clarity within the scientific context. A person is considered to be a synaesthete when they have a highly *specific*, *automatic*, and *consistent* experience in a sense that is *additional* to the sense being stimulated[4]. For example, upon hearing C-major, an experience of green is immediately evoked (sound–colour synaesthesia). This is not just any green, but a very specific shade of green (the pale, dusty green of grass dried out in the sun). Furthermore, a synaesthete is unable to turn off the green experience; the 'greenness' happens whether they like it or not. A synaesthete will experience the same shade of green each time C-major is heard, even across years. Developmental synaesthesia is furthermore characterised by the fact that synaesthetic experiences are present since early childhood. Acquired synaesthesia can happen due to brain damage, sensory loss or drug intake[5].

As you may already note, some experiences are much easier to fit inside our framework

4
Rothen, N., Tsakanikos, E., Meier, B. & Ward, J. Coloured Letters and Numbers (CLaN): A reliable factor-analysis based synaesthesia questionnaire. Conscious. Cogn. 22, 1047–1060 (2013).

5
Brogaard, B. Serotonergic hyperactivity as a potential factor in developmental, acquired and drug-induced synesthesia. Front. Hum. Neurosci. 7, 657 (2013).

for synaesthesia than others. For instance, is emotion a sense? How do we quantify an emotion? Do you ever really have the same emotion twice? Therefore, how consistent should an emotion–colour synaesthesia be, as is the case with Anna's illustration? The criterion of 'consistency' in diagnosing synaesthesia is considered the test of genuineness in the field[6,7]. Consistency in research on synaesthesia is highly valued by scientists because it is almost impossible for a non-synaesthete to reach consistency levels of 'genuine' synaesthetes. In this type of test, both synaesthetes and controls are asked to associate items (e.g. letters to colours on a colour wheel) and then retested again several weeks or months later. Only synaesthetes consistently choose the same colours (even when the controls know they will be retested and the synaesthetes do not!). The down side to this reliance on consistency is that we may be excluding variations of synaesthesia, such as emotion–colour, which by nature are bound to be less consistent while still fulfilling the other criteria for synaesthesia[8]. Anna's work wonderfully illustrates the issue when she talks about emotions evoking experiences of colour and the difference each day that she 'relives' the emotion. The criteria we employ as scientists comes from a need for specificity, not judgement or disbelief.

Many of the other experiences illustrated here by Anna are common among synaesthetes. Individuals who experience colours evoked by emotions describe the colours as evolving over

6 Baron-Cohen, S., Wyke, M. A. & Binnie, C. Hearing words and seeing colours: an experimental investigation of a case of synaesthesia. Perception 16, 761–767 (1987).

7 Asher, J. E., Aitken, M. R., Farooqi, N., Kurmani, S. & Baron-Cohen, S. Diagnosing and phenotyping visual synaesthesia: a preliminary evaluation of the revised test of genuineness (TOG-R). Cortex 42, 137–146 (2006).

8 Simner, J. Defining synaesthesia. Br. J. Psychol. 103, 1–15 (2012).

time, for example becoming more saturated in hue as a relationship deepens. Synaesthetes have described that items generally need to be familiar to evoke a synaesthetic colour. For example, in grapheme–colour synaesthesia, a language that is known will evoke colours but an unknown language will not[9]. Furthermore, synaesthetes report that real and synaesthetic experiences do not mix with each other. For example, a lexical–gustatory synaesthete (someone who tastes words) reported that when she has a conversation over dinner, the tastes of the words being spoken do not influence the taste of the food in her mouth[10]. Anna's challenges in illustrating her inner experiences in terms of 'outer reality' are those of the synaesthesia researcher as well: First, we are limited in our ability to map internal experience to a limited range (e.g. of colour) in the changing environment. It is very important to keep the environment tightly controlled so as to always minimise differences. It is often reported by colour–type synaesthetes that their experiences are more than just 2-dimensional colours, for instance, patterns of multiple colours. Often synaesthetic experiences have shape, such as 3-dimensional texture or structure[11]. Some synaesthetes report experiencing colours that they have never even seen in the 'real' world[12]!

Why is synaesthesia remarkable from a neuroscientist's perspective? The striking aspect of synaesthesia to a neuroscientist is the opportunity to study how the brain creates subjective experience in the absence of corresponding

9

Mroczko, A., Metzinger, T., Singer, W. & Nikolić, D. Immediate transfer of synesthesia to a novel inducer. J. Vis. 9, 25–25 (2009).

10

Colizoli,O., Murre, J.M. & Rouw, R. A taste for words and sounds: a case of lexical-gustatory and sound-gustatory synesthesia. Front. Psychol. 4, 775 (2013)

11

Eagleman, D. M. & Goodale, M. A. Why color synesthesia involves more than color. Trends Cogn. Sci. 13, 288–292 (2009).

12

Ramachandran, V. S. & Hubbard, E. M. The phenomenology of synaesthesia. J. Conscious. Stud. 10, 49–57 (2003).

external stimulation. A concrete example is that synaesthesia allows us to investigate how the experience of *colour* is created by the brain in the *absence of wavelengths of light* stimulating the retina. The systematic mapping between the stimulated sense (e.g. sound) and the synaesthetic sense (e.g. colour) allows for the attribution of *meaning* between the connections. These meanings are entirely subjective, the foundations of which are embedded in the individual's experience within his or her environment, as *Layers of Reality* wonderfully illustrates. This modern-day field of investigation into synaesthesia is a direct descendant of the philosophy of mind concerned with epistemology.

A common misconception is that synaesthesia is a neurological disorder. Although some individuals may have synaesthesia in addition to other psychological or neurological conditions, most synaesthetes are normal healthy individuals. In fact, synaesthetes as a group have better memories for items within the domain of their synaesthesia (e.g. a sound–colour synaesthete will have a better-than-average memory for colours)[13,14]. Developmental synaesthesia is characterised by an absence of pathology in other domains[15]. My opinion is that much misconception about synaesthesia as a disease or disorder comes from the fact that in the English language (and perhaps in most others) there is no word for a 'positive disorder'. What I mean is that there is no word to capture a biological phenomenon that is out of the range of normal experience, but that is *not* detrimental

13

Rothen, N. & Meier, B. Grapheme–colour synaesthesia yields an ordinary rather than extraordinary memory advantage: Evidence from a group study. Memory 18, 258–264 (2010).

14

Terhune, D. B., Wudarczyk, O. A., Kochuparampil, P. & Cohen Kadosh, R. Enhanced dimension-specific visual working memory in grapheme–color synesthesia. Cognition 129, 123–137 (2013).

to the individual — that does not necessarily help or hinder them. The word 'condition' is perhaps the closest in the sense that by definition it denotes something neutral, but 'condition' is too often used to denote something negative. I tend to use the term 'phenomenon,' although I do feel it implies something a bit too 'otherworldly' when in fact we are talking about the natural variation in a biological trait. Although synaesthesia has been documented since at least a few hundred years ago[16], our mode of communication has yet to catch up with this shift in paradigm in order to communicate information related to 'layers of reality' in subjective experience.

Anna takes us by the hand in order to guide us through the conscious experience of her structure of reality. Through this wonderful self-reflective experiment, we as the readers can begin to understand the connections made within another's mind. In order to accomplish this, Anna must present her personal struggles and most private thoughts. *Layers of Reality* is not just beautiful, but also incredibly brave.

15

Rouw, R., Scholte, H. S. & Colizoli, O. Brain areas involved in synaesthesia: A review. J. Neuropsychol. 5, 214–242 (2011).

16

Galton, F. Colour associations. Synaesthesia Class. Contemp. Read. 43–48 (1880).

Visualising Synaesthesia

To help visualise the experience of synaesthesia, I tried to pin down different synaesthetic experiences and come as close as possible to reproducing them. I am curious to find out how close I can come to visualising my emotional landscape.

I personally do not think that there is a structural difference between the quantity or quality of my feelings and those of others, but there seems to be an extra 'layer' in my perception that I still have trouble imagining is absent in other people's perception.

It is an impossible task to squeeze an emotion into a two-dimensional rectangle, and even more so a synaesthetic response, but I would still like to try to give an impression of how synaesthesia unfolds in my brain. Those images are what I 'see' inside of my mind, what comes up when I think of a certain place, a certain person or a situation that I was in. I do not see auras or colours around other people, as I am often asked; it is an internal phenomenon. A person's colour can change slowly over the years as my connection to that individual evolves, but also abruptly transform when my relationship to that person is suddenly altered. I remember a breakup on the phone, during which that person's welcoming, comforting warm greyish-blue turned instantly into a dirty green.

The memory of a place or a situation is mingled with emotions that are connected to that location or experience, so I guess the synaesthetic colour associations are triggered by those feelings.

This is as close as I can get to visualising my associations.

Charly

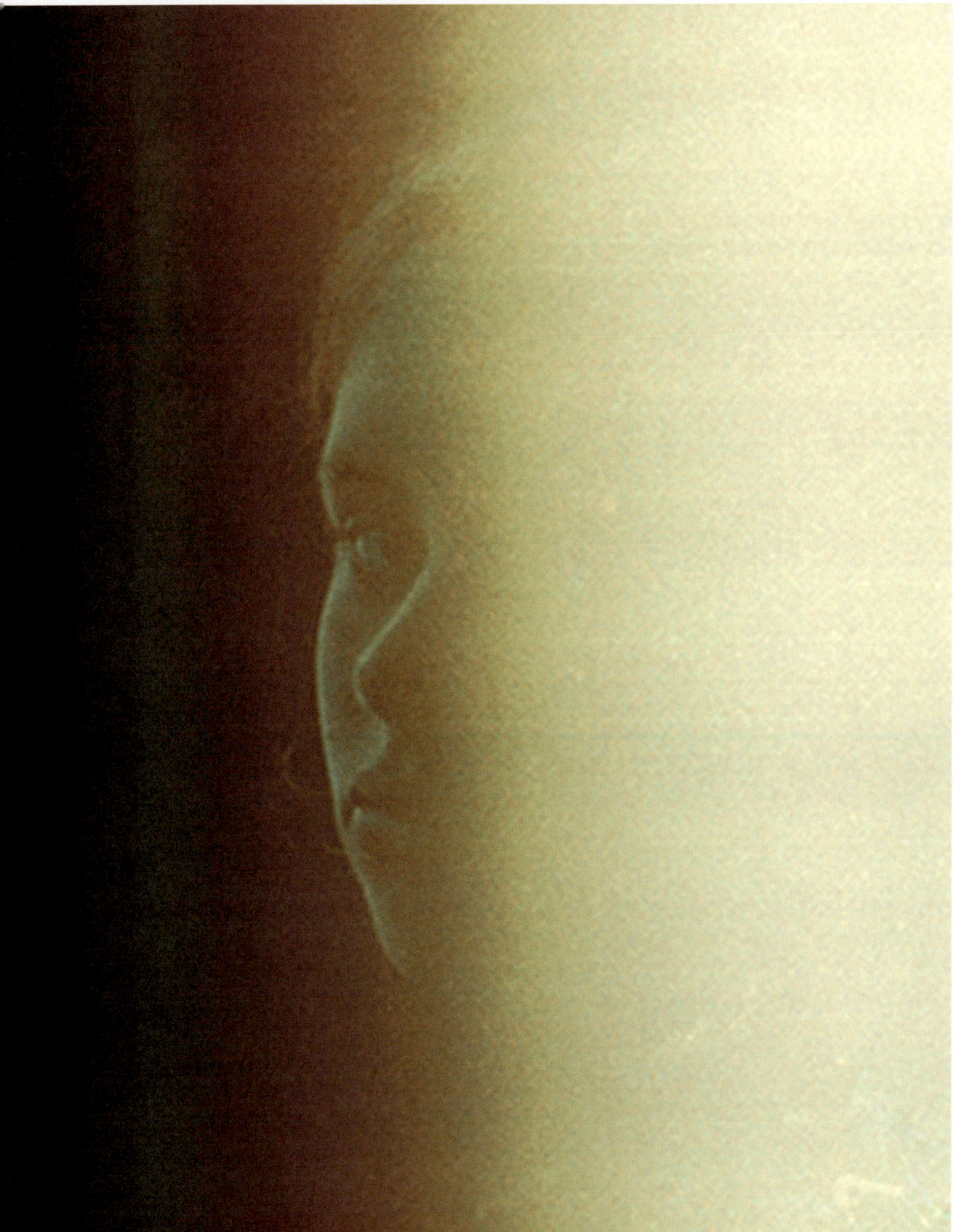

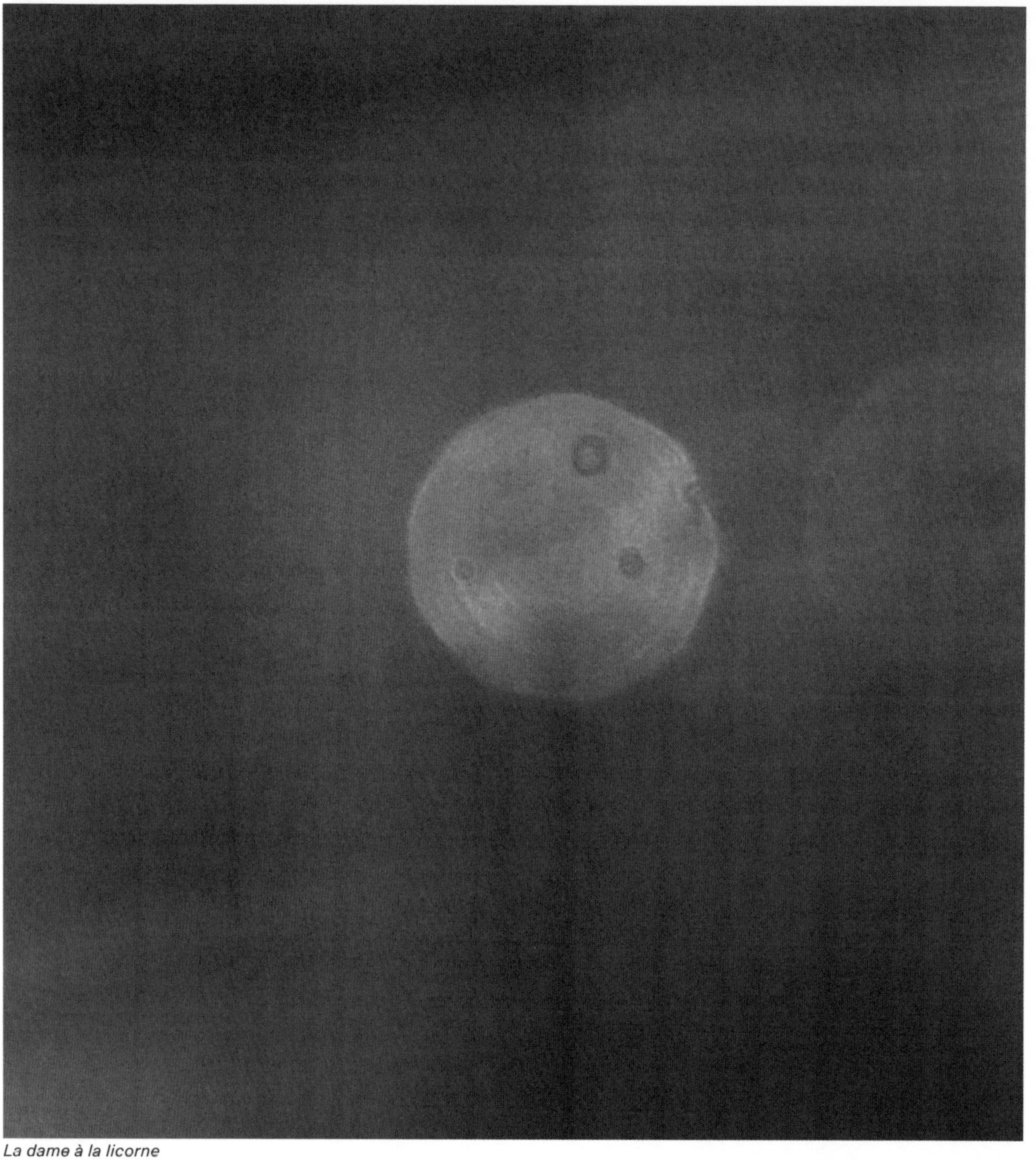

La dame à la licorne

Lange Haven, Schiedam

Finding my space

Isolation

Emilio

Beaulieu

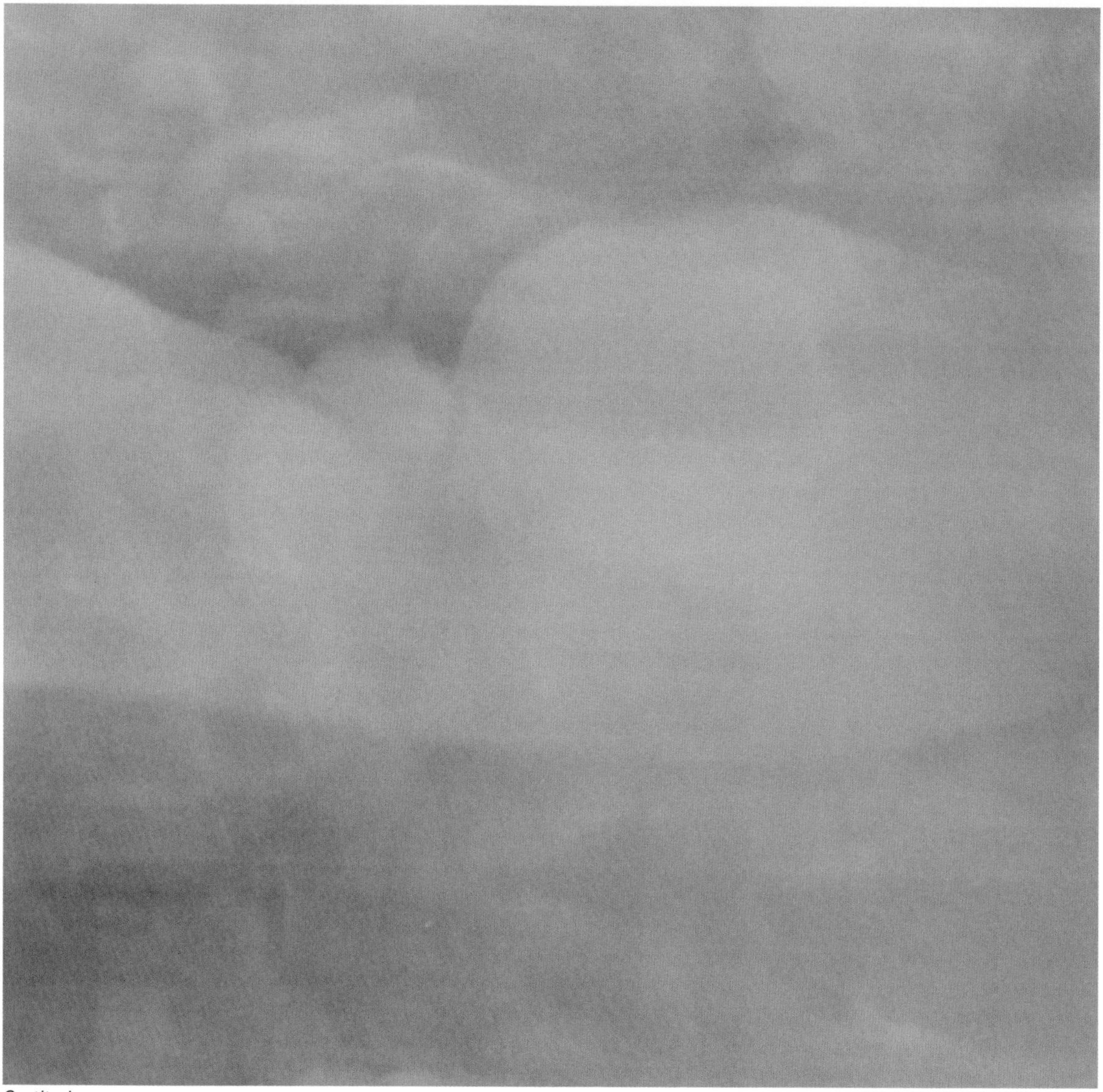

Certitude

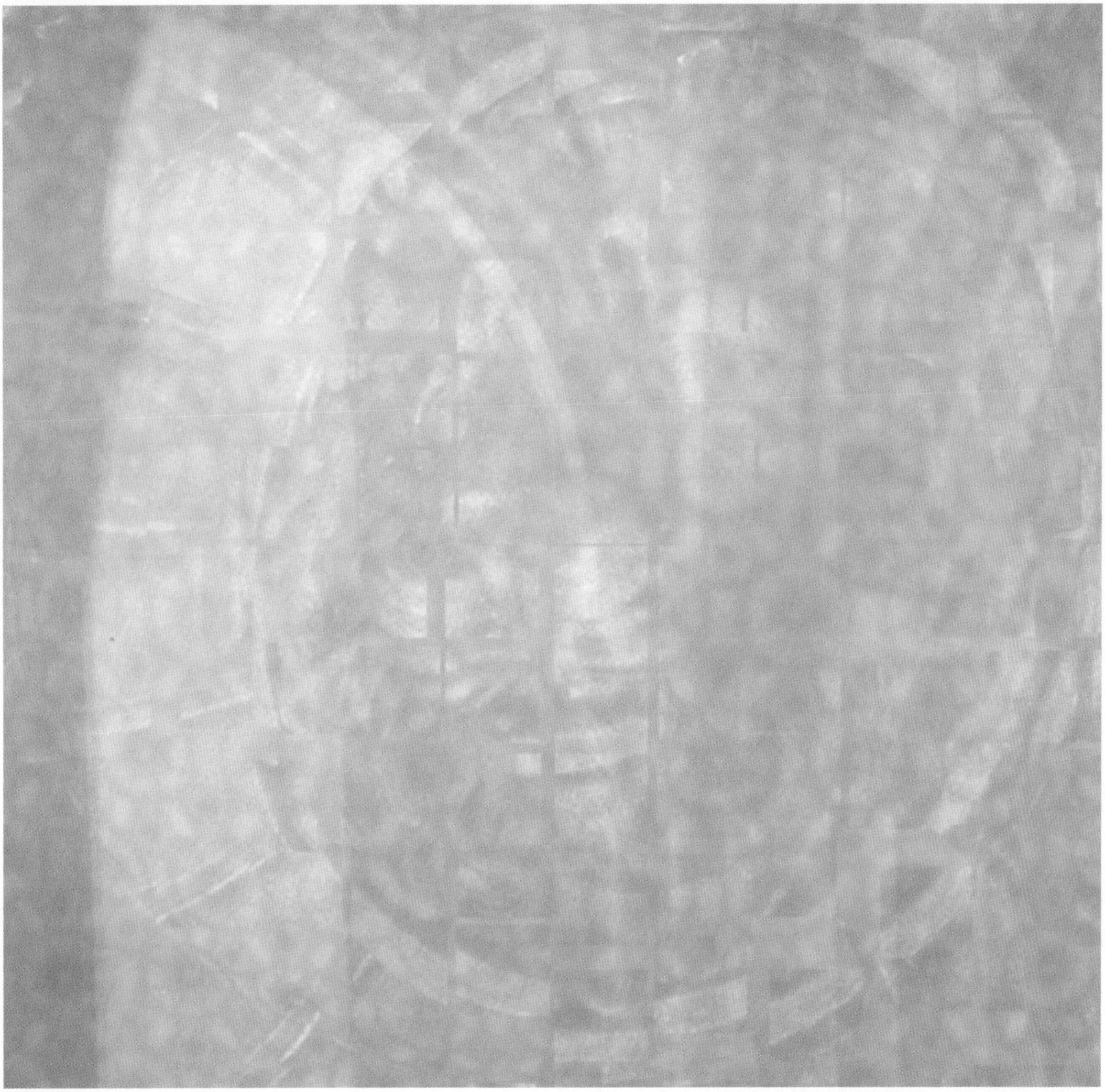

Respect

Madeleine

Coloured Emotions as Qualia

Synaesthesia from a Philosophical Point of View

Alexandra Spaeth, MA
philosopher

Colours do not exist. Simply put, they are merely sensory impressions caused by reflected light on objects. It has been known since ancient times that it would be pure illusion to assume the perceivable world to be the actual reality. Instead, we are subject to a variety of boundaries of our cognition. Related theories have changed, but this thesis is still valid.

How well our colour perceptions represent reality can most likely be assessed by the natural sciences, as these are concerned with questions such as how we perceive the physical world.[1] However, no convincing theory of synaesthesia has been developed so far, let alone an understanding of the human brain as a whole. Moreover, even if it were possible to devise such a scientific theory, it remains highly questionable whether it would be enough to decrypt the phenomenon of synaesthesia.

Although emotions, perceptions and sensations are manifested at a neurological level,[2] scientific findings do not imply that there is nothing more. At this point we should maybe look at a famous philosophical thought experiment involving a brilliant scientist called Mary,[3] specialised in the neurophysiology of vision. Mary knows all there is to know about colours. From a scientific perspective, she has all the necessary physical facts to understand what colours are.

Nevertheless, one crucial thing is missing: Mary has never seen a colour with her own eyes. For whatever reason, she has been surrounded

1
Churchland, P. S. Touching a nerve. The self as brain. W. W. Norton & Company, Inc. (2013).

2
Crick, F. & Koch, C. A framework for consciousness. Nature Neuroscience 6, 119–126 (2003).

3
Jackson, F. C. Epiphenomenal qualia. Philosophical Quarterly 32, 127-136 (1982).

by a monochromatic world for her whole life, a world of black and white. The question now is whether she would acquire any additional knowledge about colours if she experienced one herself. We tend to believe that she would.

This thought experiment is intended to illustrate that more seems to exist than just the physically experienceable reality. Apparently, there is also a phenomenal consciousness which cannot be captured within the physical world we live in.[4] This very fact is a paradigm issue in the philosophy of mind regarding the relation between mental and physical processes.

4

Alter, T. & Walter, S. Phenomenal concepts and phenomenal knowledge. New essays on consciousness and physicalism. Oxford University Press (2008).

The phenomenal character of sensory impressions is exactly what is meant by the philosophical concept of qualia. These raw feels are, paradoxically, beyond our cognitive faculties. Qualia describe the simple fact that there is obviously something which is intrinsically linked to an internal perspective which as such cannot be detached from the individual experiencing it.[5]

5

Nagel, T. What is it like to be a bat? Philosophical Review 83, 435–450 (1974).

The occurrence of synaesthesia deconstructs conventional theories of perception even more. All relevant questions arising on subjectivity and objectivity, and everything in between, remain inadequately answered by empirical sciences.[6] This also applies to the humanities such as philosophy.

6

Levine, J. Purple haze. The puzzle of consciousness. Oxford University Press (2004).

Essentially, the capability to hear colours, to taste sounds, to see temperature and the like is equally strange as seeing colours at all.

Thus, if no scientific explanation can resolve the discrepancy between the phenomenal consciousness and the physical world, this precious art book proves to be precisely the right place to visualise Anna's synaesthesia in terms of appreciating coloured emotions. She succeeds in revealing a few more layers of reality to us all—whatever reality might be.

Anna Püschel studied photography at the AKV St Joost in Breda, the Netherlands, and fine arts at LUCA in Ghent, Belgium. Her work focuses on invisible aspects of human existence. She is currently doing a Master's degree at LUCA with a project about silence.

Sam Beekhuizen earned his Master's degree in Brain & Cognition and Psychological Research at the University of Amsterdam under the supervision of Dr Romke Rouw and Stevel Scholte. His main research interests include synaesthesia, psychoactive drug effects, consciousness and music.

Erika Schippmann is a psychotherapist and psychoanalyst practicing in Bonn, Germany. Anna was her patient for a few years. She is a member of the International Psychoanalytical Association (IPA).

Olympia Colizoli holds a PhD in cognitive neuroscience and currently does research as a postdoctoral researcher at the University Medical Center Hamburg–Eppendorf, Department of Neurophysiology and Pathophysiology and at the University of Amsterdam, Department of Psychology, Amsterdam Brain and Cognition Centre.

Alexandra Spaeth studied philosophy, German studies as well as comparative literature and cultural sciences. She graduated from the University of Bonn, Germany, where she specialised in applied ethics and political philosophy. She currently works for a charitable association in Cologne, Germany, in the area of development cooperation and global sustainability.

concept, text and photography
Anna Püschel

essays
Sam Beekhuizen, MSc
Erika Schippmann
Olympia Colizoli, PhD
Alexandra Spaeth, MA

text editing
Joy Phillips

design
Rob van Hoesel
Carel Fransen

typeface
Neue Haas Unica (Monotype)

paper
Olin Smooth High White 120 grsm
Munken Print White 15 115 grsm
MCM GD2 300 grsm
Muskat Brown 100 grsm

print
Wilco Art Books, Amersfoort (NL)

screen print
Hop Zeefdruk, Bunschoten (NL)

binding
Patist, Den Dolder (NL)

edition
500

publisher
The Eriskay Connection
(eriskayconnection.com)

isbn
978-94-92051-29-5

Many thanks to all the incredible people that contributed to *Layers of Reality*:

- My parents, for all their help and support;
- Sam Bingham, the most encouraging and patient man I know;
- Machiel Botman, whose words laid the foundation for this book;
- Carel Fransen, for the amount of time he put into this project, and for (almost) never losing track;
- Rob van Hoesel, who decided to make this book before I even knew who he was;
- Louisa Püschel, for her sharp eye;
- Alexandra Spaeth, for her endless support and her contribution on epistemology;
- Olympia Colizoli, for writing this wonderful text for *Layers of Reality*;
- Sam Beekhuizen, for his introduction to my work;
- Erika Schippmann, for her view on the case;
- Regis Keuren, who gave me so much confidence in what I am doing;
- Wiebke Zander, who was always there to help me;
- Jason Edwards, for his mental and physical support;
- Mihai Gui, for his senseful feedback;
- Christian Obst, for the corrections;
- Rhodo, for introducing me to a deeper understanding of everything;
- Marte Rijsdijk, who gave me the perfect vibe on a warm day in June;
- Lilian Joosen, for her heartfelt support and uplifting words;
- Evelien Zeegers, for giving me the golden tip;
- Petra Stavast, for her feedback that meant so much to me;
- Karin Krijgsman, for her support before graduation;
- Luc Begas and Melanie Otto from brabants kenniscentrum voor kunst en cultuur, for supporting my project in many ways;
- Deutsche Synästhesie-Gesellschaft;
- IASAS, International Association of Artists and Synesthetes;

as well as every single one of you that contributed to the crowdfunding campaign for *Layers of Reality*:

Emiel Heinsbroek, Inês Fernandes, Richard Fieten, Louisa Püschel, Morry Shahrezaie, Claudine Bohère-Püschel and Paul Püschel, Sam Bingham, Jason Edwards, Monique Mutsaers, Satya Kama, Christian Obst, Chris Kierspel, Jo and Matt Bingham, Iris Sijbom, Iris Ten Bloemendal, Stéfanie Läpke and Roman Palkoska, Barry van der Rijt, Zani, Patricia and Didier Mercuer, Jostte Baggioni and Madeleine Poirier, Wiebke Zander, Marco Heijndijk, Aurélie Halsband, Caecilia Rasch, Anne Lebedelle, Maarten Tromp, Karin Krijgsman, Petra Stavast, Mayke Breukers, Marte Rijsdijk, Ernst Rijsdijk, Annet Rijsdijk-Quaadvliet, Fréderique Bruban, Isabel Appenrodt, Stephan and Danièlle Nobbe, Norbert Ticks, Joyce van Belkom, Berthold and Annick Halsband, Gabriela Cendoya, Ron de Vries, Zeffania Stokkermans, Denise van den Oosterkamp, Owen Harvey, Dominique Bohère, Norbert Ticks, Isabelle Appenrodt, Lilian Joosen, Lucy Arnink, Fabia Cnossen, Daria Scagliola and Jannes Linders, Annette Oomens, Crétien van Campen, Berry de Jong, Philippe Moroux, Matthias Feilhauer, Jasper Kloosterboer, Cassandra Gould, Sarah Lugthart, Esther van Rosmalen, Christiane and Felix Derrer, Rabea Birk, Lilian Bosten, Gisela and Andrea Paoli, Lucy Noble, Ella Werst, Marga Rotteveel, and all anonymous donors.

Thank you for your trust and support! I owe you so much.

This publication has been generously supported by VSB fonds, BKKC, Arca Printers en Supplies B.V. and Epson.